Bo Lojek

History of Semiconductor Engineering

Bo Lojek

History of Semiconductor Engineering

With 319 Figures, 1 in Color

Springer

Dr. Bo Lojek
ATMEL Corporation
1150 E. Cheyenne Mtn. Blvd.
Colorado Springs, CO 80906, USA
e-mail: blojek@atmel.com

Cover Illustration: Layout of Fairchild analog integrated circuits prepared by Dolores Talbert (Circa 1963)

Library of Congress Control Number 2006938040

ISBN-10 3-540-34257-5 Springer Berlin Heidelberg New York
ISBN-13 978-3-540-34257-1 Springer Berlin Heidelberg New York

Springer is a part of Springer Science+Business Media
springer.com

Typesetting: PTP-Berlin Protago-TEX-Production GmbH, Germany
Cover-Design: WMXDesign GmbH, Heidelberg
Production: LE-TEX Jelonek, Schmidt & Vöckler GbR, Leipzig, Germany

Printed on acid-free paper 7/3100/YL - 5 4 3 2 1 0

To the memory of my parents, Bohumila and Rudolf Lojek.
Because memory remains.

Preface – The pages that should be read first

The innovator makes enemies of all those who prospered under the old order, and only lukewarm support is forthcoming from those who would prosper under the new, because men are generally incredulous, never really trusting new things unless they have tested them by experience.

Niccolo Machiavelli, The Prince

Many years ago I was called to jury duty. During the jury selection, the plaintiffs' attorney asked my name and what I did for a living. I was very proud of my profession and what I did and I answered my name and said "*I am an engineer.*" The attorney did not ask any other questions and without hesitation said that I could go because he did not want me as a member of the jury. On the way home I was pondering what was wrong with engineers that they are not suitable for jury duty. If I had answered the attorney's question that I was a car salesperson or politician he would have considered me for jury duty – but not if I was an engineer.

My dad was an engineer; he could fix everything. He regularly took me to all kinds of junk yards and we always found something that was worth bringing home. I was taught to pick up screws or washers on the street, because after cleaning it certainly could be used again. I knew very early on that I wanted to be an engineer like my dad. I knew that being an engineer was a noble job. I studied hard and I became an engineer, and now I have faced the situation that I could not be a member of a jury because of my profession.

I decided to contact the plaintiffs' attorney and I asked if he could talk to me. We met and I told him my concern. He laughed and told me "*You cannot be a member of a jury, because you are an engineer. Engineers are too analytical and too logical. There is nothing personal.*" Good engineers approach life differently from others. They are analytical and logical, they rather trust the data and experimental evidence than loaded opinion of the attorney.

Anyone who knows anything about engineering would agree that engineers play critical, ubiquitous roles in sustaining our nation's international competitiveness, in maintaining our standard of living, in ensuring a strong national security, in improving our health, and in protecting public safety. The word "engineer" comes from the same Latin word *ingenium* as the words "genius" and "ingenious." I cannot think of any other profession that affects our lives in so many vital, significant ways. Engineers believe in numbers, in the laws of physics, laws of nature; yes, engineers are too analytical and too logical! An attorney would characterize these traits as negative or undesirable qualities, yet I believe they are essential to innovation and progress, and they are qualities of the people who contribute most to our society.

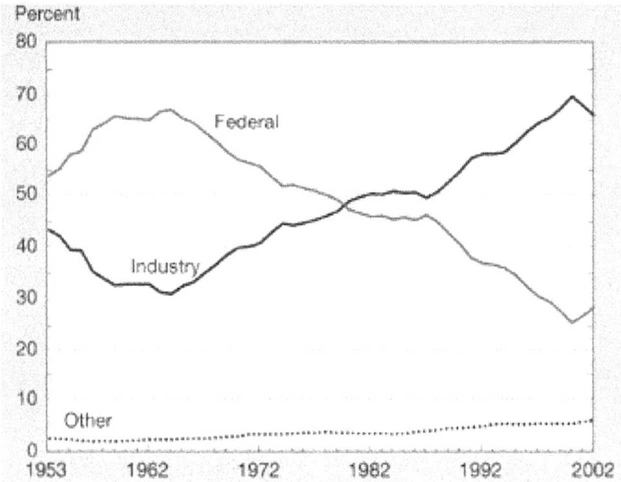

Fig. 1. National R&D expenditures 1953–2002 [Source: National Science Foundation 2004]

Creative people are sometimes seen as eccentric because they *genuinely enjoy* their work, instead of working only because they need an income. They are very seldom motivated by money. Society has changed recently so much that it is not easy to be a good engineer. Our "politically correct" society seems to delight in making it more difficult by denying resources to creative people who need them (Fig. 1.)

The growth of R&D investment in the United States has slowed steadily during the last forty years. Government data indicate that although total R&D expenditures continued to rise through 2002, industrial R&D, which fueled the growth over the prior period, failed to keep pace with inflation and experienced its first decline in real terms after 1994. This has occurred only six times in the past 49 years. The business activities of many R&D-

performing firms were curtailed following the stock market decline and the subsequent economic slowdown of 2001 and 2002.

The Federal Government was once the main source of the nation's R&D funds, funding as much as 66.7 percent of all U.S. R&D in 1964. The Federal share first fell below 50 percent in 1979, and after 1987 it fell steadily, dropping from 46.3 percent in that year to 25.1 percent in 2000 (the lowest it has ever been since 1953). Adjusting for inflation, Federal support decreased 18 percent from 1987 to 2000, although in nominal terms, Federal support grew from $58.5 billion to $66.4 billion during that period. Growth in industrial funding generally outpaced growth in Federal support, leading to the decline in Federal support as a proportion of the total.

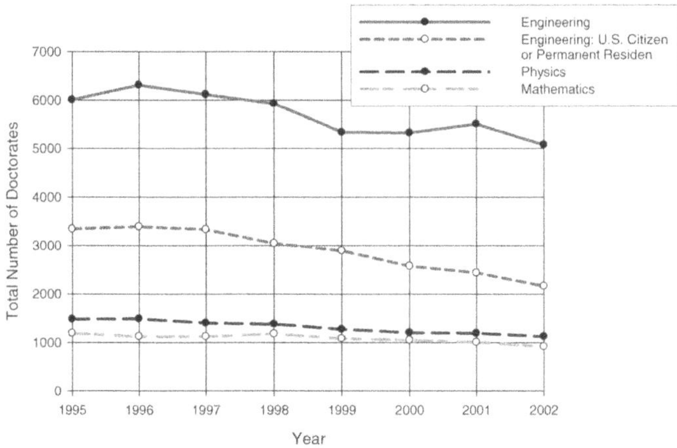

Fig. 2. Doctorates awarded in Engineering, Physics, and Mathematics: 1995–2002 [Source: National Science Foundation NSF 04–303 (October 2003)]

Figure 1 explains the most significant change in the industry which occurred in the early sixties. The industry, with pressure from Wall Street, could not finance long-range and risky basic research. The objective of basic research is to gain more comprehensive knowledge or understanding of the subject under study without specific applications in mind. Basic research advances scientific knowledge but does not have specific immediate commercial objectives. Basic research can fail and often will not bring results in a short period of time. The industry is mainly involved in the Research & Development (R&D), which is the systematic use of the knowledge or understanding gained from basic and applied research directed toward the production of useful materials, devices, systems, or methods, including the design and development of prototypes and processes.

Future American technology projects will be developed here only in the initial stages. Once the cost, rather than innovation, becomes the principal

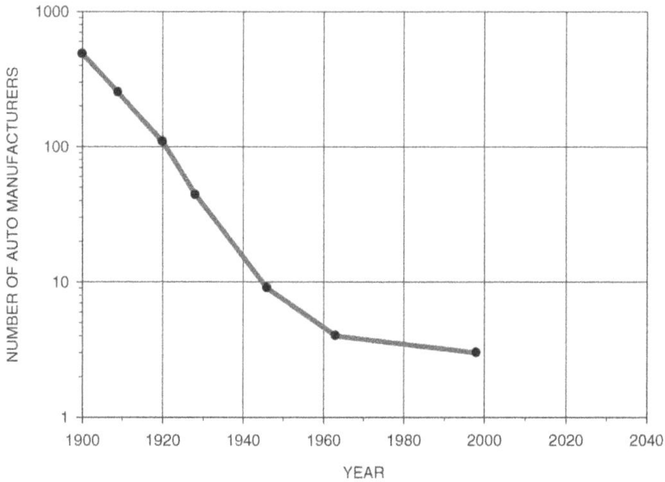

Fig. 3. Automobile Manufacturers in U.S. (1900 – Present)

factor for products, companies will ship the work to cheap-labor countries. So called "outsourcing", a result of a traditional capitalist economic development life cycle, deflects bright students away from engineering. Not surprisingly, the total number of doctorates awarded in U.S. during 1995 to 2002 in Engineering, Physics and Mathematics during the last decade steadily declined (Fig. 2.)

I was one of the fortunate engineers who had fun all of my life. My parents supported me in all I wanted to do. I enjoyed work constantly; I do not remember a time when I had a vacation. As a boy I built a crystal radio, a superheterodyne, and radio controlled airplanes. As a man, I look forward to being at work every morning. If I had a chance to live my life again, I would not change one iota. Engineering is my passion.

I wrote book about engineers. To be more explicit, this is a book about the group of engineers and scientists who invented modern transistors and integrated circuits. Historians assigned the invention of integrated circuits to Jack Kilby and Robert N. Noyce. In this book I am arguing that the group of inventors was much bigger. It happened that I know or I worked with many of the personalities described in the next chapters. I am describing the events which I lived through.

Colorado Springs, 2002–2006 Bo Lojek

Acknowledgement

Everybody who has written a book knows how difficult writing is. I know how difficult it is to write a book in a second language. I would not have been able to do it if many colleagues at Atmel Corporation where I work had not helped me. However, there are three friends who deserve special recognition:

Robert A. Pease, a distinguished engineer of National Semiconductor, helped me with proofreading and mainly encouraged me when I was in my "low." Bob is also known as "a common sense engineer" and has invented 25 integrated circuits, holds 21 patents, and has written numerous technical papers and books. He authored in Electronic Design over 240 columns and counting. Bob Pease is not only a terrific engineer and a unique personality; he is also the "Czar of Proofreading" who knows the English language, as everything that Bob does, perfectly.

The help and advice of Professor Hans-Joachim Queisser, former Shockley Semiconductor Laboratories employee and one of the founding directors of the Max Planck Institute for Solid-State Physics in Stuttgart, was critical in the final stage of book production.

Dr. Morgan Sparks, Former Bell Laboratorie's scientist, good friend of Bill Shockley, designer of the world's first junction transistor, and retired president of Sandia Laboratories guided me through many obstacles because "*my viewpoints lead to different stories and credits from those generally accepted by the media.*" During the four years of working with Morgan, I found not only my role model but I, also, recognized the nobility of engineering if carried out by a person like Dr. Sparks. I was very glad that at the end of my research I passed Morgan's criteria, and he concluded "*The book is a remarkably detailed account of accomplishments that constitute semiconductor microelectronics.*"

The majority of materials reproduced here is from my archives. I have saved for all my life newspaper clippings, photographs, vacuum tubes, silicon substrates, transistors, pieces of manufacturing equipment, wafers, integrated circuits, Rubylith foil with the mask of my first IC, etc. The paperwork alone grew during the last forty years to 672 cubic feet. When I was young, I never thought about writing a book; I was poorly organized and often, when I pulled out a page from trade magazines or newspapers, I did not record the title and date of publication. For this reason, I was not able to trace all owners of copyright material. However, the copyright holder has released the majority of images used here into the public domain, its copyright has expired, or it is ineligible for copyright.

The author is very much indebted to Ms. Alice Blanck of Springer and Ms. Steffi Hohensee of LE-TEX for their substantial contributions to the heavy task of dealing with a stubborn author, and with putting the manuscript with complicated archive materials into a finished book.

Contents

Prologue

"Who controls the past controls the future. Who controls the present controls the past."

George Orwell, 1948

On July 1, 1948 The New York Times printed, on page 48 in the "News of Radio" section, an announcement that NBC would broadcast Waltz Time on Friday night. The same section contains a brief report about a new invention, *"a device called a transistor, which has several applications in radio where a vacuum tube ordinarily is employed."*

When creative men started working on semiconductors by the late thirties and integrated circuits at the end of the fifties, they did not know that they were going to change the lives of future generations. Very few people at that time recognized the significance of perhaps the most important invention of the century. Nobody noticed that the key people behind the inventions were frequently frustrated and disappointed. Who remembers today, names such as Russell Ohl, Karl Lark-Horovitz, William Shockley, Carl Frosch, Lincoln Derick, Calvin Fuller, Kurt Lehovec, Jean Hoerni, Sheldon Roberts, Jay Last, Isy Haas, Bob Norman, Dave Allison, Jim Nall, Tom Longo, Bob Widlar, Frank Wanlass, Federico Fagin, or Dave Talbert?

In the beginning of the sixties the editors of Time-Life Books in Alexandria, V. A. published "A golden Age of Entrepreneurship" with a photograph (Fig. 1) accompanied by a legend stating "1958–1959 Robert Noyce, Jean Hoerni, Jack Kilby, and Kurt Lehovec all took part in developing the integrated circuit".

In Jack Kilby's speech to the 2000 Nobel Prize Committee the names reduced to just Robert Noyce despite the fact that Hoerni and Lehovec's ideas were so much more practical than Kilby's that even Texas Instruments adopted them.

Fig. 1. Inventors of integrated circuit as they were recognized in the early sixties

Neither Hoerni nor Lehovec had the backing of a large company. Approximately 40% of all newspaper stories originate from Press Releases prepared by Public Relation firms. Because radio and TV agencies only re-edit newspaper stories, a substantial portion of the public's "news" originates from PR releases. Naturally the connection to the PR source along with some of the people who created history are edited out.

The idea of the integrated circuit is almost as old as the transistor. For example, in October 22, 1952, Bernard M. Oliver filed a patent application with the idea to integrate several transistors onto one chip. A few months later on May 21, 1953 Harwick Johnson of Radio Corporation of America described "Semiconductor Phase Shift Oscillator and Device," now U.S. Patent # 2,816,228 (Fig. 2). To my knowledge, Johnson's patent represents the very first effort to integrate various electronic components into one piece of material.

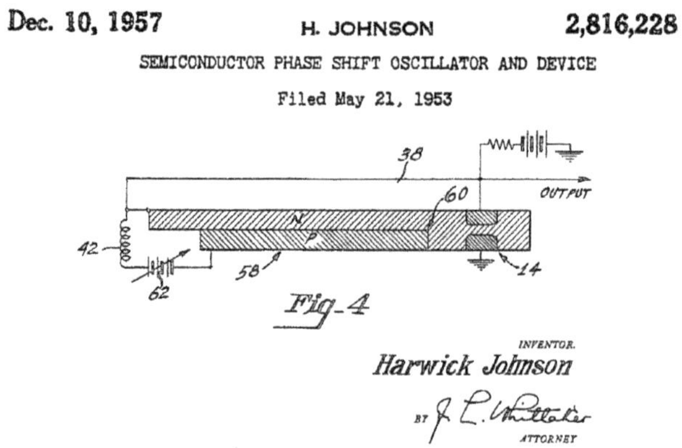

Fig. 2. Harvick Johnson Patent filed May 21, 1953

In May 1952 the British scientist G. W. A. Dummer made the following prediction at the IRE annual electronic components meeting in Washington, D.C. "With the advent of the transistor and the work in semiconductors generally, it seems now to be possible to envisage electronic equipment in a solid block with no connecting wires. The block may consist of layers of insulating, conducting, rectifying and amplifying materials, the electrical functions being connected by cutting out areas of the various layers". In the summer of 1957, Dummer described at an International Symposium on Electronics Components at Malvern, UK "a transistor flip-flop with two emitter follower outputs – a total of four transistors all contained in a chip of silicon 125 mils by 375 mils." Dummer's idea was using a loose wire to connect all circuit components. Johnson described in 1953 the same circuit as Jack Kilby of Texas Instruments many years later, and the similarity of these circuits is more than obvious from the comparison shown in Fig. 3. The natural question which probably most people would ask is, "What is the difference between the Johnson and Kilby application that allowed the invention priority to go unequivocally to Kilby?" It is not easy to answer such a question. Johnson described his circuit as a "unitary body" where much of the conventional circuitry is eliminated. Kilby used the term "circuit integrated into the body of material."

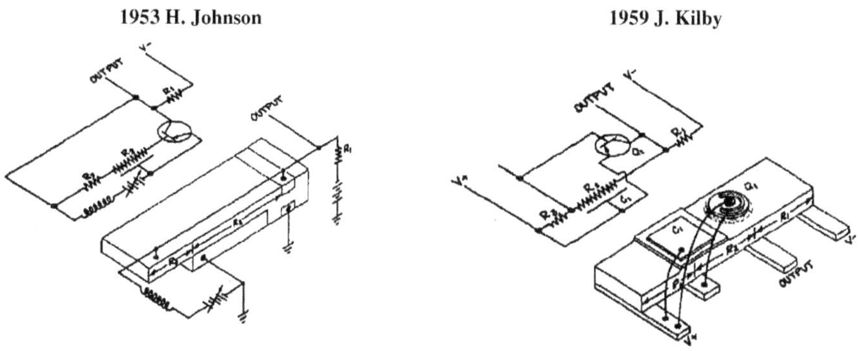

Fig. 3. Comparison of Johnson and Kilby Patents

Kilby's idea of the integrated circuit was so unpractical that it was dropped even by Texas Instruments. Kilby's patent was used only as very convenient and profitable trading material. Most likely, if Jack Kilby worked for any company other than Texas Instruments, his idea would never have been patented.

All innovative and astonishing ideas in semiconductor engineering are the products of very creative individuals. Creative individuals, not business leaders or government initiatives, are the most important factor in creating new devices. The current "politically correct" society and PR departments of large companies alter historical events, and present a development of what is today called microelectronics as a systematic effort of exceptional leadership. Many modern college business textbooks list companies such as Fairchild Semiconductor Corporation and Geophysical Service Inc. (later re-named to Texas Instruments) as examples of exceptional business wisdom.

2,816,228

SEMICONDUCTOR PHASE SHIFT OSCILLATOR AND DEVICE

Harwick Johnson, Princeton, N. J., assignor to Radio Corporation of America, a corporation of Delaware

Application May 21, 1953, Serial No. 356,407

10 Claims. (Cl. 250—36)

In the electron tube art, it is well known that a tube may be operated, in a suitable circuit, as a phase-shift oscillator. In such circuits, a resistance-capacity phase shifting network is connected between the output and the input of an amplifier tube, the circuit being proportioned to provide a 180° phase shift at the desired oscillation frequency. In accordance with the invention, a transistor may be similarly operated in such an external phase-shifting circuit. Also in accordance with a preferred embodiment of the invention, a semiconductor phase-shift oscillator is incorporated in a unitary body whereby much of the circuitry of the conventional phase-shift oscillator is eliminated.

The capacitance of a P-N junction biased in the reverse direction may be determined from the formula:

$$C=\frac{KA}{4\pi W}$$

wherein:

K=dielectric constant of the semiconductor material
W=thickness of the P-N junction
A=cross-section area of the P-N junction

The capacitance of a P-N junction biased in the forward direction may be determined from the formula:

$$C=\frac{qL^2I}{kT4D}$$

wherein

q=magnitude of electron charge
L=diffusion length of minority carriers
k=Boltzmann's constant
T=temperature in degrees Kelvin
D=diffusion constant
I=current

The resistance of a piece of semiconductor material may be determined from the formula:

$$R=\frac{rL}{A}$$

wherein:

r=resistivity of the semiconductor material
L=length of the piece
A=cross-section area of the piece

What is claimed is:

1. A semiconductor device comprising a body of semiconductor material having alternating zones of different conductivity material and a semiconductor delay line integral with one of said zones, said delay line including a plurality of series connected filaments of semiconductor material and P–N junction portions.

2. A semiconductor device comprising a body of semiconductor material having alternating zones of different conductivity material and a semiconductor delay line contiguous with one of said zones and connected to another one of said other zones and thereby providing feedback between said one zone and said other one of said zones, said delay line including a plurality of series-connected alternating filaments of semiconductor material of one type of conductivity and P–N junction portions.

3. A semiconductor device comprising a body of semiconductor material having alternating zones of different conductivity material and a semiconductor delay line connected to one of said zones, said delay line including a plurality of series connected alternating filaments of semiconductor material of one type of conductivity and P–N junction portions and bias voltage means connected to said P–N junction portions for varying the capacitance thereof.

3,138,743

MINIATURIZED ELECTRONIC CIRCUITS

Jack S. Kilby, Dallas, Tex., assignor to Texas Instruments Incorporated, Dallas, Tex., a corporation of Delaware

Filed Feb. 6, 1959, Ser. No. 791,602

25 Claims. (Cl. 317—101)

It is, therefore, a principal object of this invention to provide a novel miniaturized electronic circuit fabricated from a body of semiconductor material containing a diffused p-n junction wherein all components of the electronic circuit are completely integrated into the body of semiconductor material.

As will be apparent to one skilled in the art, ohmic connections are those which exhibit symmetry and linearity in resistance to flow of current therethrough in any available direction. If two resistors are to be connected together, it is not necessary to provide separate terminations for the common point. The resistance may be calculated from

$$R=\rho L/A$$

where L is the active length in centimeters, A is the cross sectional area, and ρ is the resistivity in ohm-cm. of the semiconductor material.

Capacitor designs may be obtained by utilizing the capacitance of a p-n junction, as shown in FIGURE 2, wherein a semiconductor wafer 15 of p-type conductivity is shown containing an n-type diffused layer 16. Ohmic contacts 17 are made to opposite faces of the wafer 15. The capacitance of a diffused junction is given by

$$C=A\epsilon\left(\frac{qa}{12\epsilon V}\right)^{1/3}$$

where A is the area of the junction in square cm., ϵ is the dielectric constant, q is electronic charge, where a is the impurity density gradient, and V is the applied voltage.

What is claimed is:

1. In an integrated circuit having a plurality of electrical circuit components in a wafer of single-crystal semiconductor material, a plurality of junction transistors defined in the wafer, each transistor including thin layers of semiconductor material of opposite conductivity-types adjacent one major face of the wafer providing a base and an emitter region which overlie a collector region, the base-emitter and base-collector junctions of each of said transistors extending wholly to said one major face, a plurality of thin elongated regions of the wafer exhibiting substantial resistance to provide semiconductor resistors, the elongated regions being spaced on said one major face from the transistors, and conductive means connecting selected ones of the elongated regions to regions of selected ones of the transistors.

2. In a semiconductor device which includes a single-crystal semiconductor wafer: a junction transistor provided adjacent one major face of the wafer by thin layers of semiconductor material of opposite conductivity types overlying one another and extending to said one major face with the emitter-base and base-collector junctions of the transistor extending wholly to said one major face; and a resistor provided in the wafer by a discrete elongated region of the semiconductor material which is spaced from the transistor on said one major face.

3. An integrated circuit comprising a wafer of semiconductor material containing a plurality of electrical circuit components including at least one active circuit component and at least one passive circuit component, the active circuit component including at least two thin layers of semiconductor material of opposite conductivity-types extending to one major face of the wafer with p-n junctions of the active circuit component extending wholly to said one major face, the passive circuit component including at least one discrete region of the semiconductor material of the wafer which is spaced on said one major face away from the thin layers of the active component, substantial electrical impedance being exhibited between the semiconductor material contiguous to the at least one discrete region of the passive component and semiconductor material immediately underlying said thin layers of the active component.

The author of this book would like to offer a different view of the history of microelectronics. My view is based on personal experience as a Diffusion Engineer for almost 40 years. I experienced, unfortunately too many times, that the company establishment was frequently one of the biggest, if not the biggest obstacle, which needed to be overcome in the introduction of new ideas.

This book is my personal story, and my recollection of events may be biased. I am not asking the reader to agree with my statements but I will be delighted if readers will exercise their own judgment. This book is addressed to creative people who think for a living but are not convinced that they already know it all. I met and worked with people I did not like, and I treated them accordingly. I am proud that I was fired by some of them and I consider it as a distinction. On the other hand, I worked with a much bigger group of individuals who made my life extraordinary and many of them became my role models. I regret that I did not save more historical materials, did not ask more questions or did not spend more time with people who were characterized as troublemakers, eccentric, whistleblowers or "difficult persons," and for whom I am using the term "creative individuals". Regretfully, many of them are not with us anymore.

Creative individuals are critical to the success of any innovative process. The common characteristic of creative individuals is their willingness to surmount obstacles and persevere. Any creative endeavor will undoubtedly face obstacles because such endeavors threaten some established or entrenched interest, or status quo.

Research of human behavior consistently reveals a significant difference between creative and uncreative people, and sheds light on why highly creative individuals frequently cause trouble. Creative individuals exhibit atypical thought processes and mental content, they are less constrained by conventional expectations, and they are less concerned with making the right impression on others. Highly creative individuals do not respect common practices. Their methods, style, authoritarian control, and temperament are frequently at odds with conventional norms.

There are also differences in how these two groups process information. Creative individuals have a wider range of attention – they can think of more things at the same time than less creative people. They are also more open to new information and willing to take a higher level of risk. Creative people usually have a very deep knowledge of their subject. This high level of expertise can very frequently lead to problems when the mind becomes "set". Once their mind is set, creative people become persistent and try to overcome any obstacle. The outcome of such activities can result in phenomenal discoveries or colossal failures.

Creative people are introverted, independent, arrogant and hostile. Creative people are over-reactive. Highly creative individuals are usually high-maintenance employees. Creative people are driven with a strong need for achievement and they have the self-belief and energy to challenge the practice of the system and their managers. An engineer usually does not tell his supervisor that he wants to realize his creative potential. Rather he expresses these needs with symptomatic behavior that may become troublesome if the needs remain frustrated. Such behavior may become increasingly disruptive. The highly competitive and ultimately detrimental interaction between strong personalities eventually results in destroying the system, the organization or, more frequently, in the separation of the creative person from the organization and creating a new spin-off or organization. This new organization is set up by rules defined by a creative person that has in the beginning top-notch knowledge of a particular problem. As the business succeeds and expands this originally creative person sometimes becomes preoccupied with business issues and becomes increasingly disconnected from creative scientific work. Unless the founder of a start up is able to adjust to new situations, then gradually, the original dynamic environment will move to the same stagnant environment from which the creative person originally separated. At this time a new generation of creative people will initiate a new cycle of unusual and unconventional thinking.

The transistor and integrated circuit was a result of the creativity of high-maintenance employees. Their creativity was colossal and, therefore, all their behavioral flaws were colossal. This, however, does not make them as a person or their accomplishments smaller.

When I was a younger engineer I was naive enough, not to look back in history. I was always bashing history and art students, who kept their class in the park in shadow of a tree and discussed Nietzsche or Shakespeare. Physics students had to be in laboratories, in front of a blackboard and had to work hard on their experiments. From history I could have learned that really almost nobody cares about the methods of polysilicon doping or channel concentration, but almost everybody cares about the quantity of money or power they have. Certain human behavior has a character of axiom, and will never change. In my naiveté, I assumed that history is something that is part of our past and no changes or editing are needed. I was wrong! To illustrate this statement I will use the example which I faced during the course of writing this book: on November 4, 2003 the CBS issued this Press Release

> *(CBS) CBS Dumps Reagan Miniseries*
> *Mounting pressure and criticism from conservatives has prompted CBS to dump "The Reagans," the network's prime-time miniseries on former President Ronald Reagan. "The Reagans" had been scheduled to air on Nov. 16 and 18. November is a ratings sweeps month of special importance to TV networks since it helps to determine how much they may charge advertisers.*

Criticism of CBS took on an official tone with a letter of complaint from Republican National Committee Chairman Ed Gillespie to CBS President Les Moonves. The New York Times, citing unidentified people close to the production of "The Reagans," reported that CBS executives had previously reviewed the script and viewed the miniseries without raising any objections to the content.

Why were the CBS executives comfortable with the miniseries, and Ed Gillespie was worried? The majority of people in this nation still remember the Reagan era, and they can easily judge if the miniseries is true to history. Obviously, to be in charge of a particular "version of the truth" is of paramount importance. We may think that this is politics, and such things cannot occur in a business environment. There could be few individuals who believe it. The majority of individuals are not that naïve.

The current "politically correct" establishment is aware of problems that creative individuals with independent minds can create so they impose rules defining what is acceptable and what is not. Of course, the threshold between acceptable and unacceptable is defined by "them." When I served in the army, our commander, whose mental ability was on a level suitable for his position, walked in front of the unit, looked into the face of each of us and then puzzled and visibly disappointed, shouted *"This is horrible. Each of you is different!"*

They want us to be the "same," to follow "them," because they know better than we do, what is good for us. If creative individuals cannot be trimmed, they need to be fired, they need to be discredited. And if this is not working, they need to be portrayed in a negative light, for example, as a racist. Of course these are the methods of fascists and communists. I have first hand experience with both of these regimes. Since the time of Plato the problem is that *"There are two kinds of people: Those who do not know, and a much bigger group of those who do not know that they do not know."* Ed Gillespie knows and he wants that we do not know.

The second example is directly related to the semiconductor industry. In the middle of the sixties several authors published predictions about the future of microelectronics. Gordon Moore of Fairchild published his prediction in Electronics Magazine [1], Jay Last of Amelco presented his view at the Teknorama Conference in February 1967 [2], and Orville R. Baker of Signetics presented his data at the National Electronics Convention [3] the same year. Last provided the source of his data and showed that between 1961 and 1966 the number of transistors increased each year by a factor of 1.6. Baker, based on Fairchild and Signetics data, showed the area in square mils per Flip-Flop circuit. The area decreased by an average of 1600 $mils^2$/year.

The details of the presentations by Baker and Last are forgotten. The "data" by Gordon Moore, became so-called Moore's Law and the ideology of the Intel Corporation. In the original paper Gordon Moore did not explain what is the definition of "number of components per integrated function." Because Moore included the single transistor data into the plot, the original figure is somewhat confusing. Intel's Public Relations Department apparently decided that transistors are not an "integrated function", and they also believed, for any base the logarithm has a singularity at zero; therefore the

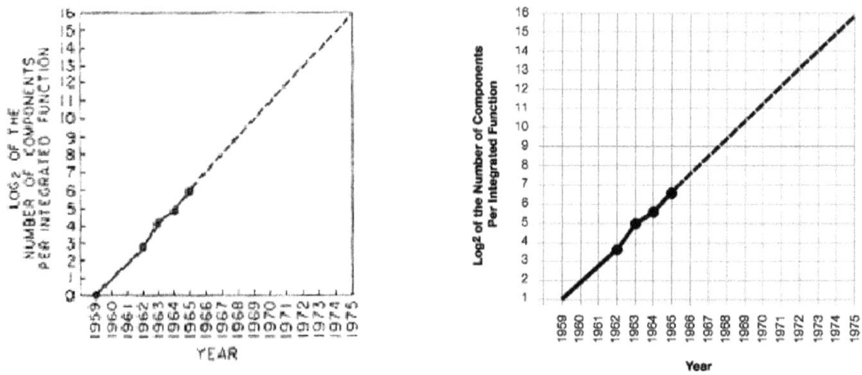

Fig. 4. Original (left) and "fabricated" (right) data defining so-called Moore law. (Original was published in Electronics magazine, April 19, 1965; the fabricated data are from Intel's Press Kit http://www.intel.com/pressroom/kits/events/moores_law_40th)

original plot was modified. In 2005, Intel's Press Kit "Moore's Law 40th Anniversary", Intel decided to create "new data" and pasted it into the original document. Because Moore's paper, contrary to Last and Baker's presentations, did not include real data, nobody really bothered to notice that the data changed the magnitude.

Moore never disclosed the source of his data. Very likely, Fairchild's Micrologic circuits, developed during 1960 and 1964 by R. Norman and R. Anderson (shown in Fig. 5) was used as a base for an observation now called "The Law."

To avoid confusion with altered data, I am trying to support my argumentation with many documents that were not previously published. We all know that human memory fades, especially in the cases of success or failure, and has a tendency to find a way to join the success, and to separate as much as possible from the failure. When success or failure is documented, none or very little memory is needed. Because fewer and fewer libraries keep old publications in their collections, for many people it is difficult to find the real historical facts. And because only a few have access to a corporation's internal documents, it is not surprising that some implanted new and "corrected" information subsequently become "truth."

This book is my personal story and it is story about engineers who refused to be "the same." I do not need to rely on historians' assessments and their research of semiconductor business; I lived this history, I was privileged enough to know many of the key figures described in this book. This claim is not the same as the claim: "I cannot be wrong." Certainly, I may not have the complete information about all events. I am more than happy to modify my statements if new facts are provided. My approach to the possible corrections, however, is the following: *"In God we trust. All others bring data."*

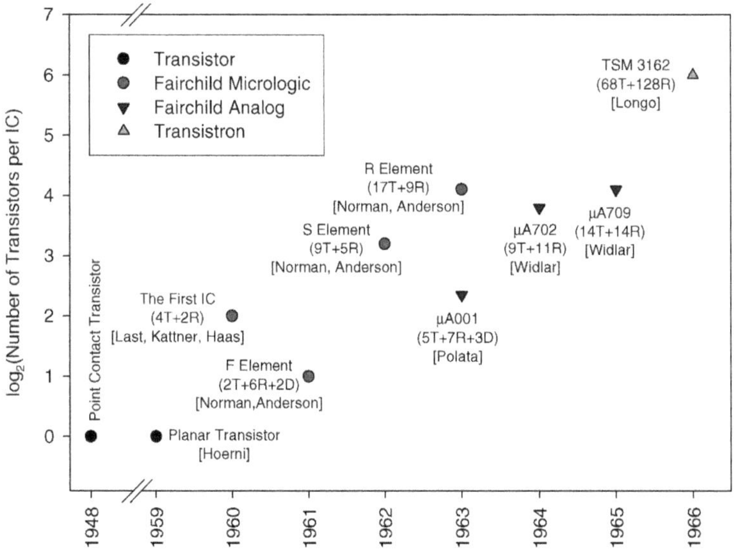

Fig. 5. Very likely source of data G. Moore used in his 1965 Electronics paper (T = Transistors, R = Resistors, D = Diodes)

I am offering a different opinion about several myths and common folklore knowledge of events, which took place almost fifty years ago when semiconductors started to be a business. During the course of this work I was relying mostly on my recollections and life-long interests in the history of semiconductor engineering. I accumulated a significant amount of samples, wafers, photographs and news clippings. I noticed that history is like a solid-state diffusion process that is evolving with time. The state-of-the-art Rapid Thermal Processing of semiconductor reduces the diffusion evolution of dopants to a minimum. Unfortunately, there is no similar process known for the treatment of history.

Historians do not own the past, but they do get to make up the rules as to what counts as history. In post literature society, where fewer and fewer books are read, every new attempt to investigate the past bears increasing responsibilities to be chroniclers of history. The main obstacle is that there is no absolute objectivity in history. You can pick virtually any topic and find twenty, a hundred, or a thousand histories, all of which will be different. I can go down the checklist of all the historical facts, be totally faithful to it – and manipulate the hell out of people through iconography and symbolism. I do not know of a way to be truthful in my book except to be emotionally truthful.

Of course, many of these fictional recreations may or may not be true to history, but my objectives are to present my knowledge and let the reader conclude with his own opinion. You do not censor things. History can put us in relation with the past in a way that tells us something about ourselves.

With this book I want also to pay a tribute to the memory of William B. Shockley. In my entire life I have not met a more creative and resilient person than William B. Shockley. I found between my friends only a few persons who would say a friendly word about Shockley. Shockley was very complex, difficult and extraordinary in many ways. He was a brilliant physicist who could not pretend anything. His social skills were close to none. If you were stupid he would tell you and it would not be wrapped in politically correct nonsense.

Niccolo Machiavelli wrote in “The Prince”: *The Innovator makes enemies of all those who prospered under the old order, and only lukewarm support is forthcoming from those who would prosper under the new.* Shockley made many of his enemies because in ten cases he was right nine times. A majority of people, do not appreciate when someone constantly demonstrates that he is sharper.

Interestingly enough, the majority of those who sharply criticize Shockley never forget to put into their resumes or biographies a note that they worked with and were trained by Bill Shockley.

Shockley created not only Silicon Valley and a new industry, but he changed the way we live. If the atomic bomb had not been invented, or if we did not reach the Moon, the life of the majority of people would not be affected. However, I cannot imagine my life without the transistor, even though I do not use a mobile phone.

The fundamental theory of the PN junction as used today was formulated by Shockley in a very short period of time. Shockley became the most respected man by many at the age of forty. There was, and still is a bigger group of others who envy him. What was the source of his genius or what some call evil genius? My answer is that Shockley was a man that Nature rarely produces and who only appears on Earth at intervals of centuries.

It is too sad that Shockley never shared in the rewards that so many Silicon Valley pioneers have reaped.

References

[1] G. E. Moore, “Cramming more components onto integrated circuits,” Electronics, Vol. 38 (1965), April 15

[2] J. Last, Presentation at Teknorama Conference, Stockholm, February 1967, (published as “Integrerad elektronisk kretsteknik av idag”, Elektronik, Vol. 4 (1967), p. 40

[3] O. R. Baker, “Aspects of Large Scale Integration,” NEC Record 1967, p. 56

1

Research Organization: Bell Telephone Laboratories

"Our species is the only creative species, and it has only one creative instrument, the individual mind and spirit of man. There are no good collaborations, whether in music, in poetry, in mathematics, in philosophy. Once the miracle of creation has taken place, the group can build and extend it, but the group never invents anything. The preciousness lies in the lonely mind of man."

John Steinbeck, East of Eden

Bell Laboratories were jointly owned by the American Telephone & Telegraph Company and Western Electric, AT&T's production subsidiary. At the end of World War II the Laboratories employed about 11,000 people, of whom about one-third were professional scientist and engineers, about one-third technical aides, and about one-third clerical and support personnel. Approximately 85 percent of the laboratory staff was engaged in the development of specific devices and systems for use in telephone systems or by the military. About 15 percent of the professional staff under the Vice-President of Research, William O. Baker, was involved in research which was not related to any specific objective.

During World War II, Bell Laboratories undertook more than 2000 research projects for the Army, Navy, and the National Defense Research Council. Between 1949–1959, the U.S. Government funded more than $600 million of research at Western Electric and Bell Laboratories (approximately 50% of total Research Budget of Bell Laboratories.) During this period the Department of Defense allocated between $1 million and $2 million annually to over one hundred doctoral candidates working on basic research of solid-state physics. Support of scientific research began to pay important dividends very quickly.

The idea to develop a solid-state switching device originate from Bell Laboratories' brilliant Director of Research, Mervin J. Kelly. Kelly was very exceptional, extraordinarily keen, alert and practical. Visionary Kelly was motivated by a desire to replace mechanical relays in telephone exchanges with solid state devices.

On November 2, 1956 when William B. Shockley won the Nobel Prize and responded to Kelly's congratulation, Shockley wrote:

> *"Dear Mervin, It is hard for me to see as a research director and vice president in your position could have proceeded more effectively to get a transistor out of a solid state physicist like myself. The background of experience I had in the vacuum tube area and some talks you once gave me on the importance of electronic switching stimulated me to be alert to such possibilities. This was then followed by the freedom to work on subjects of my own choosing in the solid state physics area. I hope that we shall have an opportunity to pat each other on the back over a drink before too long."*

Shockley obtained his doctorate from MIT in 1936 where he became the protégé of Prof. P. M. Morse. His doctoral thesis was entitled "Calculations of wave functions for electrons in Sodium Chloride crystals" and was supervised by Professor John C. Slater. Shockley turned down several offers (General Electric, Yale University) and joined Bell Laboratories because he wanted to work with C. J. Davisson (who later won a Nobel Prize for his work on electron diffraction). Shockley was assigned to the Vacuum Tube Department previously headed by Mervin J. Kelly.

When Kelly became the research director of Bell Laboratories he put William Shockley in charge of the Solid State project. Shockley first considered materials which were investigated earlier by the Pohl group in Germany, and Davydov and Joffe in Russia and which was reasonably well understood, Copper Oxide, for example. Shockley envisioned a device (Fig. 1.1) which later failed to behave as predicted by theory. However, the experience learned during the course of this work firmly established motivation and desires to pursue the idea.

The research work was interrupted by World War II. In 1940 Shockley worked on the J. B. Fisk project "Uranium as a source of power." In 1942 Shockley left for assignment to the newly established Office of Scientific Research and Development (OSRD.) This was headed by MIT engineer and legend Vannevar Bush with James B. Connant, a chemist and president of Harvard University and other prestigious members of the scientific community. OSRD was a federally-funded civilian organization with the main goal to coordinate the war effort between the science community, business and the government.

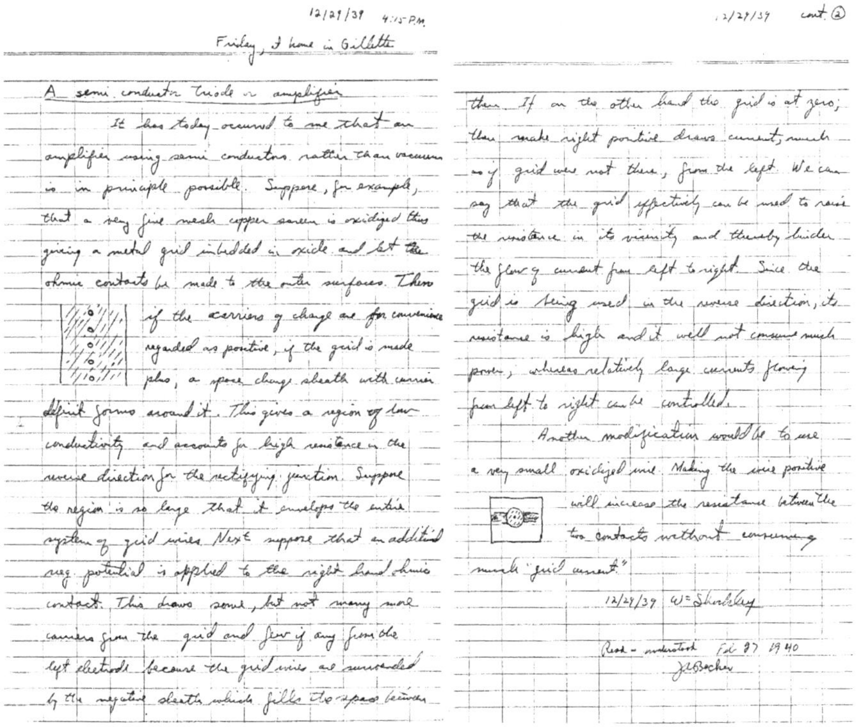

12/29/39 4:15 P.M.

Friday, I home in Gillette

A semi conductor triode or amplifier

It has today occurred to me that an amplifier using semi conductors rather than vacuum is in principle possible. Suppose, for example, that a very fine mesh copper screen is oxidized thus giving a metal grid imbedded in oxide and let the ohmic contacts be made to the outer surfaces. Then if the carriers of charge are for convenience regarded as positive, if the grid is made plus, a space charge sheath with carrier deficit forms around it. This gives a region of low conductivity and accounts for high resistance in the reverse direction for the rectifying junction. Suppose the region is so large that it envelops the entire system of grid wires. Next suppose that an additional neg. potential is applied to the right hand ohmic contact. This draws some, but not many more carriers from the grid and few if any from the left electrode because the grid wires are surrounded by the negative sheath which fills the space between

12/29/39 cont. ②

them. If on the other hand the grid is at zero; then make right positive draws current, much as if grid were not there, from the left. We can say that the grid effectively can be used to raise the resistance in its vicinity and thereby hinder the flow of current from left to right. Since the grid is being used in the reverse direction, its resistance is high and it will not consume much power, whereas relatively large currents flowing from left to right can be controlled.

Another modification would be to use a very small oxidized wire. Making the wire positive will increase the resistance between the two contacts without consuming much "grid current."

12/29/39 W. Shockley

Read - understood Feb 27 1940
J.A. Becker

Fig. 1.1. Copper Oxide Solid State Amplifier envisioned by W. Shockley on December 29, 1939, Bell Telephone Laboratories Notebook # 17006 assigned to William Shockley (Dept. 328-3), September 1, 1939

Under the direction of V. Bush, Shockley become Director of Research of the Antisubmarine Warfare Operations Research Group[1] and in 1944 he became an Expert Consultant in The Office of the Secretary of War working on deployment of radar in the B-29 program.

The small group at Bell Laboratories continued solid state research under the direction of MIT Radiation Laboratory to purify the semiconductor material for microwave detector used in radar. Russell Shoemaker Ohl (1898–1987) who was trained in electrochemistry and a graduate of Penn State in 1918, discovered during the course of investigations of the properties of crystal detectors for radar, the first p-n junction device when he accidentally cut a section of sample across an (invisible) boundary between p and n regions of a silicon ingot solidifying from a doped melt.

On March 6, 1940 Ohl showed his sample to Mervin Kelly. Kelly called Walter Brattain and Joseph Becker. Brattain immediately suggested that the

[1] Shockley actually invented and coined the phrase "*Operations Research*" during his work on antisubmarine warfare

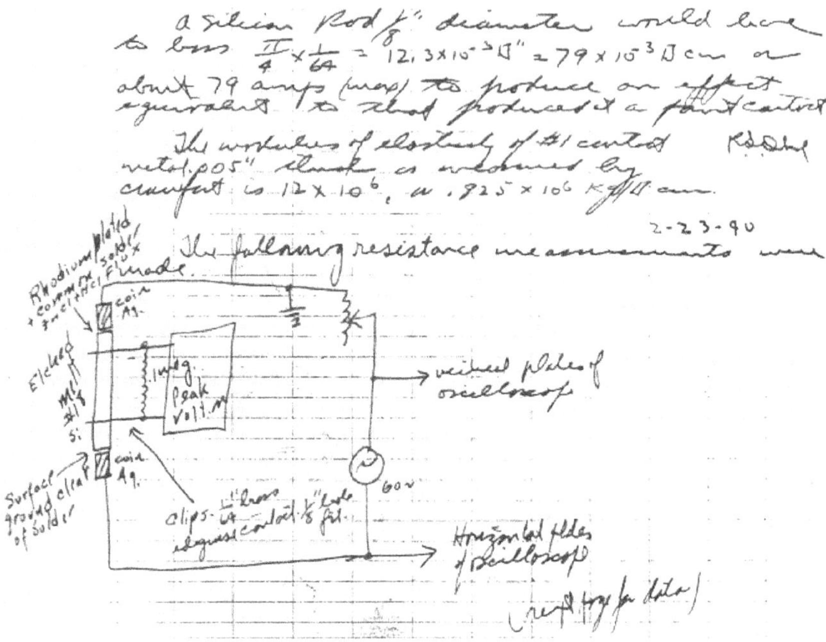

Fig. 1.2. Circuit used by R. Ohl to measure the resistance of a silicon rod. The unusual behavior of the current passing through the sample let to the discovery of the P-N Junction. [R. S. Ohl Laboratory Notebook, February 23, 1940]

electrical current must be due to "some barrier being formed in the crystal" and nothing else happened.

Mervin J. Kelly in the spring of 1945 was promoted to vice-president in charge of research. He immediately, in June 1945, organized goal-oriented research and signed the Authorization for Work requesting "*new knowledge that can be used in the development of completely new and improved components.*" (Fig. 1.3.) Kelly established several new research groups. William Shockley and physical chemist Stanley Morgan headed Solid State Physics Department. Physical chemist Addison H. White was in charge of Electronic Materials Department, and Jack A. Morton directed Basic Research. Shockley was still heavily involved in military projects (Policy Council Joint Research) and often in Washington D.C.; S. Morgan substituted as group leader during Shockley's absence.

The department originally included experimentalist Walter Brattain, John Bardeen, a theoretician who joined Bell Laboratories from the Naval Ordnance Laboratory in late 1945, experimentalist Gerald Pearson, physical chemist Robert Gibney, and electrical engineer Hilbert R. Moore.

Shockley and Morgan's Solid-State Department, reported to Harvey Fletcher, the Director of Physical Research. Fletcher reported to Ralph Bown, the Director of Research.

President Harry Truman honored Shockley with a Medal of Merit on July 19, 1946 with the citation *"By his tireless efforts, initiative and skilful application of scientific techniques to the problems confronting the army, he made an exceptional contribution to the war effort."*

On September 6, 1945 Shockley and Morgan visited Karl Lark-Horovitz group at Purdue University. Prof. Lark-Horovitz with small group of students and scientists (V. A. Johnson, S. Benzer, R. Bray, R. E. Davis, L. G. Dowell and W. W. Scanlon) conducted since March 1942 research funded by OSRD on "Preparation of Semi-Conductors and Development of Crystal Rectifiers." Based on the work of Purdue group Shockley reach very important conclusion – the only semiconductor materials at that time with good prospect, were elemental Germanium and Silicon.

In January 1946, Shockley predicted, based on the existing theories for Germanium and Silicon, that a significant modulation of conductivity of thin layers of semiconductor should be produced by inducing a surface charge by strong electric field. The proposed form of modulation became known as

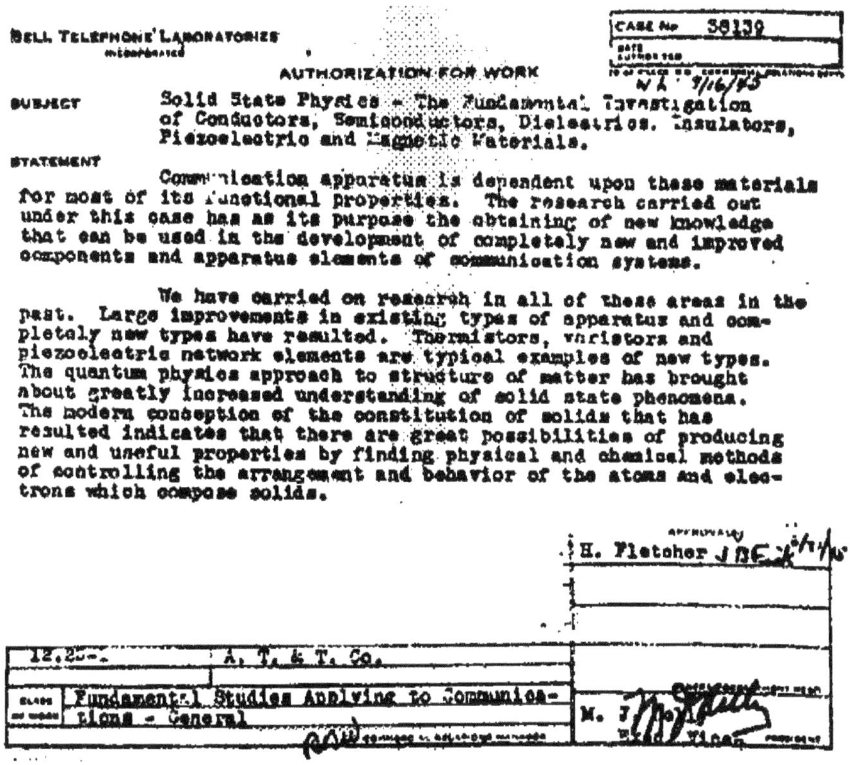

BELL TELEPHONE LABORATORIES
INCORPORATED

CASE No 38139

AUTHORIZATION FOR WORK

SUBJECT Solid State Physics - The Fundamental Investigation of Conductors, Semiconductors, Dielectrics, Insulators, Piezoelectric and Magnetic Materials.

STATEMENT

Communication apparatus is dependent upon these materials for most of its functional properties. The research carried out under this case has as its purpose the obtaining of new knowledge that can be used in the development of completely new and improved components and apparatus elements of communication systems.

We have carried on research in all of these areas in the past. Large improvements in existing types of apparatus and completely new types have resulted. Thermistors, varistors and piezoelectric network elements are typical examples of new types. The quantum physics approach to structure of matter has brought about greatly increased understanding of solid state phenomena. The modern conception of the constitution of solids that has resulted indicates that there are great possibilities of producing new and useful properties by finding physical and chemical methods of controlling the arrangement and behavior of the atoms and electrons which compose solids.

H. Fletcher

12.25-. A. T. & T. Co.

Fundamental Studies Applying to Communications - General

M. J. Kelly

Fig. 1.3. Authorization for Work to begin Solid State Physics research in Bell Laboratories signed by Mervin Kelly in June 1945.

the "field effect." Realizing the practical implication of such a possibility, Shockley proposed experiments to test his hypothesis.

A number of experiments were carried out by J. R. Haynes, H. J. McSkimin, W. A. Yager and R. S. Ohl. However, the degree of modulation had been considerably less than predicted by theory. Those results lead to a re-examination of the theory and the postulation of surface states by John Bardeen. Bardeen's idea resulted in additional speculation about the presence of a space charge region that may exist at the surface of a semiconductor. According to Pearson and Brattain [1], having postulated a space charge region at the free surface of a semiconductor, the question arose how to verify experimentally its existence. W. Shockley pointed out that "according to this picture the contact potential between n and p type samples should increase with doping." Experiments performed by Pearson and Brattain proved that this was the case.

In the fall of 1947 Brattain and Gibney experimentally studied properties of Bardeen's surface states. There was little or no theory explaining the unusual experimental behavior observed on measured samples. In November 1947, R. B. Gibney made a key suggestion which influenced all future experiments. Gibney suggested that voltage be applied between the metal plate and semiconductor (Fig. 1.4.) Gibney proposed a structure of semiconductor with contact at the periphery, and with a second contact in the center of the structure formed by electrolyte. When these connections were made, a current flowed through the sample and its magnitude was mainly determined by sample resistivity. When the potential of the electrolyte was modulated the current in the external circuit was accordingly modulated.

Brattain and Gibney had overcome the blocking effect of the surface states – the practical problem that had caused the failure of the original "field effect" experiment. They proposed amplifiers using the field effect with electrolyte to obtain the desired high electric field.

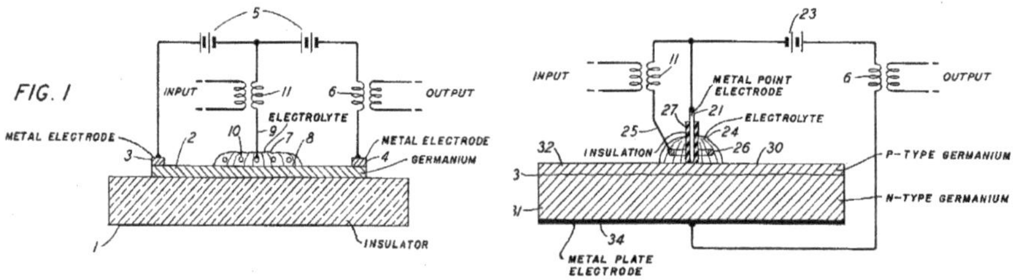

Fig. 1.4. A drawing from the Brattain and Gibney patent application experimentally verified on November 20, 1947 [U.S. Patent 2,524,034].

On November 23, 1947 (Sunday) Bardeen referred to observations by Brattain and Gibney and suggested a modified structure (Fig. 1.5) where liquid

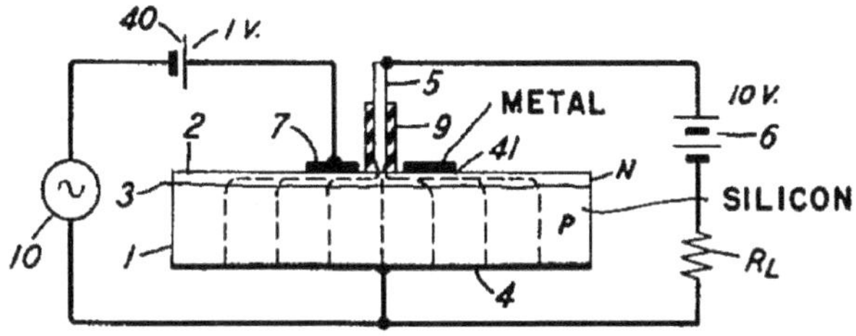

Fig. 1.5. Bardeen's disclosure of "Three-Electrode Circuit Element Utilizing Semiconductor Materials" dated November 23, 1947 [U.S. Patent 2, 524,033]

electrolyte was replaced with metal forming a rectifying contact with semiconductor. The suggestion become U.S. Patent 2,524,033 where Bardeen wrote:

> *"the current, in making its way through the block from the source electrode to the collector, first spreads out laterally in the surface layer in all directions from the source electrode before crossing the barrier. In accordance with the invention in one of its aspects, a third electrode, denoted the control electrode, is disposed to exert its influence on this spreading resistance.*
>
> *The result is a substantial modification of a substantial part of the whole internal resistance of the device, and so a substantial alteration of the current in the external circuit."*

Bardeen's patent application referred to Gibney's previous work which later became U.S. Patent 2, 560,792 where Gibney suggested the structure which leads to transistor version as demonstrated in December 23, 1947. Gibney wrote:

> *"thin surface layer of P-type material containing fixed negative charges and mobile positive charges, and high resistance barrier which separates this thin surface layer from the main body of the block which has N-type characteristics containing fixed positive charges and mobile negative charges.*
>
> *Positively biased metallic electrode placed on the P-type surface layer serves as emitter and positive charges "holes" tend to flow away from the emitter electrode in all direction before crossing the barrier. Some of them flow in the neighborhood of the negatively biased electrode which may be termed collector.*
>
> *Evidently the portion of the emitter current which is collected by the collector depends on the distance which separates these two electrodes."*

Brattain's notebook from December 8, 1947 reports a very important change. The Gibney device (Fig. 1.4) with the drop of electrolyte and the point-contact structure exhibited voltage and power gain; however, the device had a new feature: so-called "high back voltage" N-type germanium. "High back voltage" germanium is high resistivity material – a central feature necessary for achieving the voltage gain of the point contact transistor. Brattain's entry also contained a note about a luncheon discussion with Shockley and Bardeen where Bardeen suggested use "high back voltage" germanium studied by Lark-Horovitz group at Purdue. The reasoning behind this suggestion was to get a better rectifying contact of high resistivity material in comparison with low-resistivity silicon or germanium samples they used before.

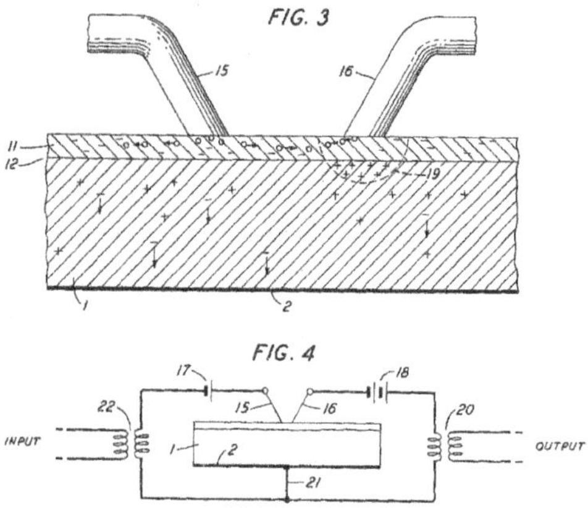

Fig. 1.6. Gibney patent # 2,560,792 which for the first time introduced terms "Emitter" and "Collector"

Then Brattain, sometime before December 16, 1947, got a brilliant idea to apply gold on a wedge and then separate the gold at the point of the wedge with a razor blade to make two closely spaced contacts as shown in Fig. 1.7. Twenty years later Brattain in interview for IEEE Spectrum magazine recalled his experiment: *"I accomplished it by getting my technical aide to cut me a polystyrene triangle which had a smart, narrow, flat edge and I cemented a piece of gold foil on it. After I got the gold on the triangle, very firmly, and dried, and we made contact to both ends of the gold, I took a razor and very carefully cut the gold in two at the apex of the triangle. I could tell when I had separated the gold. That's all I did. I cut carefully with the razor until the circuit opened and put it on a spring and put it down on the same piece of germanium that had been anodized but standing around the room now pretty near a week probably. I found that if I wiggled it just right so that I had contact with both ends of the gold that I could make one contact an emitter and the other a collector, and that I had an amplifier with the order of magnitude of 100 amplification*[2]*, clear up to the audio range."*

[2] Human memory is imperfect, and later accounts are often subject to "Retrospective Realism." Documented amplification of the Brattain device, during tests performed on December 16, 1947, had a voltage gain of fifteen.

Fig. 1.7. The triangular wedge is made of plastic and is covered with gold foil slit in half at wedge's tip. One side of the wedge serves as the emitter, the other as the collector. The piece of germanium acts as the base. Actual size of the wedge is approximately 30 mm

The wedge assembly was completed on December 16, 1947 and the testing was completed that same afternoon. The transistor was born.

On the afternoon of December 23, 1947 H. R. Moore connected the input of the transistor to a 1 kHz signal and the output to an oscilloscope. R. B. Gibney, J. Bardeen, G. L. Pearson, W. Shockley, W. Brattain, H. Fletcher and R. Bown witnessed the test (Fig. 1.8 and 1.9.) The power gain was 1.3 and the voltage gain fifteen. The next morning, on December 24, 1947 Brattain and Moore demonstrated to M. Kelly, Bell Laboratories Vice-President, Harvey Fletcher, the Director of Physical Research, and Ralph Bown, the Director of research, device operating as an oscillator.

Bell Laboratories immediately declared the invention as "BTL Confidential" and added more people to "Surface States Project," Among them were John Shive, Jack Scaff, William Pfann, and J. A. Becker. Bell Laboratories filed five patents in February 26, 1948 covering the basic principle of the transistor. Gibney's name appears on two patent applications. Although his contribution was crucial to the discovery of the transistor, his name disappeared from history.

Gibney was born in Wilmington, DE on August 30, 1911. His undergraduate degree was in Metallurgy from the University of Delaware, and his Ph.D. was in Physical Chemistry from Northwestern University. He began working at Bell Labs right out of graduate school in 1936. He worked in the chemistry

Inventors	Case No.
John Bardeen - Walter H. Brattain	3-9

Conception of Any Complete Form of Invention

1. Date. (Approximately if unable to fix exactly.)	1 December 15, 1947
2. Circumstances attending conception	2 Investigation of effects of electric fields on surfaces of semiconductors

Drawings Showing Features of Invention

3. Nature of drawing? (Pencil sketch, marked blueprint, etc.)	3 Pen and ink drawing
4. Date and by whom could the date be proved?	4 December 19, 1947
5. Where filed?	5 Notebook 21780, page 6
6. Subsequent drawings, if any?	6 Notebook 20780, page 71

Disclosure
(Explanation of Invention Made to Another)

7. To whom? Written or oral?	7 W. Shockley, H. R. Moore - oral
8. Date?	8 December 16, 1947
9. If recorded, where?	9 Not recorded

Written Description

10. What was it? (Memorandum, "circuit description", notes on a drawing, etc.)	10 Circuit description
11. Date?	11 December 16, 1947
12. Where is original filed?	12 Notebook 18194, page 192

Reduction to Practice
(Physical Embodiment Constructed and Tested)

13. What was constructed?	13 Voice frequency amplifier
14. How tested?	14 Talking over circuit
15. Did it operate successfully?	15 Yes
16. Who constructed it and date?	16 W. H. Brattain, H. R. Moore, December 23, 1947
17. Who witnessed test and date?	17 R. Bown, H. Fletcher, W. Shockley, G. L. Pearson, December 23, 1947
18. Where recorded?	18 Notebook 21780, pages 7,8
19. Is the original model or circuit preserved? If so, where?	19 Yes. Room 1E455 Murray Hill
20. Subsequent tests, if any?	20 December 29, 1947-1E355 and later.

Fig. 1.8. Bell Telephone Laboratories *History of Invention* as recorded by H. C. Hart [HCH:EM 6-4-48]

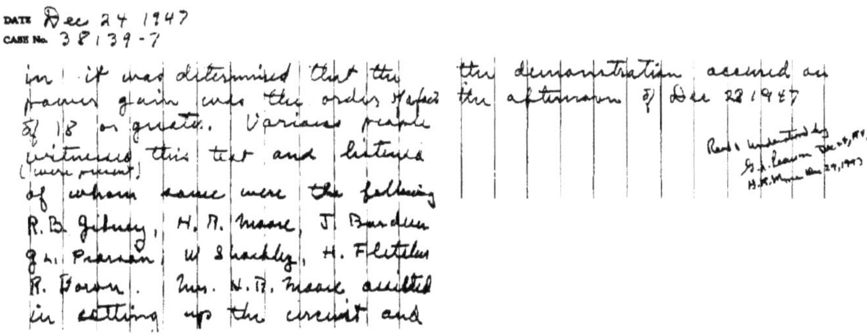

Fig. 1.9. Recorded note in Brattain notebook from December 24, 1947 *(Various people were present and witnessed this test and listened of whom some were the following: R. B. Gibney, M. H. Moore, J. Bardeen, G. L. Pearson, W. Shockley, H. Fletcher, and R. Brown. Mr. H. R. Moore assisted in setting up the circuit and the demonstration occurred on the afternoon of December 23, 1947)*

department on storage batteries until 1945. Then he was transferred into a newly-formed lab for solid state science.

Gibney's wife was ill with asthma and "dissatisfied with the cold weather on the East Coast." At the time of the transistor demonstrations in December 1947 Gibney was at Los Alamos for an interview. Since March 1948 he managed the Physical Metallurgy Group in Los Alamos Laboratories. Not unexpectedly, the bitterness developed between Gibney and his former colleagues Bardeen and Brattain, and since he left Bell Laboratories they never spoke since, mainly because Bardeen's claimed that the biasing the electrolyte was his idea.

Success was achieved – but there was still poor understanding what physical mechanism was behind the transistor action. As almost everybody knows today, the transistor action can be very simplistically described as following (pnp case): the emitter-base junction is forward biased, therefore holes (majority carriers in the emitter) are injected across the junction into the base region. Injected holes create an excess concentration of minority carriers in the base. Due to a large gradient of minority carriers in the base, some of them will diffuse into reverse-biased collector-base junction where they become majority carriers.

The title of the project suggests what Bardeen and Brattain considered at that time – surface effect as a basis for "transistor action." The imperfect level of understanding of the transistor effect is also quite obvious from papers they submitted to Physical Review in December 27, 1948 [6].

Bardeen and Brattain wrote:

> *"the fact that the collector current may actually change more than the emitter current is believed to result from an alteration of the space charge in the barrier layer at the collector by the hole current flowing*

into the junction. The increase in density of space charge and in field strength make it easier for electrons to flow out from the collector, so that there is an increase in electron current. It is better to think of the hole current from the emitter as modifying the current-voltage characteristic of the collector, rather than as simply adding to the current flowing to the collector."

The explanation of "transistor effect" is even more confusing in Bardeen and Brattain patents # 2,524,035 which was filed on February 26, 1948 with a continuation filed on June 17, 1948. The application states *"The some sort of barrier layer that Schottky found for rectifying contacts may exist beneath the free surface of a semiconductor, the space charge of the barrier layer being balanced by a charge of opposite sign on the surface atoms."*

The term "minority carriers" or "injection" is never used in any Bardeen and Brattain publication disclosed before 1949.

The roles of minority carriers were known from the work of Soviet scientist. Boris Davydov, a student of Abram Joffe, who is considered to be founder of Soviet semiconductor physics, developed, in 1938, a very advanced theory of the rectification in copper oxide using the continuity equation in the form as is used today. Davydov extended previous work of Walter Schottky performed in Siemens & Halske Laboratories in Berlin, and one decade before Bardeen correctly recognized the importance of surface states. Davydov wrote:

"Even in the absence of any electrode, these surface charges cause a rearrangement of the electrons in the surface layer of the semi-conductor which results in the appearance of space charges equal in magnitude and opposite in sign to the surface charges. A double layer extending to a definite depth is thus obtained which influences the value of the work function. If an electrode is now brought into contact with the semi-conductor, these space charges will change depending on the contact potential difference between semi-conductor and electrode (Fig. 1.10.)"

In working out the theory of this effect, Davydov recognized the importance of minority carriers which he included into his theoretical model.

Until Shockley completed his theory of junction transistors in early 1948, "transistor effect" was not well understood. Shockley built on the Davydov junction theory and expanded it for transistors.

Recently several individuals suggested that Shockley's contribution was not critical to the transistor discovery and various versions of history were presented indicating that Bardeen was aware of the importance of minority carriers.[3] Others claim that Shockley was not even presented during the

[3] In the recent book "True genius," L. Hoddeson and V. Daitch suggested that it is implausible that Bardeen would not recognize "hole injection." They argue that Bardeen was very likely familiar with what they call an "abstruse mathematical

transistor discovery in BTL. In reality, Shockley was very "hands-on" type of project leader. He discussed on daily basis and in a great length the experimental results and progress of the project. He was not good experimentalist, but he was excellent in the interpretation of observed data.

Although the date December 23, 1947 is considered and accepted as the date of the birth of the transistor, the December 16, 1947 is actually the date when "wedge" point contact transistor worked for the first time. During the period of the time of December 16 to December 23, 1947 many demonstration and tweaking of the device had been performed in the expectation of the show on December 23, 1947.

The not often emphasized fact is that Bardeen's, Brattain's, and Gibney's experiments were conducted with samples prepared by W. G. Pfann, J. H. Scaff, and H. C. Theuerer. It is difficult to understand that there was a time when good quality semiconductor material was not available. All samples were grown using the method of growing single crystals of silicon and germanium developed by G. K. Teal and J. B. Little. The crystal purification was performed by zone refining method developed by W. G. Pfann. Without Pfann's, Scaff's and Theuerer's hard work and inventions the transistor could not be accomplished.

The other key factor of the transistor discovery was research philosophy at Bell Telephone Laboratories. The endeavor to probe deeply into the logical consequences of the fundamental theory, to reduce these consequences to pictorial terms, to find experimental counterparts to the theoretical concepts and patience, are factors of research work which must sooner or later bare fruit. The fundamental scientific knowledge is a self-contained good and, in any case, it will in due course lead to all manner of practical consequences serving the varied interest of men. The invention of the transistor occurred in connection with research programs based on this philosophy and the creativity of individual men.

The newly developed device was first presented to the BTL Research Department Technical Staff on June 22, 1948. It was demonstrated to the press on June 30, 1948 in the West Street Auditorium in New York. An oscillator, pulse amplifier and TV amplifier was demonstrated during the first

---

paper" by Russian Boris Davydov [2,3,4] who in 1938 described rectification effect in copper oxide and points out the importance of the minority carriers. Bardeen indeed was familiar with Davydov's work and referenced Davydov's work on page 718 of his article submitted to Physical Review on February 13, 1947 [6]. Shockley referenced in his work the Joffe summary which includes the Davydov paper.

Brattain during the 1960 Physics Conference in Prague, where he, Bardeen, Shockley and A. Joffe met, acknowledged that he was familiar with Davydov's work, however, with emphasis on the success of the space charge theory of rectification they did not consider Davydov work as a feasible alternative. Brattain explained this mistake in his recollection [5] where he stated *"the minority carriers were, after all, present in too small concentrations in most semiconductors to matter very much."*

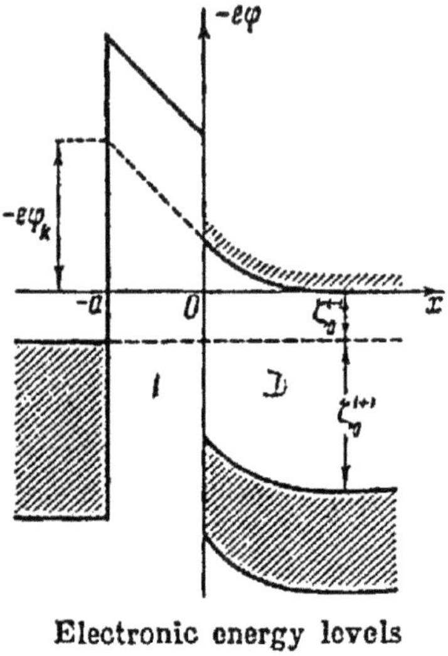

Fig. 1.10. Band diagram including the surface states as introduced by Boris Davydov in 1938 [3]

public announcement of the transistor. Timing was selected to coincide with the appearance of the articles in Physical Review and Bell System Technical Journal.

The announcement received little public notice. The New York Times carried nothing more than a small description five days later:

> *"A device called a transistor, which has several applications in radio where a vacuum tube ordinarily is employed, was demonstrated for the first time yesterday at Bell Telephone Laboratories....In the shape of a small metal cylinder about a half-inch long, the transistor contains no vacuum, grid, plate or glass envelope to keep the air away. Its action is instantaneous, there being no warm-up delay since no heat is developed as in a vacuum tube. The working parts of the device consist solely of two fine wires that run down to a pinhead of solid semi-conductive material soldered to a metal base. The substance on the metal base amplified the current carried to it by one wire and the other wire carries away the amplified current."*

To make a few laboratory point-contact transistors to demonstrate feasibility was not difficult. To make them in reproducible in the manufacturing environment was a new big challenge. In mid 1948 Mervin J. Kelly appointed human dynamo Jack A. Morton head of the effort to develop a manufacturable transistor. Morton was an electrical engineer with degree from the University of Michigan and joined BTL in 1936. Morton was strong outspoken personality. He was never neutral and can never be ignored.

Fig. 1.11. June 30, 1948 press conference announcing the transistor invention. Ralph Bown address reporters in the West Street auditorium

PRESS RELEASE *from* Bell Telephone Laboratories
463 West Street, New York 14 · CHelsea 3-1000

Release: A.M. Papers of Thursday, July 1, 1948.

An amazingly simple device, capable of performing efficiently nearly all the functions of an ordinary vacuum tube, was demonstrated for the first time yesterday at Bell Telephone Laboratories where it was invented.

Known as the Transistor, the device works on an entirely new physical principle discovered by the Laboratories in the course of fundamental research into the electrical properties of solids. Although the device is still in the laboratory stage, Bell scientists and engineers expect it may have far-reaching significance in electronics and electrical communication.

Fig. 1.12. Front page of Bell Telephone Laboratories Press Release from July 1, 1948

Morton set up a group about of thirty scientist which between 1945 and 1950 was supported by a budget $500,000 a year. There was a hope that the transistor would quickly replace the vacuum tube for its great potentials. But trouble arose immediately. No two transistors worked in the same way.

W. G. Pfann modified the structure of the 1N26 shielded microwave point-contact rectifier which he developed with R. Ohl and suggested a "plug-point" contact transistor. The axial nickel pin rectifier was replaced by two parallel pins with an offset C-spring welded to each pin. The Ge was N-type. The emitter contact wire was made of beryllium-copper wire with 180 μm in diameter. A phosphor-bronze wire was used for collector contact; later it was found that phosphorus was the crucial substance. The mechanical and electrical stability of A-Type Transistor depend on mechanical spring pressure. The pressure at the point contact, on the order 0.07 kg/cm^2, is sufficient to produce a fairly reliable contact. A special method of "electrically" formed collector contact was developed. An electrical overload was applied as a pulse to the wire terminal. The forming changes a small volume of N-type base material to P-type. Within the small P-type insert the change in conductivity may be caused by diffusion of Phosphorus during the thermal shock. The general form of structure is P-N–P-N and for this reason some point contact transistors may exhibit a current multiplication factor αs much greater than the theoretical value of unity, and negative resistance characteristics. However, this understanding was learned much later when E. A. Anderson worked on switching circuits in the late 1950's.

It was very quickly learned that it is relatively easy to produce a point-contact transistor which exhibits transistor action, but very difficult to produce one with specific characteristics. The combination of the forming process, which was an art, and the unknown quantities of the theory and structure properties forced the development and manufacture into empirical approaches. This transistor was the prototype of the Type A Transistor, which was produced by Western Electric for almost ten years.

Shockley naturally felt somewhat disappointed by "not being one of the inventors." Shockley was also puzzled by the fact, that the "transistor principle" was not clear. The point-contact transistor exhibits a current gain in excess of unity. Typical value is in the range 2 to 3. Various explanation were

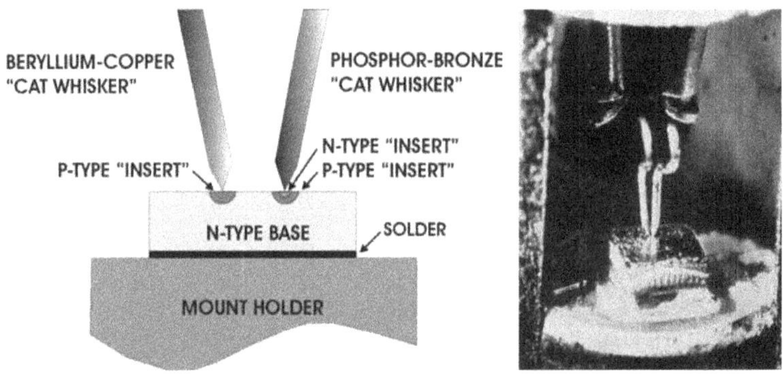

Fig. 1.13. W. G. Pfann's improved Point Contact Transistor (1948)

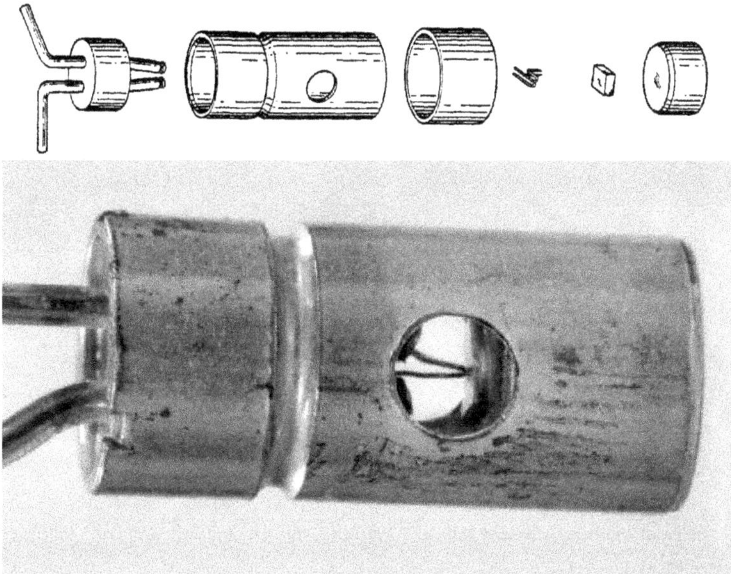

Fig. 1.14. The prototype of the point-contact transistor type M 1698 and parts of its assembly

offered, but as matter of fact, a good solid physical theory of point-contact transistor does not exist and even today we do not know, for example, why point-contact transistors occasionally exhibit negative resistance behavior.

In the process of the forming of the collector contact, a complicated three-dimensional doping pattern is produced. Shockley proposed the so-called "hook collector" with complex n-p-n structure. When working on this analysis, the theory returned him back to his work in the spring of 1947 (Fig. 1.19) and on January 23, 1948, he completed the P-N junction theory which included injection of minority carriers and introduced quasi-Fermi (imref) concept.

Shockley worked out the fundamental studies of carrier injection in germanium in January 1948. This work and better understanding of the function of point contacts, laid the groundwork for understanding of "transistor effect."

On January 28, 1947, one month after the point-contact demonstration, Shockley in his notebook described three key concepts of his new innovation – the junction transistor:

1. exponentially-increasing minority carrier injection across the base-emitter junction
2. reverse bias on the collector-base junction
3. appropriate device geometrical dimension and doping profile

The patent application for the junction transistor was completed on June 26, 1948. Application details includes heavy doping near contacts, hetero-

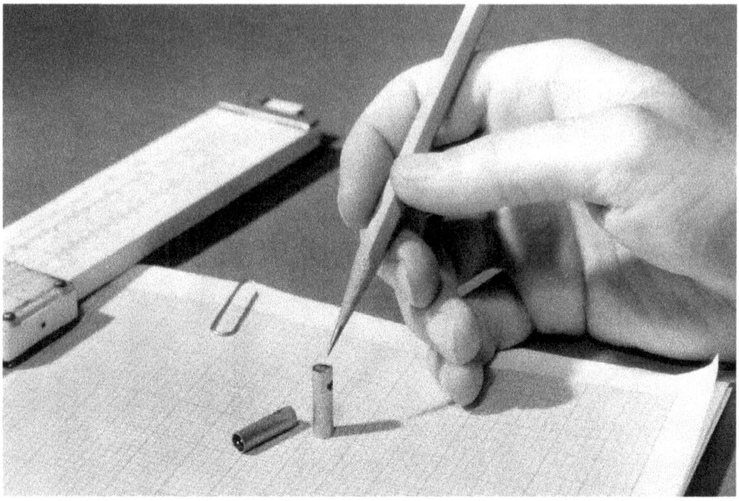

Fig. 1.15. One of the publicity photographs of the point contact transistor, released for news media (1948)

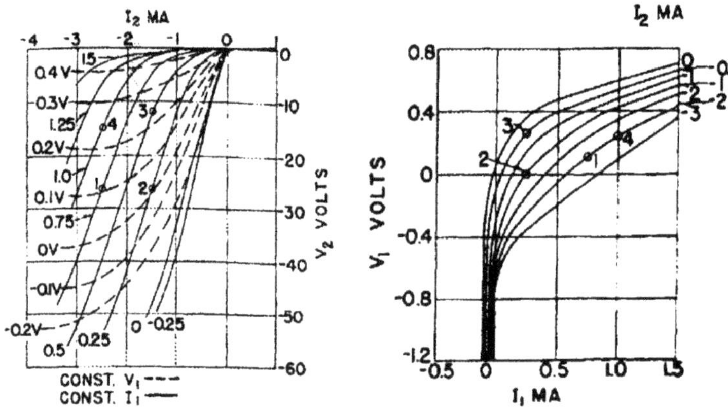

Fig. 1.16. Collector and emitter characteristics of Type A Transistor

junctions with wide energy gap to increase emitter efficiency, and detailed minority carrier analysis.

The first preliminary information about this new device was presented at the Cambridge Meeting of the American Physical Society in June 16–18, 1949 as a joint presentation of W. Shockley, G. L. Pearson, M. Sparks and W. H. Brattain.

Shockley's achievement is important for several reasons: First, the theory of the junction transistor and its operation was developed before the device was made. Bardeen's and Brattain's point-contact transistor was more a discovery than an invention – and has never been understood.

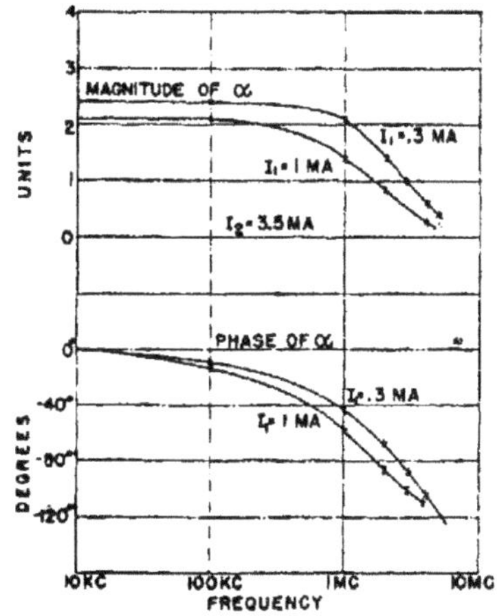

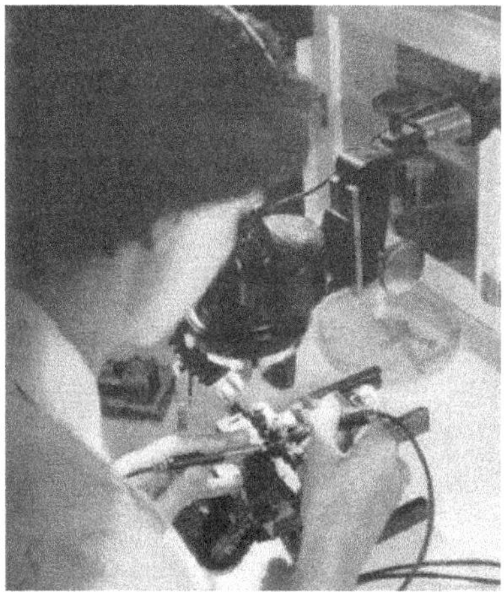

Fig. 1.17. Current gain α versus frequency for the Type A Transistor, and labor-intensive transistor assembly in Allentown's Western Electric facility

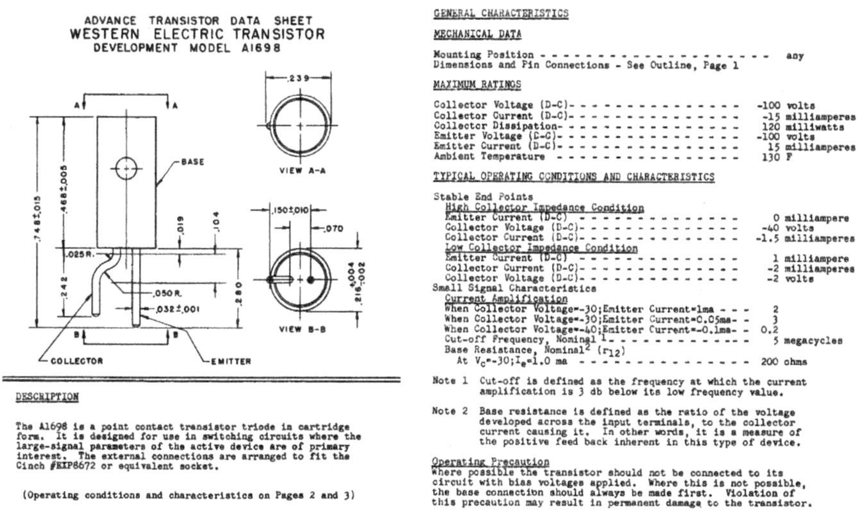

ADVANCE TRANSISTOR DATA SHEET
WESTERN ELECTRIC TRANSISTOR
DEVELOPMENT MODEL A1698

DESCRIPTION

The A1698 is a point contact transistor triode in cartridge form. It is designed for use in switching circuits where the large-signal parameters of the active device are of primary interest. The external connections are arranged to fit the Cinch #EIP8672 or equivalent socket.

(Operating conditions and characteristics on Pages 2 and 3)

GENERAL CHARACTERISTICS

MECHANICAL DATA

Mounting Position - - - - - - - - - - - - - - - - - - - any
Dimensions and Pin Connections - See Outline, Page 1

MAXIMUM RATINGS

Collector Voltage (D-C) - - - - - - - - - - - - - - -100 volts
Collector Current (D-C) - - - - - - - - - - - - - - -15 milliamperes
Collector Dissipation - - - - - - - - - - - - - - 120 milliwatts
Emitter Voltage (D-C) - - - - - - - - - - - - - - -100 volts
Emitter Current (D-C) - - - - - - - - - - - - - - 15 milliamperes
Ambient Temperature - - - - - - - - - - - - - - 130 F

TYPICAL OPERATING CONDITIONS AND CHARACTERISTICS

Stable End Points
High Collector Impedance Condition
Emitter Current (D-C) - - - - - - - - - - - - - - 0 milliampere
Collector Voltage (D-C) - - - - - - - - - - - - - - -40 volts
Collector Current (D-C) - - - - - - - - - - - - - - -1.5 milliamperes
Low Collector Impedance Condition
Emitter Current (D-C) - - - - - - - - - - - - - - 1 milliampere
Collector Current (D-C) - - - - - - - - - - - - - - -2 milliamperes
Collector Voltage (D-C) - - - - - - - - - - - - - - -2 volts
Small Signal Characteristics
Current Amplification
When Collector Voltage=-30;Emitter Current=1ma - - - 2
When Collector Voltage=-30;Emitter Current=0.05ma - - 3
When Collector Voltage=-40;Emitter Current=-0.1ma - - 0.2
Cut-off Frequency, Nominal 1 - - - - - - - - - - - - - 5 megacycles
Base Resistance, Nominal 2 (r_{12})
At V_c=-30;I_e=1.0 ma - - - - - - - - - - - - - - 200 ohms

Note 1 Cut-off is defined as the frequency at which the current amplification is 3 db below its low frequency value.

Note 2 Base resistance is defined as the ratio of the voltage developed across the input terminals, to the collector current causing it. In other words, it is a measure of the positive feed back inherent in this type of device.

Operating Precaution
Where possible the transistor should not be connected to its circuit with bias voltages applied. Where this is not possible, the base connection should always be made first. Violation of this precaution may result in permanent damage to the transistor.

Fig. 1.18. The A1698 Data Sheet issued on September 7, 1951

In spite of the pressure and frustration that accompanied the work in Shockley's group in the nineteen fifties, a majority of scientists characterized the environment as exciting. At the Device Research Conference, held at the University of Colorado campus in 1957, the "Murray Hill people" offered a folk song with the refrain *"And Uncle Bill Shockley and all, and all."*

The vast amount of the work on transistors has been a result of Shockley's influence. Shockley was not only *"The Authority"* he also had an exceptional ability to interest others in what interested him.

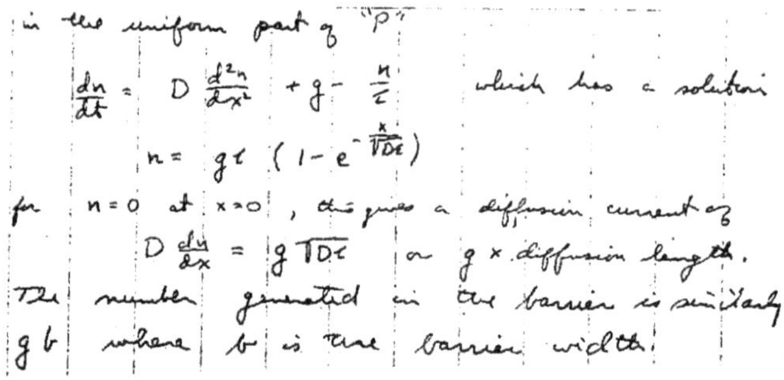

in the uniform part of "p"

$$\frac{dn}{dt} = D\frac{d^2n}{dx^2} + g - \frac{n}{\tau}$$ which has a solution

$$n = g\tau\left(1 - e^{-\frac{x}{\sqrt{D\tau}}}\right)$$

for $n = 0$ at $x = 0$, this gives a diffusion current of

$$D\frac{dn}{dx} = g\sqrt{D\tau}$$ or $g \times$ diffusion length.

The number generated in the barrier is similarly gb where b is the barrier width.

Fig. 1.19. W. Shockley's Laboratory Notebook from April 24, 1947 describing the first record of diffusion theory for minority carriers in a p-type layer of a reverse-biased p-n junction

After practical realization of the junction transistor in April 1950, it became clear that the point-contact transistor had no future. CBS-Hytron was the only company who, beside Western Electric, put into production point-contact transistors (Fig.1.20.)

It became also clear that the original transistor theory, as described in Bell Laboratories' five patents from February 1948, were incomplete. Shockley put, with his work on PN junctions, transistor theory on solid physics foundation. With publishing of Shockley's book *Electrons and Holes in Semiconductor*[4] at the end of 1950, everybody recognized Shockley as a transistor guru.

Bardeen felt pushed away, and put himself into a difficult situation. In a written statement to M. Kelly in May, 1951 he demanded that Bell Laboratories establish Bardeen's group, because he did not want to work for Shockley in the Solid State Physics Department. The language used in the letter was quite unusual, especially the part describing Shockley's attitude. M. Kelly, J. Morton, and Ralph Bown knew that transistor discovery was a result of Shockley's thought, which he wanted to investigate experimentally. In most proposed experimental cases Shockley already did theoretical work[5].

Bardeen's letter was not well received by the Bell Laboratories establishment; his demands were rejected, and by June 1951 Bardeen was no longer

[4] Shockley had great difficulties to get title accepted. Shockley wanted "Holes and Electrons in Semiconductors", which van Nostrand editors deemed vulgar, but the Shockley idea of a "hole" was central – although it was not at all accepted in the community at first, even experts preferred thinking of electrons only.

[5] When asked about Shockley Jack A. Morton points out *"During the course of my own career, I have had opportunity to study under, or work with, quite number of Nobel prize winners in physics. I think I can say without equivocation that of them all Shockley stands out in combining outstanding strengths in teaching, in leadership, and in independent contributions"* [7].

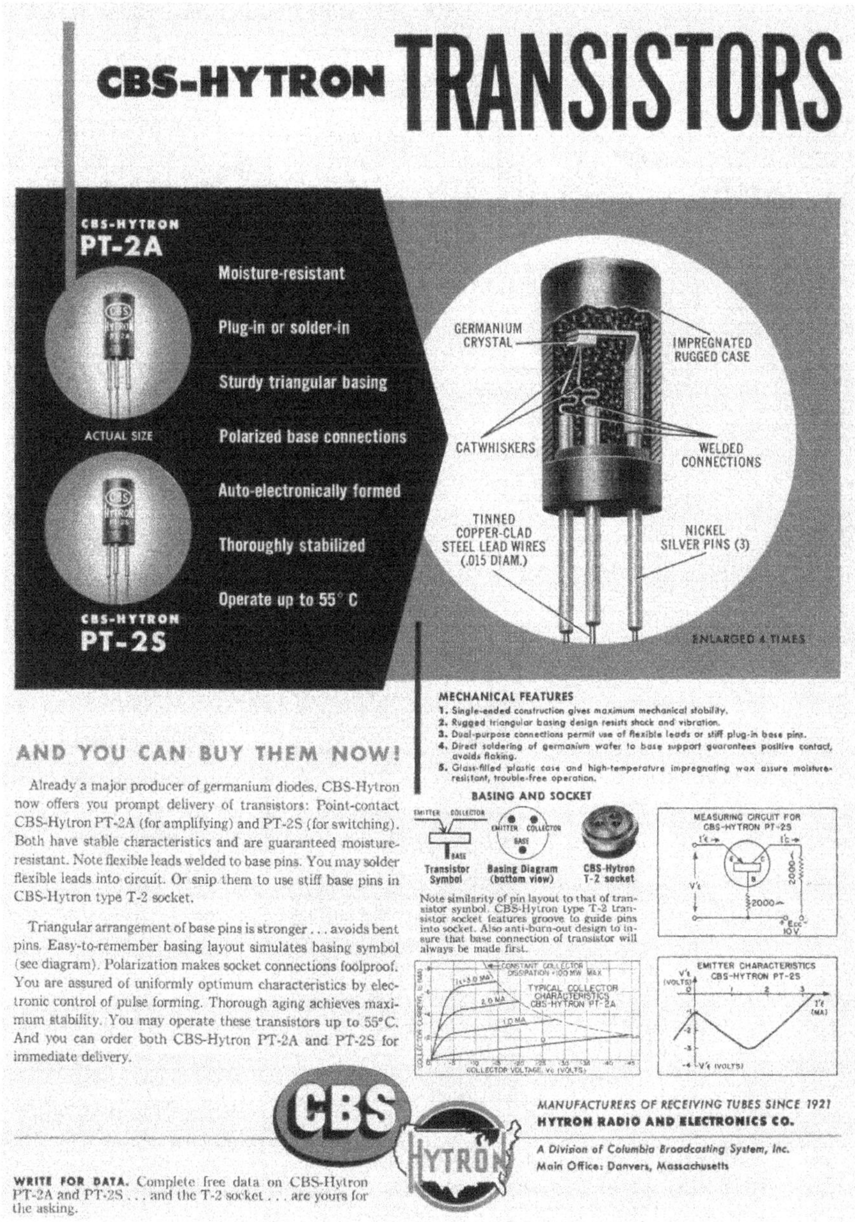

Fig. 1.20. The CBS-Hytron point-contact transistor PT-2A was produced in 1950 to 1951. Note the emitter characteristics which are typical for the point-contact transistors

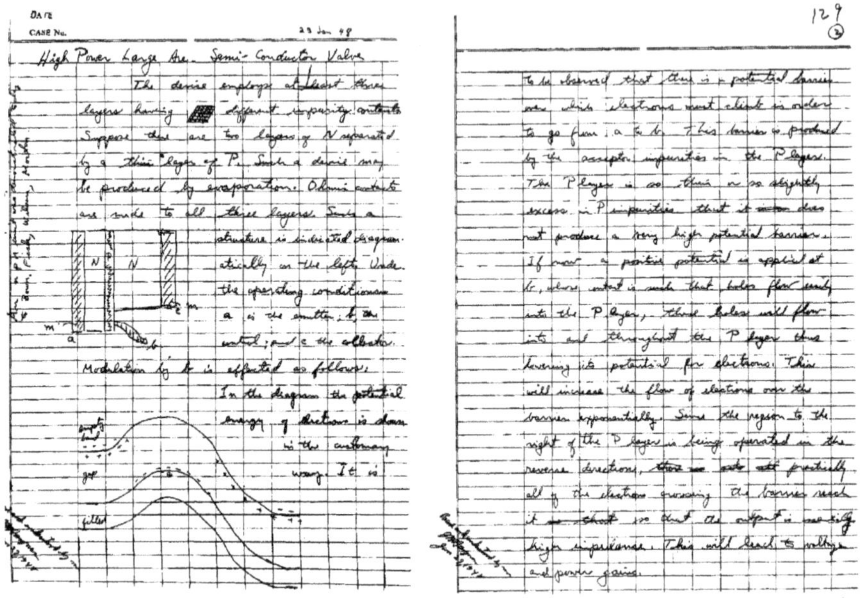

Fig. 1.21. The conception of the junction transistor (Shockley's notebook dated January 23, 1948)

at Bell Laboratories. Both men managed to talk to each other when they met on various meetings, but that was all they could do. Bardeen changed his position slightly after the Nobel Prize nomination. In the Western Union cable to Shockley on November 1, 1956 Bardeen cabled:

> *"Heartiest congratulations and thanks for your contributions which made award possible. Hope to see you in Stockholm."*[6]

However, shortly after Shockley's death in 1989, Bardeen refused to write a note about Shockley to the IEEE Spectrum magazine.

Morgan Sparks, who worked long enough with Bardeen, Brattain, and Shockley to know them, told me

> *"there is no doubt that all these men were brilliant physicists, but they were conventional. Shockley was unconventional."*

Mervin Kelly knew that Shockley was the catalyst; the highly competitive and hard-driving leader Shockley knew how to challenge people and how put things together.

William Shockley never took credit for Bardeen's and Brattain's work and the discovery of the point-contact transistor. Contrary, he very often

[6] Brattain's congratulation to Shockley cabled the same day says *"Trusting you still sober and coherent at this stage of the game congratulations. Brattain, Bell-TelLabs."*

took extra steps and explained his contribution. In Fig. 1.22 is shown draft of the letter which Shockley mailed to Newsweek editor Malcolm Muir after Newsveek published an article about the transistor. The draft says:

> "*You state that I "came upon the principle while investigating the behavior of semi-conductors;" to amplify words instead of electricity may I add that I come on it only after it was found and displayed by Drs. John Bardeen and Walter Brattain to whom credit for the invention is due.*"

The Department of Justice initiated in 1949 an antitrust suit against AT&T which influenced the companys policy of disseminating its new technology. The litigation resulted in a consent decree requiring Western Electric to license all existing Western Electric and Bell Laboratories patents to domestic companies, royalty free, and to provide licenses of future patents at "reasonable rates." AT&T was permitted to maintain its vertically integrated structure but was prohibited from entering into any product markets other than basic phone carrier provider.

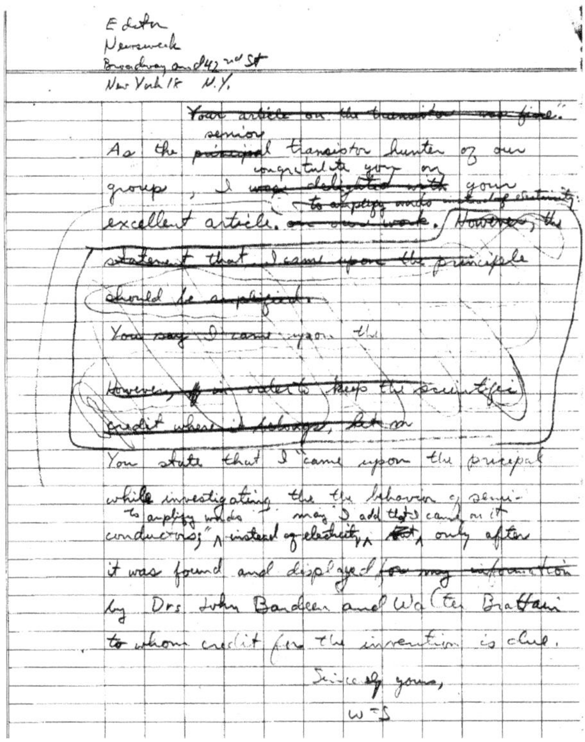
Editor
Newsweek
Broadway and 42nd St
New York 18 N.Y.

As the senior transistor hunter of our group, I congratulate you on your excellent article.

You state that I "came upon the principle while investigating the behavior of semi-conductors;" to amplify words instead of electricity may I add that I came on it only after it was found and displayed by Drs. John Bardeen and Walter Brattain to whom credit for the invention is due.

Sincerely yours,
W S

Fig. 1.22. Shockley's letter to Malcolm Muir, editor of Newsweek [September 9, 1949]

The Bell Laboratories organized in November 1951 the first "Transistor Symposium" strictly for military and government officials. During this one week long symposium, thirty-five papers were presented to some 300 attendees mostly from the Army, Navy, Air Force and government agencies. All participants were either United States citizen or specifically cleared by the military services. In opening the symposium, Dr. Kelly stated that improved point-contact transistors are becoming available in small quantities through Western Electric, and the new junction transistors ware expected to be available in 1952. Topics covered basic transistor physics and mainly transistor characteristics and circuitry.

Later, the second symposium was attended by more than 80 industrial firms. All presentations focused on the properties of the transistor with very little information about manufacturing technology. Actually, the first two symposia dwelt largely on the point-contact transistors.

Third symposium for Western Electric licensees was held in April 1952, with 26 domestic and 14 foreign organization attended intensive sessions lasting eight days. For a price of $25,000, all licensees received the so-called "Twenty-Five Thousand Dollar Book," bound in 792 pages (Fig. 1.23), called "The Transistor – Selected Reference Material on Characteristics & Application."

The licensees were:

Arnold Engineering
Globe Union
Hanovia Chemical and Manufacturing Comp.
T.R. Mallory Company
Microwaves Associates
Minneapolis Honeywell
Raytheon Manufacturing
Radio Receptor Company
Sprague Electric Company
Texas Instruments
Tung-Sol Electric Company
Automatic Electric
Automatic Telephone and Electric Comp.
The Baldwin Company
Bowser
British Thomson-Houston Company
Bulova Watch Company
Crane Company
L.M. Ericsson
Felton and Guilleaume Carlfswerk
General Electric Company
Hughes Tool Company
IBM Corporation
IT&T Corporation
Lenkurt Electric Company
National Cash Register Company
National Fabricated Fabrics
N.V. Philips
Pye
Radio Development and Research Corp.
Siemens and Halske
Telefunken Gesellschaft
Transistor Products
English Electric Company

The symposium was held in the BTL Murray Hill auditorium under H. A. Affel's chairmanship. Point- contact and limited junction transistor samples were available to all qualified users.

Jack Morton presented a tutorial session. Gerald Pearson presented a semiconductor theory. Morgan Sparks discussed transistor theory, and made some previously mysterious phenomena more understandable. R. M. Ryder described characteristics of point-contact and junction transistors, and described the effective characterization of these devices. Then he discussed the

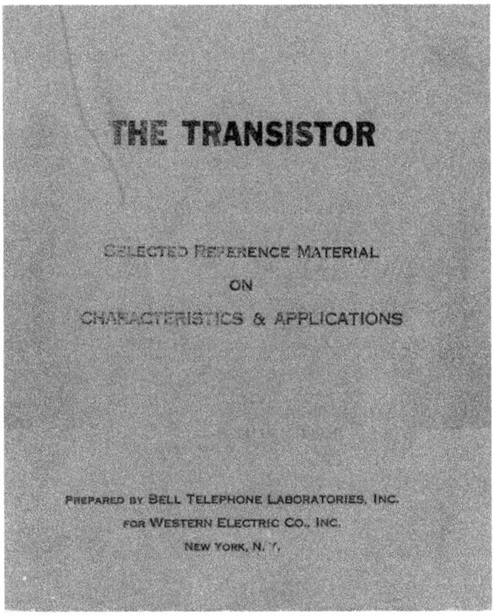

Fig. 1.23. Author's copy of "Twenty-Five Thousand Dollar Book"

point-contact transistor amplifier behavior. He showed that apparently unstable devices could be properly harnessed by proper circuit design. John N. Shive discussed phototransistor and described the generation of base region minority carriers by light. Gordon Raisbeck discussed the duality concept, based on the fact that point-contact transistor circuit theory behaves a great deal like a vacuum tube theory if you just interchange the roles of voltage and current in the circuit matrix equations. R. L. Wallace presented the temperature sensitivities of junction transistors. R. S. Caruthers discussed system application of transistors. Gordon Raisbeck covered practical oscillator circuit principles and modulator circuits. J. H. Felker fascinated attendees with the idea of application of transistors in high-speed digital computers. A. E. Anderson and J. R. Dacey discussed the applications of point-contact transistors in pulse and switching circuits.

In the closing presentation Jack Morton noted that there were three primary early limitations in transistors. These were poor reproducibility, reliability and poor understanding of circuitry with transistors. Circuit design limitations were due to not-yet understood phenomena in gain, noise, frequency and power. Morton, with some bravado, stated that these limitations had been overcome, at least to the level needed for practical systems applications at moderate temperatures. It was apparent that small size, low power drain, small heat generation, innately long life and high shock/vibration capabilities spelled the doom of vacuum tubes in many system applications. Transistors had their own unique capabilities and one must start with the systems requirements and design around transistors, rather than trying to substitute transistors one-for-one for vacuum tubes. In summary talk, Jack Morton supercharged attendees to take a BTL transistor technology to the outside world.

The first very aggressive firm which appreciated the potential of semiconductors for advanced electronic system was Hughes Aircraft Company. Si Ramo, Dean Wooldridge and Burt Miller of Hughes hired Harper Q. North, who was previously with MIT Radiation Laboratory and General Electric to start Hughes Advanced Electronic Development Laboratory. In April 1949 North hired a young engineer Sanford H. Barnes to duplicate Bell Laboratories' point-contact transistor. Barnes worked for six months; he had no germanium crystals so he was using large crystalline grain from polycrystalline germanium which he polished. This was a very difficult process and Barnes was able to demonstrate only some feasibility of a new device; however no device really worked as a transistor. The project was scrubbed by the end of 1949.

The first point-contact transistors built successfully outside Bell Laboratories (and before Bell released details about technology) were designed by Helmar Frank and Jan Tauc in Prague. They had only limited information published by Bell scientists in Physical Review, but they had germanium crystals of very good quality which the Germans used for microwave diodes in their radar research. Professor Frank actually developed a more advanced method of "contact sharpening" than the method developed by Pfann. Frank and Tauc transistors did not need any contact adjusting and there was no window to access point contacts. (Fig. 1.24). Some of these devices survived until now and they are still working[7].

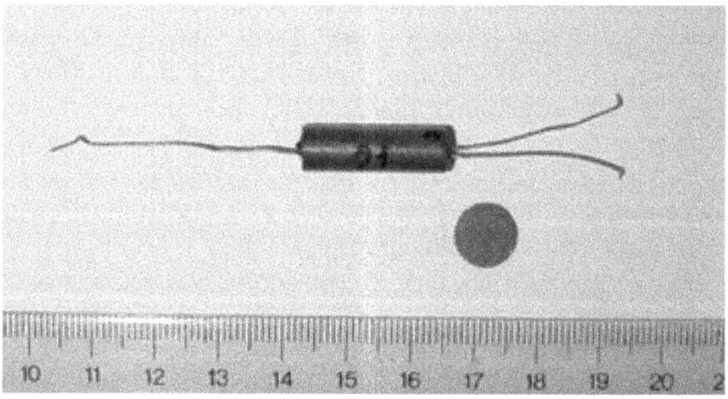

Fig. 1.24. The first European point contact transistor designed by H. Frank and J. Tauc in 1949. "dot" is unpolished Germanium sample

[7] The author "inherited" germanium samples from Prof. Frank, and learned how to build point-contact transistor from scratch. The whole procedure takes about sixty minutes. No really special tools are needed.

In May 1952 the British scientist G. W. A. Dummer made the following prediction at the IRE (Institute of Radio Engineers) annual electronic components meeting in Washington, D.C.:

> *"With the advent of the transistor and the work in semiconductors generally, it seems now to be possible to envisage electronic equipment in a solid block with no connecting wires. The block may consist of layers of insulating, conducting, rectifying and amplifying materials, the electrical functions being connected by cutting out areas of the various layers".*

Except for a few individuals, nobody was really concerned about the potential of microelectronics. The few concerned, hard headed and hard working scientists and engineers transformed the vision into reality – into reality that exceeded their vision.

Shockley commented on Dummer's presentation and said: *"We will build these transistors for a nickel a piece."* He was overly pessimistic.

Shockley decided to establish his own company and left Bell Telephone Laboratories in 1955 (Fig. 1.25.) Mervin J. Kelly retired in 1959 (Fig. 1.26); Jack Morton who became the vice-president of technology at BTL was murdered in a senseless and bizarre incident in 1971. The feminist movement tried to show that men were useless[8]. The new business establishment found easier ways to make a "profit." Rather then organize and finance good research, profit could be "massaged" through mega mergers of companies. Uncontrolled spending and raising national debt forced the government to make cuts. As always in history, the easiest way to cut government spending that would not concern the general public was first to cut first government-support of art and libraries programs, and second, to cut government-supported science and research programs. Down spiral movement of fundamental science research in U.S. began.

By 1990, AT&T's revenues were \$43.7 billion and its R&D budget was \$2.4 billion with more than 24,000 people at Bell Laboratories. AT&T formed in 1996 the spin-off of Lucent Technologies. AT&T was unable to generate

---

[8] Through history, for some reason, the terms Brilliance and Genius is associated with the image of a man, rarely with a woman. Politically incorrect individuals argue that men are more aggressive than women, owing to testosterone. Author is intrigued by the observation that women are much more common in the police and military occupations that involve violence, than in physics or mathematics, which are safe, clean, and creative occupations. Similarly, many attorneys who successfully litigate cases are female, more proof that women can succeed in a profession that requires aggression and stamina. So I am baffled by the absence of women from physics and mathematics, particularly when one considers the number of women in police, military, and litigation.

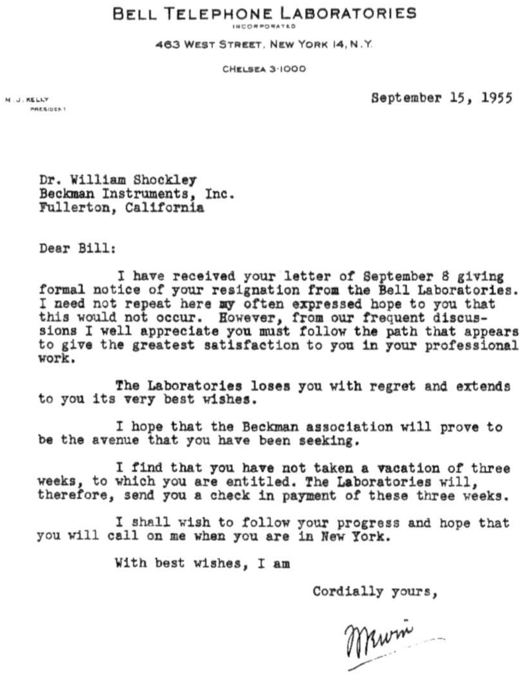

BELL TELEPHONE LABORATORIES
INCORPORATED
463 WEST STREET, NEW YORK 14, N.Y.
CHELSEA 3-1000

M. J. KELLY
PRESIDENT

September 15, 1955

Dr. William Shockley
Beckman Instruments, Inc.
Fullerton, California

Dear Bill:

I have received your letter of September 8 giving formal notice of your resignation from the Bell Laboratories. I need not repeat here my often expressed hope to you that this would not occur. However, from our frequent discussions I well appreciate you must follow the path that appears to give the greatest satisfaction to you in your professional work.

The Laboratories loses you with regret and extends to you its very best wishes.

I hope that the Beckman association will prove to be the avenue that you have been seeking.

I find that you have not taken a vacation of three weeks, to which you are entitled. The Laboratories will, therefore, send you a check in payment of these three weeks.

I shall wish to follow your progress and hope that you will call on me when you are in New York.

With best wishes, I am

Cordially yours,

Mervin

Fig. 1.25. Letter to W. Shockley from Mervin Kelly dated September 15, 1955

marketable disruptive technologies. For decades the business of the manufacturing divisions of AT&T was phone equipment. But a history of supplying the country's phone monopoly made Lucent a bureaucratic organization with many arrogant personalities not able to recognize any good engineering except their own. Lucent engineers tended to work more on their personal careers rather than as a "champion of technology" overcoming the difficulties of the transfer of scientific ideas into reality.

The market migration to a universal network instead of traditional circuit-based switching increased the competition to provide the equipment, software and service that would serve as the infrastructure of universal networks. In 1998 William O'Hea, President of Lucent's Communication Group admits:

> *"Quite frankly, we missed the last generation of data networking while a part of AT&T. We had all the technology to compete, but it was buried under a huge organization"* [8]

The change of Lucent Technology management in 1998 from strictly internal R&D strategy to acquisition started continued spiral downturn. In 1998 Lucent acquired 11 companies and 17 in 1999. Lucent Technologies made a decision to buy technology that had not yet been commercialized, rather than develop their own technology internally. The major problem with this

NEW YORK, March 1. — Dr. Mervin J. Kelly, chairman of the board of Bell Telephone Laboratories, retires today after 41 years with the Bell Telephone System.

He will serve as a research consultant at Bausch & Lomb Optical Co., Rochester, N.Y., also effective today. The appointment was disclosed in the firm's annual report.

He has also been engaged by International Business Machines Corp., New York, as consultant on research and engineering matters starting today.

KELLY

Fig. 1.26. Electronic News, March 2, 1959 announcing Dr. Mervin J. Kelly's retirement

approach is that the acquisition strategy required a specific managerial leadership and careful organizational coordination. With each acquisition a mix of people was transferred from the mother company to control and govern the transition phase of acquisition. As in the case of vacuum tube companies that wanted to go into the semiconductor business, the people governing the transition phase knew very little about the specific issues of technology that the mother company was acquiring. The vacuum tube technology was very different from that of the semiconductor, and just the ability to run a vacuum tube business does not guarantee that people will be able to run a semiconductor business.

In the case of Lucent, rather than acquiring more expensive mature technology, they bought cheaper, less mature technology. The price was lower but the investment is much more risky and many of the acquisitions just become money losing enterprises. The additional problem was that many of the acquired companies were set up and funded by venture capital, focusing more on the profitability of investment than on the potential transferability of technology to the large company.

After a series of AT&T's many wrong managerial decisions, Bell Laboratories, the jewel of American innovation, virtually disappeared by 2000.

References

[1] G. L. Pearson, W. H. Brattain, History of Semiconductor Research, Proc. IRE, Vol. 43 (1955), p. 1794

[2] B. Davydov, On the photo-electromotive force in semiconductors, Technical Physics of the USSR, Vol. 5 (1938) , p. 79
[3] B. Davydov, The rectifying action in semiconductors, Technical Physics of the USSR, Vol. 5 (1938), p. 87
[4] B. Davydov, On the contact resistance of semiconductors, Journal of Physics of the USSR, Vol. 2 (1939), p. 167
[5] J. Bardeen, Surface States and rectification at a Metal Semi-Conductor Contact, Phys. Rev. Vol. 71 (1947), p. 717
[6] J. Bardeen, W. H. Brattain, Physical Principles Involved in Transistor Action, Phys. Rev. Vol. 75 (1949), p. 1208
[7] Interview with Shirley Thomas, 1960
[8] Lucent Technologies, Red Herring, August 1998, p. 41

2

Grown Junction and Diffused Transistors

When we look at how long it took us to get certain ideas, we are impressed with how dumb we were – on how long it took us, and how stupid we were. But we have learned to live with this stupidity, and to find from it what relationships we should have seen in the first place. This recognition that we are not perfect but that persistence pays is a very important factor, in giving one the will to think – you do not need to be worry so much about the mistakes you make.

William B. Shockley, July 1976

Shockley's P-N junction theory, which he completed in January 1948, suggested that the injection of minority carriers in forward biased, and their collection at a reverse-biased junction, was probably the dominant physical mechanism of the "transistor effect." The challenging problem was that there was not an appropriate device which might prove or disprove his theory.

William Shockley knew from his work at the National Defense Research Committee during the war, another physical chemist, Morgan Sparks. Sparks joined Bell Laboratories in 1943 after his graduation from the University of Illinois, working on electrolytic capacitors and batteries. Morgan Sparks was getting along well with Shockley, and he replaced Gibney. Morgan Sparks' initial work with Shockley's group was related to the investigation of P-N junctions cut from cast ingots, where a junction was occasionally formed during solidification.

Shockley visualized a new structure in which the processes would not be restricted to the surface but would occur inside a single crystal. If his ideas were correct, the performance of the unit would be highly calculable and could be varied by controlling the composition of the crystal.

In December 1949 a single-growth technique, first used by J. Czochralski in 1917, was adapted and improved by J. B. Little and Gordon K. Teal.

They met by pure chance on a bus when returning from work at Murray Hill, and while talking about project goals, they agreed to use a new mechanical design for growth of a single crystal of germanium by slowly pulling a seed of crystal from a melt of high-purity germanium. Czochralski's technique, as improved by Gordon Teal and John Little, was suitable for the type of experiments Sparks was doing. Teal suggested to Sparks that they should explore the new material he was able to prepare[1]. Ernie Buehler was assigned to a new project and helped Teal and Sparks to modify the original system for growth of a single crystal by adding capabilities to control donor or acceptor impurity concentration in the melt during crystal growth [1]. In this system the germanium was melted inside a graphite crucible surrounded by a heating coil. The seed crystal was partially immersed into a melt (approximately 1 mm), and when surface tension between the seed and melt was established, the seed was gradually pulled up with a speed of approximately 20 mm/min, inside a bell jar under a constant flow of hydrogen. The diameter of the rod of germanium was determined by the flow rate of hydrogen and the temperature of the melt. A higher temperature of the melt results in a smaller diameter of the pulled rod. If the N-type seed is immersed into a P-type melt, a P-N junction may be formed. After crystal pulling, the rod containing the P-N junction is machined as desired. After sandblasting, the processed structure is etched and polished to remove surface defects.

There was no defined diameter of the pulled crystals. A typical diameter ranged from 3 to 10 mm, the crystal diameter varied along the length of the crystal rod due to poor control of the temperature profile. Very often the crystals varied also in the resistivity and other physical parameters. The length of crystal rods was typically in the range of 1 to 2 inches.

The research effort resulted in a single crystal P-N junction. This junction was not only a revolutionary device in 1949, but also an important scientific contribution, which confirmed Shockley's elegant theory describing electronic behavior of the P-N junction he developed, and which he published in June 1949 [3]. The typical I-V characteristics of P-N grown junction with cross section 0.6×0.6 mm are shown in Fig. 2.2 and they correlated with theory very well.

The same system was used to produce the NPN transistor. A crystal of germanium or silicon was grown from molten material by immersion of the single-crystal material into the melt and gradually withdrawing it. Impurities necessary to form the base and emitter region was introduced into the molten material as the crystal was being withdrawn. Typical resistivity were ~ 8 Ω-cm in the emitter region, ~ 2 Ω-cm in the base region, and ~ 0.001 Ω-cm in the collector region.

[1] Hans J. Queisser remembers from the discussion with Shockley that Shockley admitted in several discussions that he did not really see the significance of single crystals substrate materials. Shockley thought this would be by far too expensive.

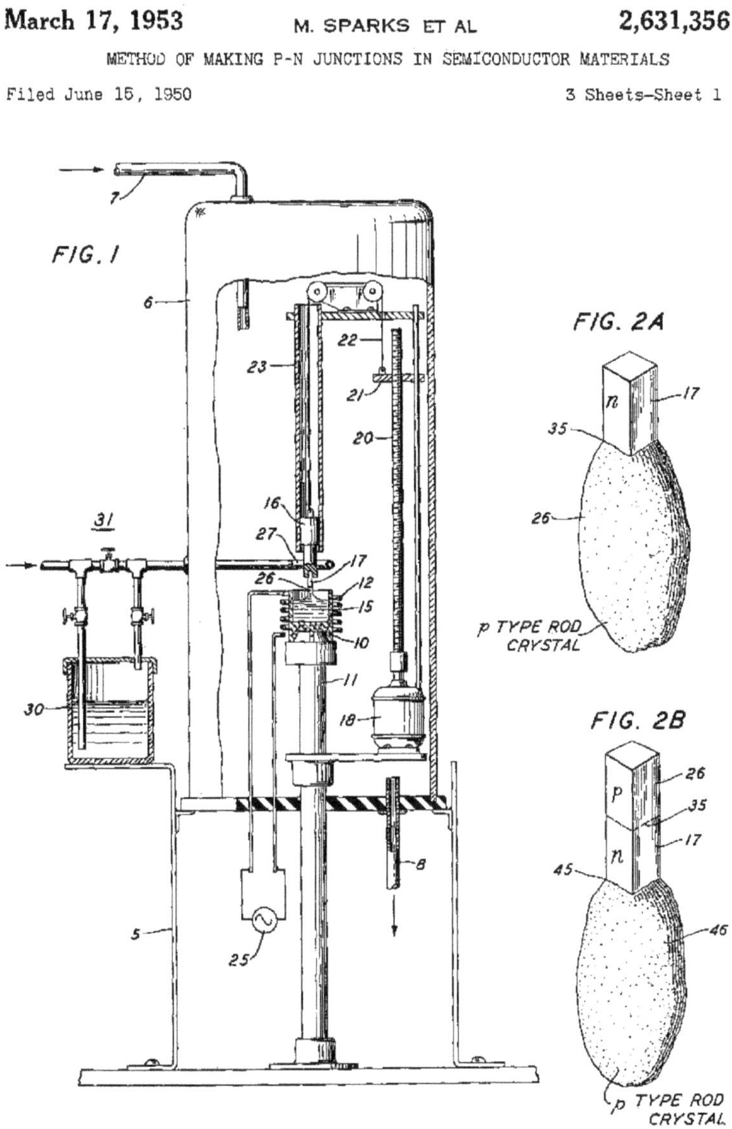

Fig. 2.1. Sparks' and Teal's system used for making P-N junctions

The construction of an NPN grown-junction transistors began with the preparation of a single crystal of germanium which has two P-N junctions grown into it. If in the growing process, a seed crystal was slowly withdrawn from a bath of molten germanium in such a way that more germanium solidifies onto the seed, it then grows into a rod-sized crystal. The molten germanium is N-type at the beginning; at the moment of withdrawing a small pellet of

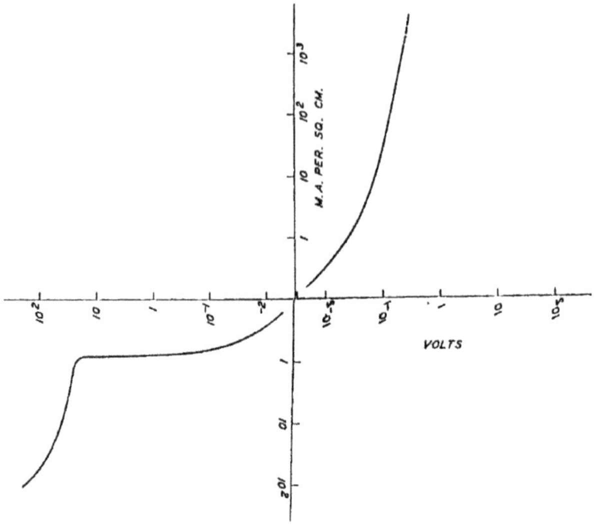

Fig. 2.2. I-V characteristic of a P-N grown junction prepared by Morgan Sparks at the end of 1949

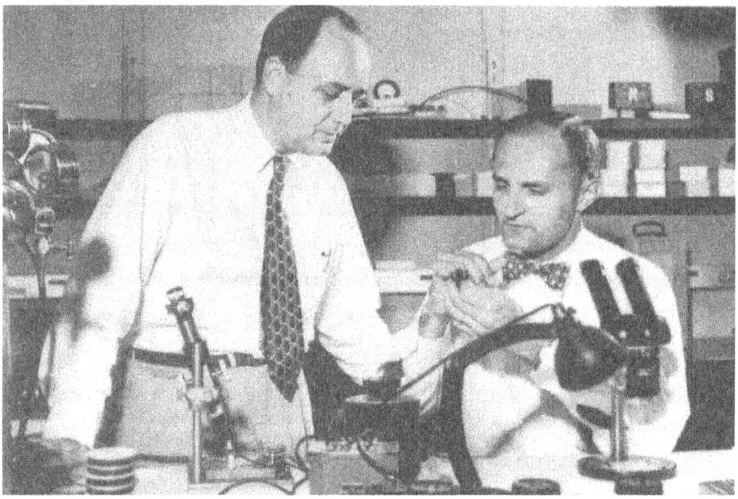

Fig. 2.3. Gordon Teal, who developed the crystal growing technique, and Morgan Sparks, who fabricated the first good junction transistor [1950]

an acceptor impurity is added, to be followed by a somewhat larger pellet of donor impurity. These impurities dissolve in the melt and control the type of the germanium subsequently grown. The resulting crystal then has a thin layer of P-type material between sections of N-type material. This structure is sketched in the left part of Fig. 2.4. Initially, the P-type layer had a thickness over 100 microns; latter experiments resulted in reduction of the thickness down to 25 microns. In the next step the crystal is sliced, leaving the P layer in the center of the slice, which is then cut and sandblasted into rectangular bars. The germanium bar is etched to reduce undesirable electrical effects caused by irregular surfaces.

In January 1950, Morgan Sparks succeeded in growing crystal rods with a thin-base layer. On April 12, 1950, Morgan Sparks entered into his laboratory notebook, data for a chemically-etched sample made by this grown-junction technique which was working as a large-area transistor (Fig. 2.4.) The etch recessed the emitter and collector regions formed a base "bump" which could be contacted. The very first working transistor had a thick base that operated only up to 20 kHz, but this was not important. The proof of the concept was the most important achievement. The transistor era was launched. The grown-junction transistor was an important milestone in transistor science which explained the basic physics of transistor effect and established a solid foundation for future work.

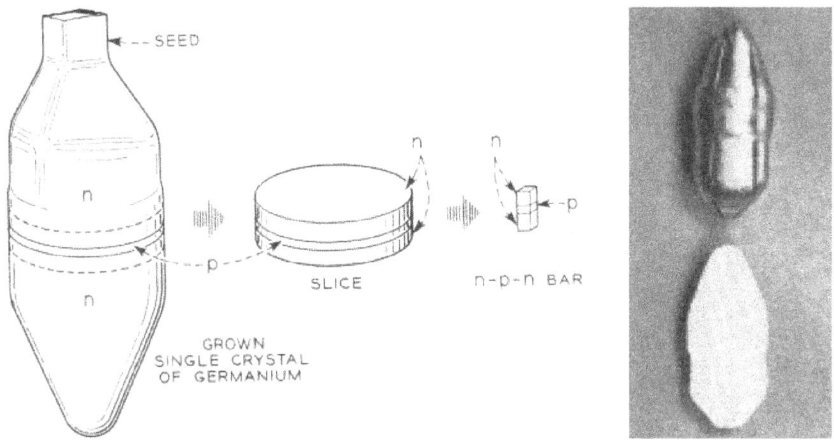

Fig. 2.4. Processing steps in the fabrication of grown junction transistors. The germanium rod in the photographs is about 50 mm long

The big problem, which involved art, was to weld the gold wire with a diameter of 50 µm to the P-type base region which was thinner ($\sim$ 25 µm) and which could not be seen. The patience of Morgan Sparks and Robert M. Mikulyak was tested to the limit. They spent countless hours to verify several different approaches before they found one which worked.

The end of the gold wire was flattened, and the wire was mounted in a special micromanipulator (shown in Fig. 2.6) which was designed by W. L. Bond to allow accurate positioning of the contact. The actual weld was made by passing a current of about two amperes for a few milliseconds through the gold wire. The current heated the contact region sufficiently to form a little "puddle" of gold-germanium alloy. To avoid a short- circuit be-

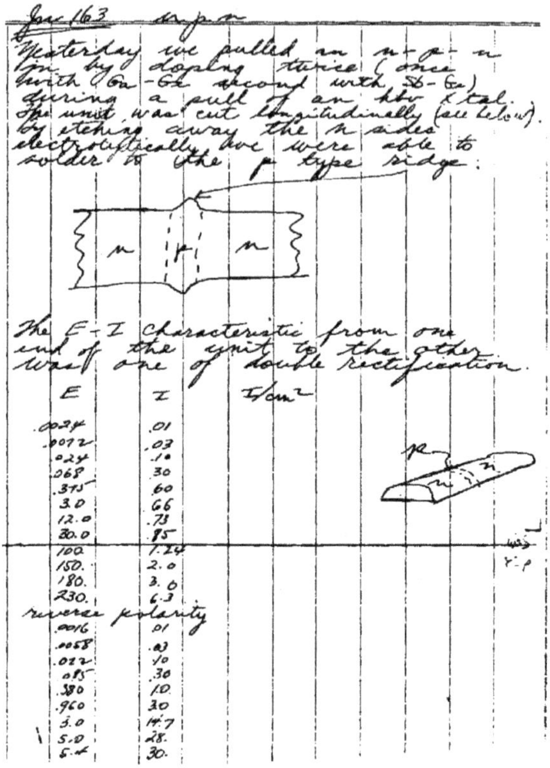

Fig. 2.5. Page of M. Sparks' laboratory notebook dated April 12, 1950

tween the collector region and emitter regions at the weld, the gold wire was "doped" with a small percentage of a P-type impurity (gallium). This insures that the germanium-gold alloy region will be strongly P-type and will act as part of the base region, even though it extended into what was N-type germanium before the bond was made.

The base location was determined from the potential gradient measurement across the NPN structure (Fig.2.7.) In the construction of such P-N structures, it was important that the junctions were planar and parallel.

The performance of the grown-junction transistor was first discussed at the June 1950 IRE conference on Electron Devices at the University of Michigan. Typical electrical parameters of Sparks' transistor, which was later manufactured in high volume as Western Electric M-1752 (Fig.2.9) and A-1858[2] type, at $V_c = 5$ V and $I_e = 1$ mA are: $\alpha = 0.97$, collector leakage current ~ 7 µA, base resistance $\sim 1\text{k}\Omega$, and $f_T \sim 1.5$ MHz. The cross section of the contacted base, and assembled transistor are shown in Fig. 2.8a and 2.8b.

[2] Transistors developed in BTL Murray Hill laboratory had prefix M. Devices developed at Allentown facility had prefix A.

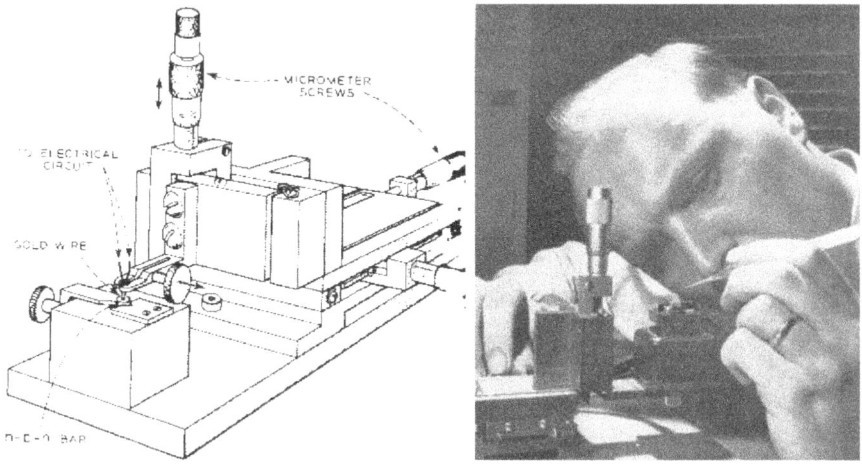

Fig. 2.6. Positioning micromanipulator for bonding of wires to the base region of grown-junction transistors [W. L. Bond 1949] (Left) and Robert M. Mikulyak working on contacting of the transistor base (Right)

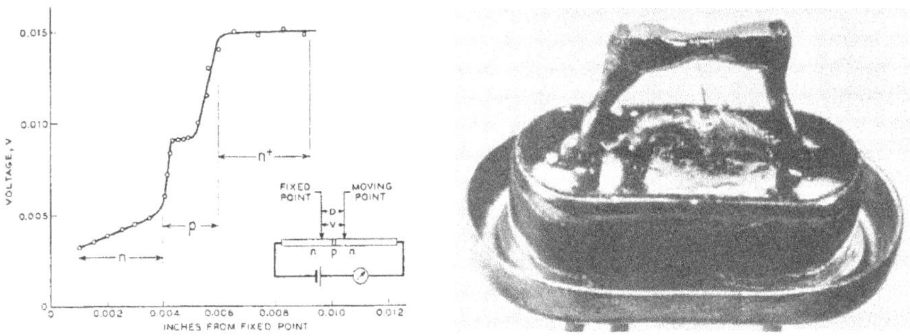

Fig. 2.7. The oscilloscope was used to find the location of the base for bonding of the contacting wire

Early exploratory work was done at BTL with the M-1752 transistor. The 1752 grown-junction transistor was encapsulated in plastic. The second transistor type, the M-1852, was enclosed in a hermetically sealed can and the semiconductor body was covered by red-lead polyethylene-polyisubutylene to reduce some of the detrimental surface effects.

The transistor type 1852 was used in so called "rural carrier system" Americus manufactured by Western Electric Company in 1954–55. In one respect, however, performance of the 1852 fell below expectations – during the initial aging period (about six month), the current gain (alpha) decreased significantly.

There is no existing sample of the original junction transistor as designed by Sparks. For a very long time, an exhibit in the entrance hall of Bell Tele-

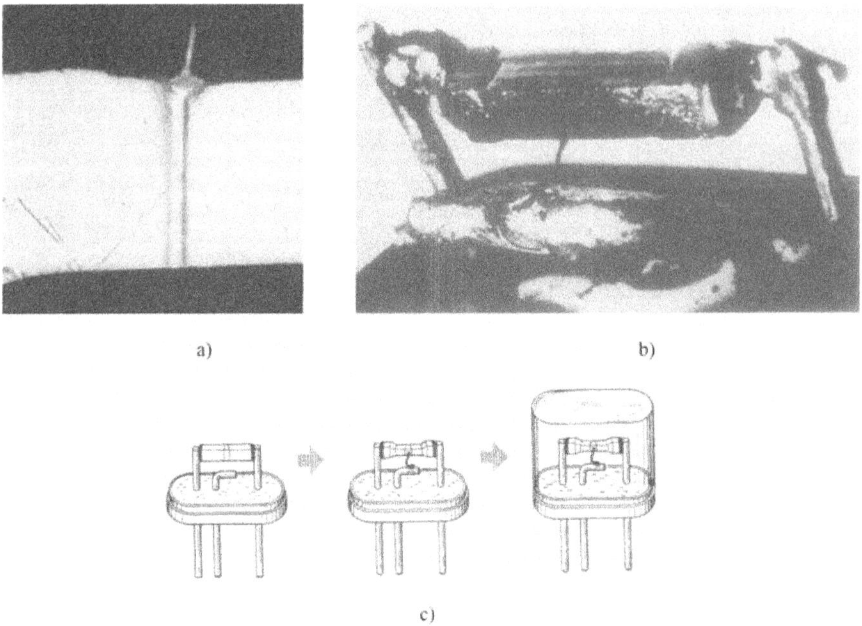

Fig. 2.8. Contact to the base with thickness ~ 25 µm (dark line in a). The wire diameter at the top of the base is 50 µm. In b) is shown the assembled transistor with contacted base, emitter and collector. The cross-section of the bulk of germanium is approximately 0.6 × 0.6 mm. Figure c) shows a typical package assembly

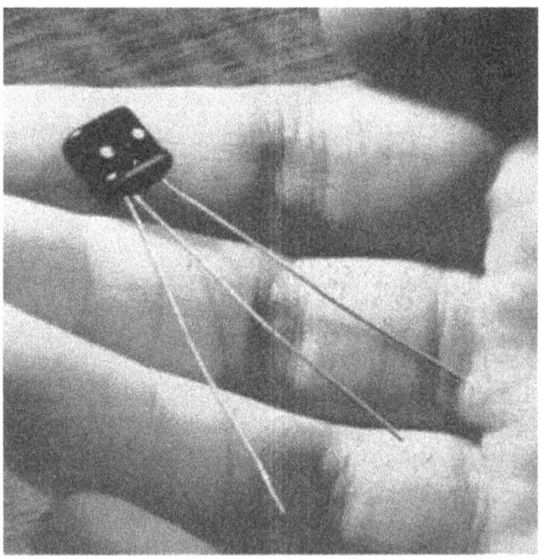

Fig. 2.9. The Western Electric M-1852 grown-junction transistor. This transistor was in production during the years 1952 to 1956

phone Laboratories in Murray Hill displayed a junction transistor as shown in Fig.2.10 with the legend "*First Junction Grown Transistor*". Several publications used this photograph to illustrate the appearance of the first transistor. In reality this sample was just a replica which did not even use the same materials as the original sample. The "copper block" is a piece of plastic coated with paint and in a pile of wax there was no semiconductor material; the thin base wire had a diameter about 120 μm instead 50 μm.

The M-1752 was the very first design of a junction transistor ever made. All units had serial numbers in order to track individual performance characteristics of each batch produced. The earliest units had tags, while later units had the serial number hand painted on one side of the case.

Western Electric was not printing the transistor type number on the case. The transistor type was coded with three color dots. The M-1752 color code was for example violet-green-red. The transistor shown in Fig. 2.9. was quite large, approximately $12\times10\times5$ mm and the color-code was very bright. In the early fifties it was quite fashionable for ladies in the Murray Hill neighborhood to use transistors as colorful jewelry.

The performance of the grown-junction transistors was discussed at the June 1950 IRE Conference on Electron Devices at the University of Michigan.

The grown-junction transistors were not the only major accomplishment of 1950. William B. Shockley completed and submitted for publishing his Magnum Opus: The Book "*Electrons and Holes in Semiconductors.*" The book was published in November 1950. Shockley worked on the book for about eighteen months. This elaborate and exquisite book is still one of the best books about semiconductors and transistors after more than fifty years. Many books published later imitated Shockley's text frequently, but with one difference: they made the explanation hazier or even wrong.

One must appreciate Shockley's work discipline. A large part of this book was based on original, not previously published material, and, yes, he bothered to use units in demonstrated calculations. The "Energy Bands in Crystals" and "Application to Transistor Electronics" chapters were published previously in the Bell Systems Technical Journal. Everyone who wanted to work with semiconductors had to be familiar with Shockley's book. Not surprisingly Shockley dedicated his book to M.B.S., which stood for May B. Shockley, Stanford University educated lady and Shockley's mother, who not only understood every detail in a car's carburetor system, but became Bill Shockley's trusted advisor.

The book was typed by young Miss Bette MacEvoy who joined Bell Laboratory's Transcript Department shortly after the war and was assigned to work mostly with William Shockley. Bette worked closely with Shockley and says that he was a wonderful person to work with – though very meticulous.

> "*Working for him, I had to have a separate, special set of typewriter keys that were actually nothing but the Roman numerals, the equa-*

Fig. 2.10. Replica of the Grown Junction Transistor

> *tions, the little square root signs, and everything – the craziest looking symbols. And I had to learn, out of a big old-fashioned Underwood typewriter, how to change those keys. If I came across something in the text, I would have to remove a certain key – they all had a number on them – and replace it. And then I would type the equation.*"

While Shockley wanted Bette to focus on his work, Morgan Sparks had a very different plan for her. He asked Bette if she would join him to see a production of Cyrano De Bergerac in New York City. "*Of course not!*" was Bette's reply, as most girls might have replied at that time. Morgan Sparks overcame bigger obstacles in his life than the date debacle. He well knew that success would not come free. Success is a synonym for hard work and attitude. If you always shoot for the moon, even if you miss you will end up among stars. Bette and Morgan married two years after the failed date.

When I was very young my father told me "*Under certain circumstance the water will flow uphill. But this will be a very rare situation.*" To find a modest and humble person through the history of the semiconductor engineering is such an uphill effort. Morgan Sparks was one of the very few and rare exceptions to this rule.

Morgan Sparks strove to be better than what was expected, to do what is commonly regarded as 'right', regardless of personal expense. His nobility was striving towards the 'good,' but not the personal one. His integrity is a matter of personal honor and consistency in applying personal values to every action. Those who are lucky enough to be close to Morgan know that his belief in the ideal 'rightness' is what gives us strength even with a complete absence of evidence.

Morgan Sparks is firmly in charge of his ego. He knew that humility is not flashy; it gains less respect as a weapon of virtue than does courage, loyalty, largesse or fidelity. Sincerity is the key to humility. To seek sincerity requires the onerous duty of peering inside yourself to see both the light and the dark, the good and the bad, the excellent and the poor. To accept these things as truths is a daunting, yet ennobling task. The noble man Morgan Sparks looks first within himself, then to those around him for clues as to how the battle progresses. This quality of sincere humility enables him to hear rings from that quiet bell of truth that resides within our own conscience. Sometimes the bell rings with a 'rightness' that is at once comforting and fulfilling, but there are other times when it rings with discord. Morgan Sparks listens most carefully at these times.

When Public TV prepared a program about the fiftieth anniversary of the transistor invention, they recorded several hours of conversation with Morgan and Bette Sparks. When you talk to Morgan Sparks, you always know where he stands. Perhaps for this reason PBS did not use any of the recording in the final version of the broadcasted program.

The original point contact transistor does not require thermal processing of the semiconductor. The manufacturing of the point contact and grown transistors required extensive manual labor that was performed in a similar way to the assembly of vacuum tubes. Many vacuum tube companies such as RCA, General Electric, and Sylvania in the East started production of transistors. Assembly of the early transistors was difficult, if not impossible, to automate. In the early 'fifties, thermal processing of semiconductors caused a breakthrough in semiconductor device fabrication. Diffusion processing enabled high-volume semiconductor manufacturing. At that time, diffusion engineering and wet etch was all that was needed to produce a transistor. At that time, being a diffusion engineer was the dream of many young scientists, contrary to the low appreciation of diffusion engineering today. Diffusion and thermal processing now play a less important role in state-of-the-art semiconductor manufacturing than they did fifty years ago. At that time the technique of diffusing impurities from a gaseous state into the semiconductor crystal opened up the possibilities of device performance never before possible.

In 1947 Bell Telephone Laboratories divided the responsibility for the research and development of materials between Chemical Laboratories and Apparatus Development Department. Calvin S. Fuller was in charge of all work on plastics including polymers at Chemical Laboratories. After the merger of both groups C. S. Fuller characterized the new situation; *"J. R. Townsend was put in the same position with me and also was assigned as head of the new organization; my fate was not at all happy with the arrangement".*

Fuller was gradually pushed away from his successful work on polymers and eventually he was offered two choices: 1) to go along with direction of the group, or 2) leave. By 1950, Calvin Fuller decided to switch to solid state chemistry. He immediately began experimental work relating to the surface properties of germanium. At that time there was a problem with germanium that nobody seemed to understand, and which was called "thermal conversion." Calvin Fuller described the problem as the following: "*The germanium crystal was so sensitive that if you went into the laboratory and grabbed the doorknob, and then happened to lightly touched the crystal, it would convert; that is it would change type from N to P on subsequent heating above about 500° C. If the sample was etched and rinsed in "highly purified water" conversion would not occur.*" With the help of J. D. Struthers and Kathy Wolfstrin, Fuller proved that copper was the rapidly diffusing impurity that created "acceptors" in the crystal and converted it from N-type to P-type. This work on copper led to investigation of other fast-diffusing species and to an examination of diffusion in semiconductors in general. Fuller, with the help of John Ditzenberger, investigated diffusion of Group III and Group V elements, during 1954.

When Calvin Fuller was developing the diffusion method, he felt that the BTL management was not fully appreciative of the possibilities of diffusion. He wrote a letter to the Lab management [letter dated March 7, 1951, case 37939, Bell Laboratories], but received neither response nor bigger support for his work. When in 1954 Fuller's work resulted in the invention of the solar cell, which Bell Labs management highly publicized, he noted "*It was very interesting because people like to speak of teamwork when they talk about research. The solar cell just sort of happened and had none of the aspects of team work.*" The apparatus used for diffusion from gaseous compounds experiments in Bell Telephone Laboratories at the beginning 1950's is shown in Fig. 2.11.

Most of the works reported by C. S. Fuller, J. A. Ditzenberger and C. J. Frosch were done in this system. The essential components were elongated fused quartz tubes with an inner diameter of approximately 25 mm extending through a high-temperature controlled "Glo-bar" furnace. The "Glo-bar" furnace provided a constant temperature zone approximately 100 mm long. The semiconductor samples were placed vertically inside the tube with faces parallel to the gas flow during processing. The zone temperature was controlled to $\approx \pm 2°C$. A photograph of the actual system (schematically shown in Fig. 2.11) is shown in Fig. 2.12.

In the mid 'fifties the first generation of bench-top furnaces was improved and heating coils were added to the system. A very simple on-off controller and bubbler were added to the system. Temperature was sensed by a thermocouple inserted into a cavity in the middle of the wafer boat. All gas

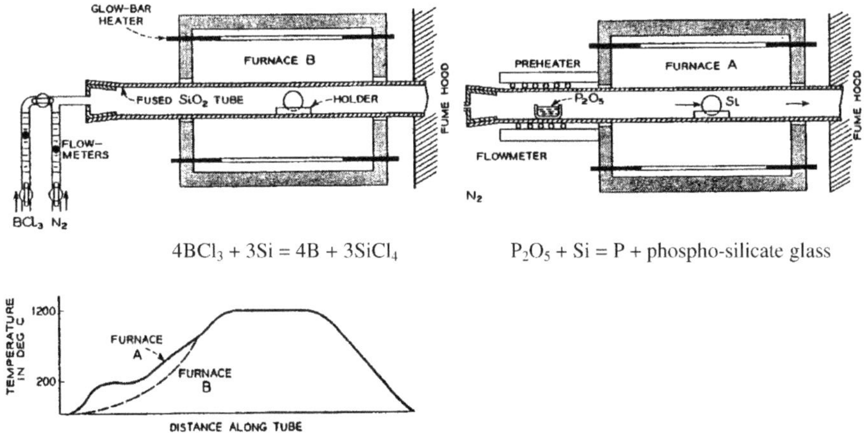

Fig. 2.11. Diffusion furnace used in early 1950's in BTL

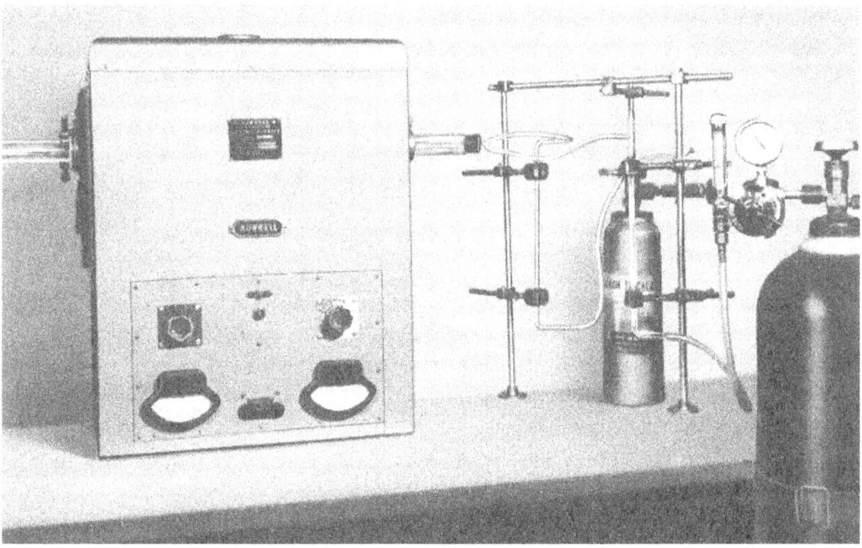

Fig. 2.12. Diffusion furnace used at Bell Telephone Laboratories during early fifties. In this type of furnace Carl Frosch and Lincoln Derick discovered by accident the oxidation of silicon

distribution was accomplished through the quartz tubing and quartz valves. No specialized gases for the semiconductor industry existed at that time.

Regular technical-quality gases were commonly used. The H_2, N_2, O_2 and CO_2 or their mixture was used as carrier gases. After the reaction in the tube, carrier gases were vented through the hood into the New Jersey atmosphere.

The recipe sequence started with the stabilization of required temperature (no temperature ramping was available). The processing gases were turned on

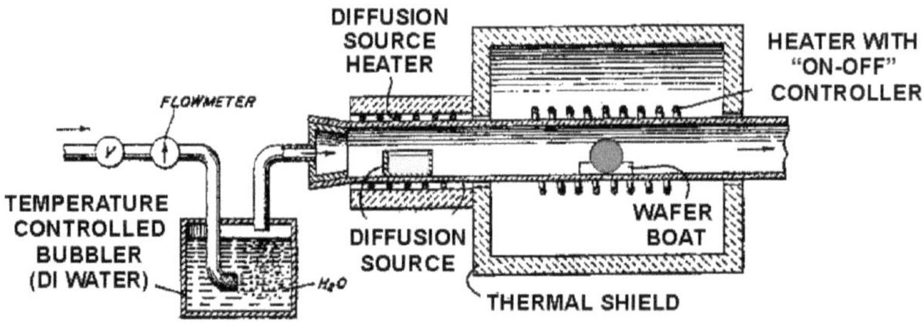

Fig. 2.13. Oxidation system used by Carl J. Frosch and Lincoln Derick at BTL in early 1950's

after the temperature stabilized and then the boat with vertically positioned wafers was manually inserted into the tube.

Fuller's investigation provided the necessary information about the high temperature diffusion parameters required for controlling the impurity concentrations and depths of the diffused layers.

There were, however, major problems with the application of diffusion to silicon:

- The drastic decrease in lifetime, which generally occurred when silicon samples were heated to the high temperatures.
- Pitting of the semiconductor surface during the diffusion process when non-oxidizing gases were used. The surface pitting resulted in non-uniform junction depth.

Fig. 2.14. shows an example of surface erosion of silicon <111> surface after annealing at 1000°C in hydrogen ambient. Fuller identified the decrease of lifetime as a consequence of metallic contamination.

Frosch and Derick discovered that the use of an oxidizing atmosphere provided surface protection of silicon surface. The discovery of the wet oxidation of silicon happened when Carl Frosch forgot to close a hydrogen valve during one of the experiments.

In early 1954, based on the work of Calvin S. Fuller, and J. A. Ditzenberger, W. Shockley suggested to Charles A. Lee, G. C. Dacey and P. W. Foy, that diffusion of impurities into Ge or Si from elemental impurities could be used to form semiconductors junction with well-controlled junction depth.

In March 1955 they filed a patent application describing the mesa diffusion transistor. However, the theoretical concept in this patent was still far away from the practical realization because there was no available process or equipment. The significance of this patent lay in the suggested method of manufacturing which enabled a high volume production.

The main obstacle which had to be overcome was non-existent manufacturing equipment. In a series of experiments P. W. Foy demonstrated that the relatively high vapor pressure of elemental dopants could be reduced to a desirable range (about 10^{-4} mm Hg) by diluting them in germanium.

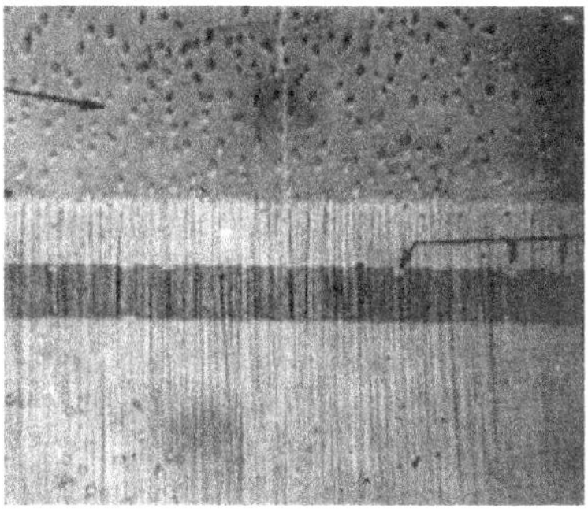

Fig. 2.14. The cross section of <111> Silicon surface after annealing at 1000°C in a hydrogen ambient

Fig. 2.15. Silicon <111> surface after annealing in hydrogen ambient [C. J. Frosch, 1955]

C. A. Lee demonstrated in 1955 a PNP Germanium transistor with diffused base. The structure of Lee's device is shown in Fig. 2.20. The 250 μm thick P-type Ge sample with dimension 5×1.5 mm was placed into the evacuated molybdenum capsule shown in Fig. 2.17 with Arsenic-doped Germanium and heated by a tungsten coil.

The typical sheet resistivity of the Arsenic base diffused region was in the order $\sim 200\ \Omega/\mathrm{sq}$ and had a junction depth $\sim 1.5\ \mu\mathrm{m} \pm 0.3\ \mu\mathrm{m}$. The emitter was formed by alloying of Aluminum film approximately 1000 Å thick. This film was evaporated to the base surface through a mask with emitter size 25×50 μm. The sample was then placed on a strip heater in a hydrogen atmosphere and heated to a temperature sufficient to alloy aluminum.

The contact to the emitter was formed from 3000 to 4000 Å thick film of gold-antimony evaporated and alloyed onto the surface of the emitter.

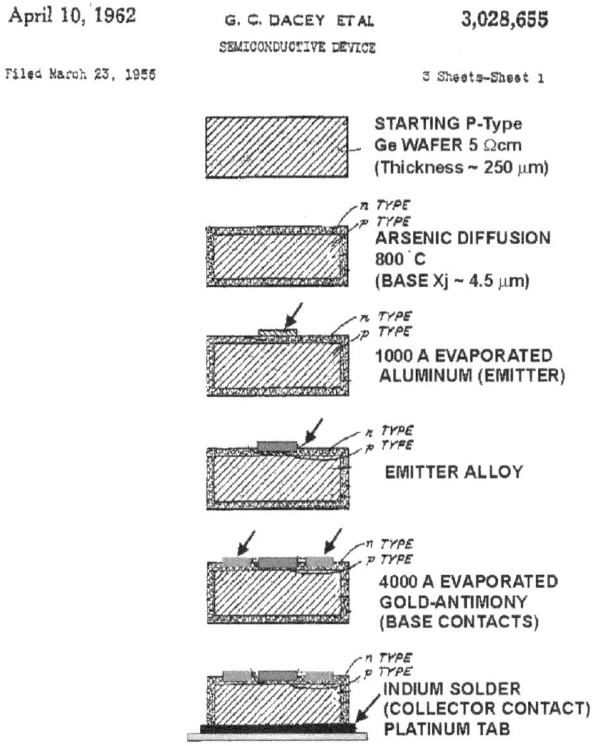

Fig. 2.16. Mesa diffused-base transistor as suggested by Dacey, Lee, and Shockley in March 23, 1955

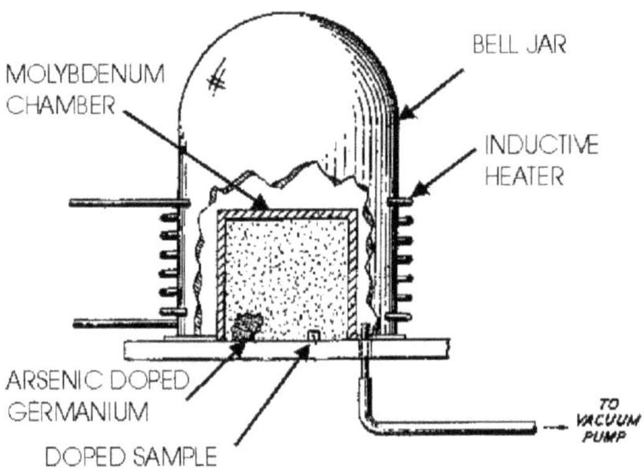

Fig. 2.17. Apparatus for the diffusion of elemental dopants into semiconductor. (BTL 1955)

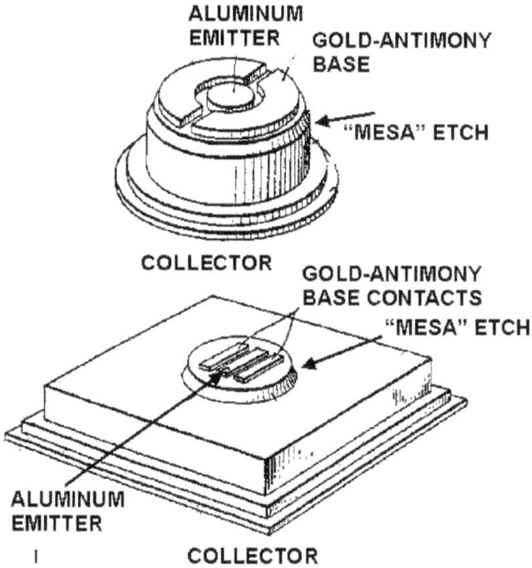

Fig. 2.18. Ring-Dot and Stripe configuration of Mesa transistor suggested by Dacey, Lee, and Shockley

Fig. 2.19. Bell Laboratories metal evaporating chamber. The mask and semiconductor sample located in the upper part of the chamber

The main goal of Shockley and Lee was to produce an impurity gradient in the base region of the transistor. The doping gradient produced a "built-in" electric field enhancing the transport of minority carriers (The Drift Transistor). This transistor was described earlier in 1954 by John M. Early and John L. Moll. John Moll was frequently pondering about the new device so

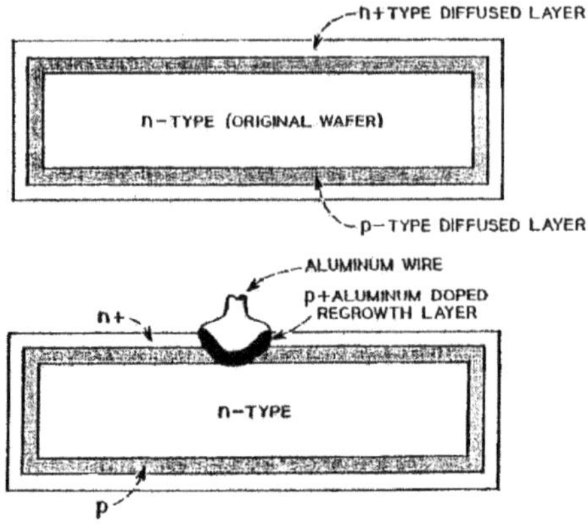

Fig. 2.20. Direct alloying method of forming the base contact to the diffused base

much, that one morning when he was car-pooling with Ray Warner, Jr., he suddenly discovered that he forgot his shoes.

At almost the same time when Lee demonstrated his transistor, M. Tanenbaum and D. E. Thomas demonstrated a silicon NPN transistor with the base and emitter regions produced by simultaneously diffusing impurities of Aluminum and Antimony.

The starting material was 3–5 Ω-cm N-Type Silicon. After diffusion the entire surface of the silicon sample was covered with the diffused N and P diffusion layer. The thickness of the P type base was approximately 2 μm. After etching and removing unwanted layers, the electrical contact to the base was formed on the sample surface by alloying an aluminum layer evaporated through a 50 × 150 μm mask. An Aluminum layer was alloyed through the emitter layer as shown in Fig. 2.20. Contact to the emitter layer was achieved by alloying a film of gold containing a small amount of antimony. Since this alloy produces an N-type regrowth layer, the gold-antimony film does not alloy through the P-base layer. Ohmic contact to the collector was produced by alloying the silicon wafer to an inert metal tab plated with a gold-antimony alloy. Electrical contact to the base was made with tungsten point-pressure contact to the alloyed aluminum. Contact to the emitter was made by bringing a gold-antimony plated tungsten point into pressure contact with emitter layer. A photograph of a mounted unit is shown in Fig. 2.21.

The laboratory works of Lee, Tanenbaum and Thomas resulted in a transistor that was not easy to manufacture. However, very good transistor performance with diffused layers put into question the main deterrent to the application of diffusion to transistor technology – the drastic decrease of lifetime (especially for silicon) when the sample was exposed to the high temperatures required for diffusion.

An additional major achievement was a theory of the diffused transistor developed by John Moll and Ian Ross [5]. It is difficult to improve the

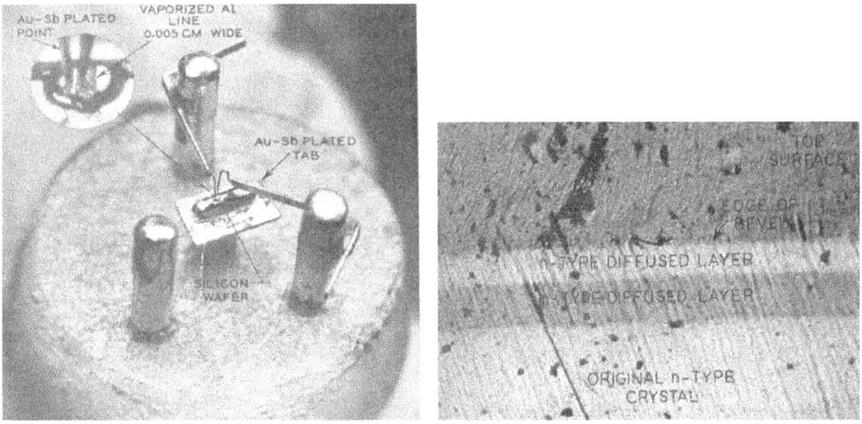

Fig. 2.21. An experimental Bell Laboratories NPN diffused transistor and the cross-section of the diffused areas

performance of any electronic device unless detailed understandings of all phenomena involved in the device behavior are understood. Moll and Ross identified two factors of primary importance:

1) The magnitude of the built-in fields
2) The distance over which the built-in field extends

The double-diffused transistor developed by M. Tannenbaum, D. E. Thomas, C. S. Fuller, J. A. Ditzenberger (who were assisted by P. W. Foy, G. Kaminski, J. M. Klein, F. Maier, C. A. Lee, and G. Weinreich) showed high speed, low saturation current, and satisfactory operation at high temperatures. The concept of this device became the work horse of semiconductor industry for more than three decades.

It is interesting to note that General Electric engineers discovered the importance of diffusion earlier than Bell Laboratories when they developed the alloyed junction transistor. R. N. Hall, W. C. Dunlap and J. S. Saby of the GE Research Laboratories, developed the P-N junction produced by an "alloying process" in 1950. A piece of antimony, or a lead containing Antimony, is soldered to P-type germanium, and then exposed to a temperature of several hundred degrees. Part of the metal will dissolve in a molten puddle; after cooling, the recrystallized germanium will retain in solid solution enough of the antimony to convert P germanium into N-type. The remaining part of the metal will be rejected and form the "button" firmly adhering to the germanium surface. At the elevated temperature some of the antimony atoms penetrate farther into the germanium by solid diffusion and thus producing the junction. By soldering lead wires onto the two metal buttons an NPN transistor structure may be obtained. An NPN transistor structure obtained by alloying junctions in the opposite surfaces of a semiconductor wafer is shown in Fig. 2.24.

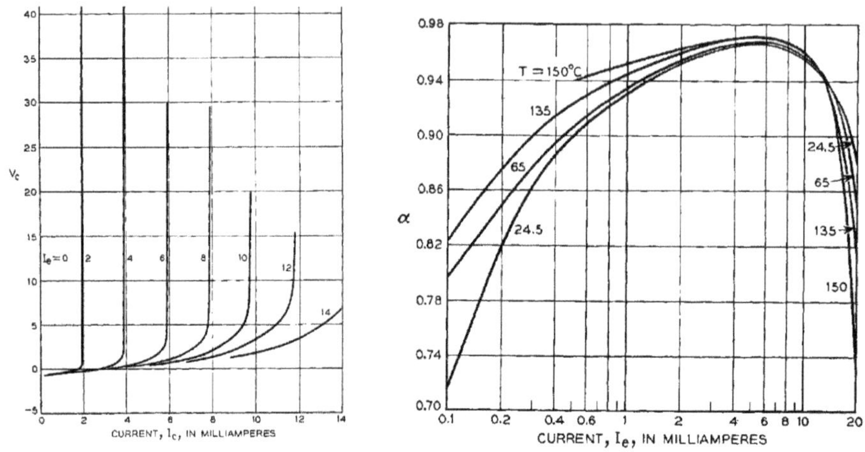

Fig. 2.22. Collector characteristics and current gain vs. emitter current for the double diffused Si NPN transistor

Fig. 2.23. Bell Labs Transistor with simultaneously diffused base and emitter and alloyed contacts

The significant dimensions of alloyed junction transistor are:

1. Germanium wafer 2×2 mm, thickness 1500 μm
2. Diameter of alloyed N-type regions 350–700 μm
3. Junction depth of alloying 500–700 μm

An alloyed junction transistor has electrical characteristics generally similar to those for grown-junction transistors having comparable junction areas. The alloying process results in a thinner base region; therefore collector breakdown voltages for an alloy device are lower.

A photograph in Fig. 2.24 illustrates the cross-section of GE devices from the mid 1950s, which were manufactured using the "alloy-diffusion" process. However, GE did not follow up on the diffusion work except for John Saby who extended basic alloying technology to create a "control device" in 1951 which later become known as the alloy diffusion junction transistor. The mounting of his device to the header is shown in Fig. 2.25.

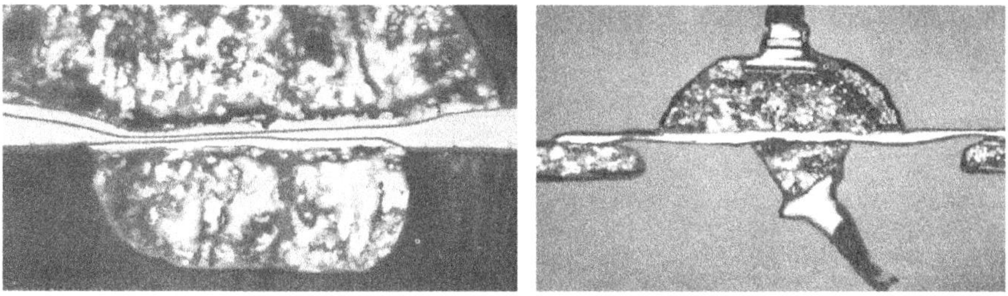

Fig. 2.24. Cross-section of General Electric alloyed junction transistors

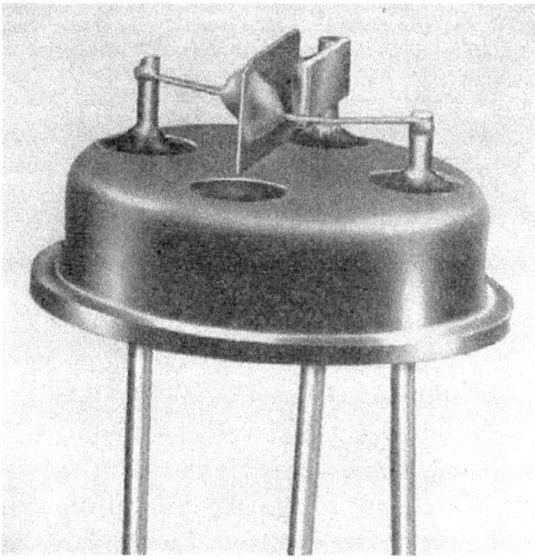

Fig. 2.25. Alloyed junction transistors (1951)

In the end of nineteen fifties IBM established Component Engineering Department located in IBM's Poughkeepsie facility. At that time IBM was one of the biggest customers of Texas Instruments and did very limited work on transistor technology. This situation changed in 1959 when Thomas J. Leach was put in the charge of the new project "Automated assembly of Ge alloy-junction transistors." Leach's group developed fully automated manufacturing process for assembly of germanium alloyed-junction transistors. The equipment called "Transistor Assembler" was completed in February 1960. Equipment fabricated the transistors from pre-formed components – emitter and collector dots, the germanium disks, the base washer, the whisker wires, and the mounting base. The process started with alloying of dots with the germanium disk. In the next step the base washer and the emitter lead and

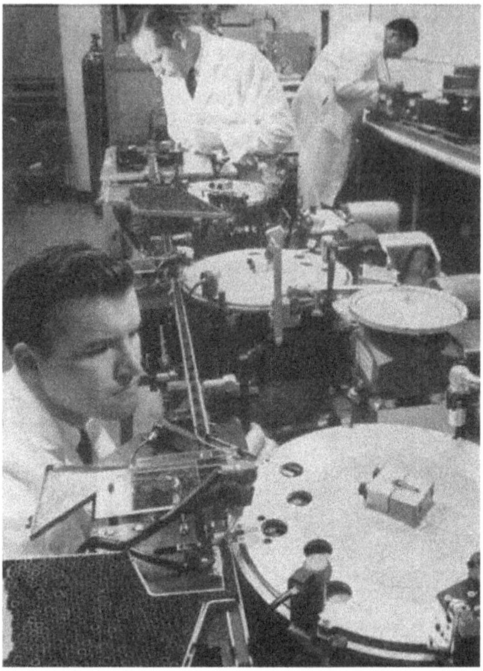

Fig. 2.26. IBM fully automated "Transistor Assembler" of Ge alloy-junction transistors [1960]

bonds were assembled. Finally, the welding of the base washer to the transistor mounting base was performed and then the emitter and collector wires were welded to the mounting base posts. The individual operations are summarized in Fig. 2.27. The complete system consists from nine major operating units (six assembly station, two ovens and the welding unit), integrated by the series of conveyors. Unfortunately system was never fully functional and the project was terminated in mid of 1960'.

Bell Laboratories did not follow up on GE's, TI's nad IBM's alloyed transistor technology. Instead L. E. Miller of Western Electric in Laureldale, PA further advanced Tanenbaum's silicon transistor with several refinements in fabrication techniques. These refinements included an independent emitter and base diffusion, and a high-vacuum evaporation technique for producing the ohmic contacts with thermocompression bonding of lead wires.

Lew Miller designed the silicon NPN transistor with mesa "ring-dot" structure shown in Fig. 2.28 which was manufactured by Western Electric as the 2N560 for military logic applications.

The base diffusion was performed in an open tube used in Frosch and Derick experiments with the gallium sesquioxide (Ga_2O_3) as the diffusion source in a wet hydrogen atmosphere. The surface concentration of the diffusion layer was approximately 10^{18} cm^{-3} and it was 0.5 μm thick.

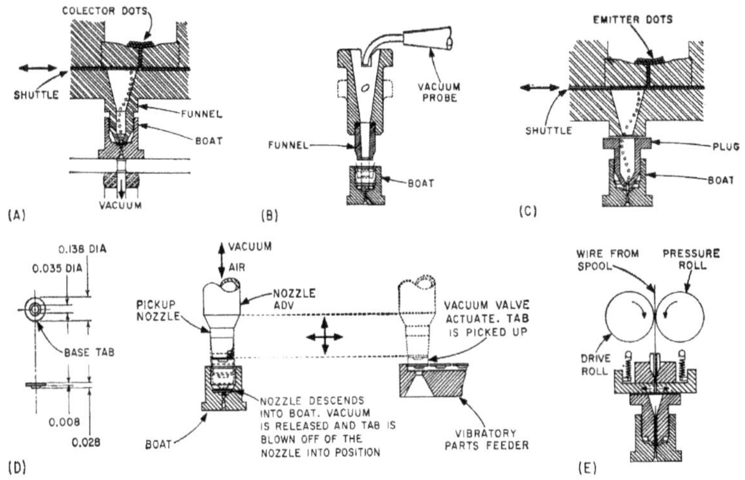

Fig. 2.27. Transistor assembly operations: a) collector dot loading; b) germanium base disk loading station; c) emitter dot loading; d) base tab insertion; e) emitter wire insertion

Fig. 2.28. Beveled section of 2N560 ring-dot mesa silicon transistor

The N-layer with a surface concentration of 10^{20} cm^{-3} and approximately 0.25 μm thick was processed in the same type of open tube diffusion furnace using P_2O_5 with dry oxygen as the carrier gas.

An Aluminum contact to the ring base is evaporated through the mask and alloyed. The silver is alloyed to the N-type emitter layer. The individual transistor dice were bonded to the gold-plated header. For the first time, the Bell Laboratories' newly-developed thermocompression bonding of gold wires was used to wire the electrical contacts.

Miller's design of the transistor enabled for the first time production of a large number of transistors simultaneously. However, the 2N560 did not use photoresist masks. Instead, the metal "shadowing masks" were used for:

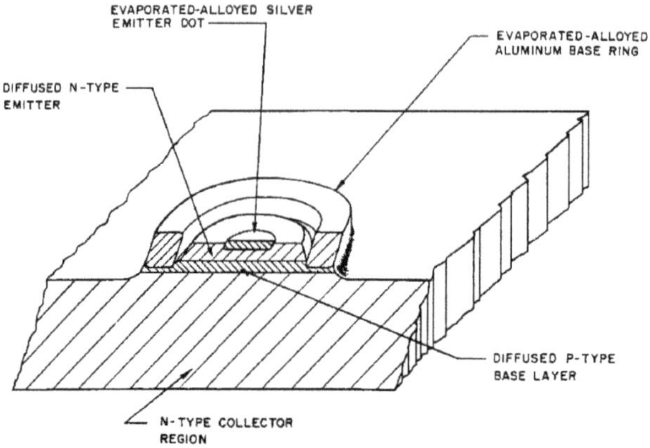

Fig. 2.29. Cross-section of 2N560 ring-dot mesa silicon transistor

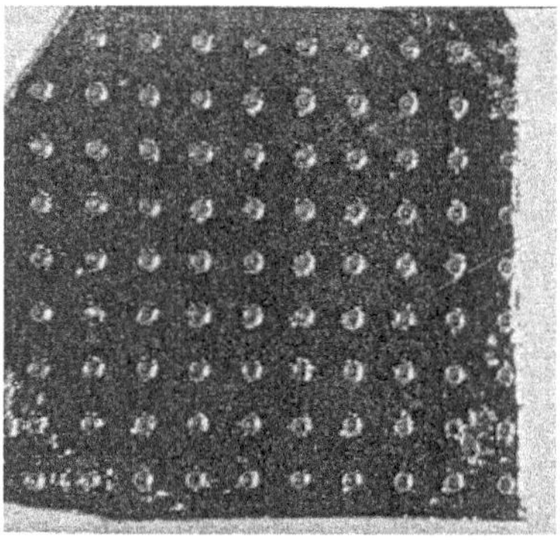

Fig. 2.30. Wafer with 2N560 transistors ready for dicing and assembly

a) emitter diffusion,
b) base contact
c) emitter contact

The aligned (at that time so called "indexed") mask was used to etch the mesa structure. In Fig. 2.30. is shown a slice of silicon with transistors ready for assembly.

Bell Laboratories announced the 2N560 at the WESCON Convention in August 1958 in Los Angeles in the same session where Victor Grinich and Robert Noyce announced the Fairchild mesa transistor 2N696.

To put transistor invention into perspective, Bardeen and Brattain had only observed transistor action using a point-contact in polycrystaline Ge in December 1947. Shockley invented the junction transistor on paper in January 1948, but he had to wait for Teal and Little's development of single-crystal Ge before he could reduce it to practice in 1950. In fact, it was Morgan Sparks, along with another co-worker, Gordon Teal, who actually made the

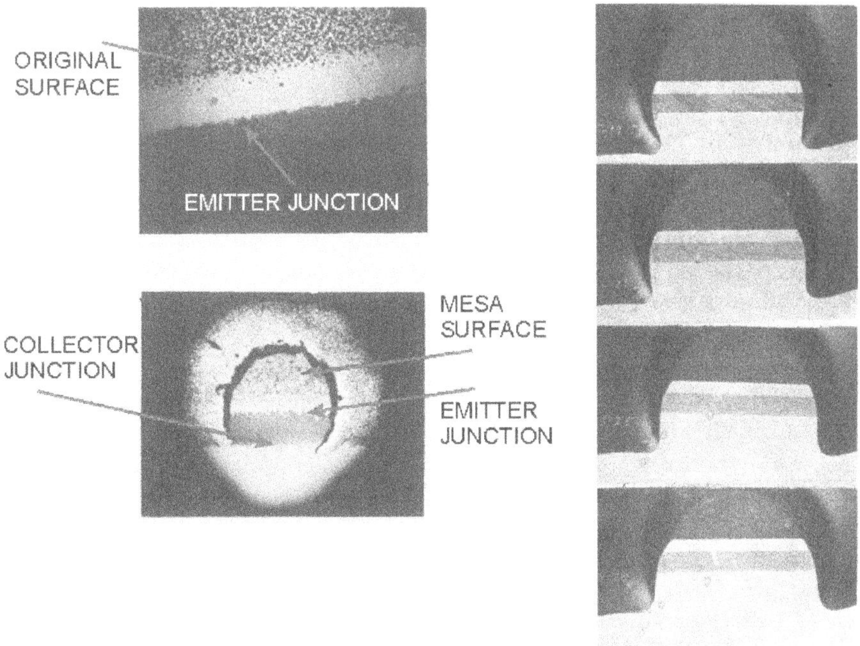

Fig. 2.31. The 2N560 Western Electric Transistor – collector and emitter junctions (Beveled section)

first working junction transistor. L. E. Miller further added improvements and made transistor manufacturable. Looking today on the crude construction of these transistors from the nineteen fifties it is hard to believe that such devices would work. But there was a magic and they worked.

Shortly after the announcement of the transistor discovery in June 30, 1948, Ralph Bown asked W. S. Gorton, an assistant of the director of research, to prepare a "Memorandum for Record" summarizing the events which resulted in the transistor discovery.

Gorton's memo was completed on December 27, 1949 and included a list of 12 persons who did the crucial work.

> "*Group was directed by W. Shockley and S. O. Morgan, J. Bardeen worked on the theory, G. L. Pearson conducted the experiments on the bulk properties and with H. R. Moore experimentally confirmed Bardeen's surface state theory; W. H. Brattain worked on surface properties. J. R. McSkimin, W. A. Yager, R. S. Ohl, and R. B. Gibney contributed by a number of experimental tests. J. H. Scaff and H. C. Theuerer prepared most of the semiconductor material, both germanium and silicon, and recognized the first P-type and N-type behavior of semiconductors.*"

During 1947 transistor research was done in the new Bell Laboratories building complex in Murray Hill; only Russell Ohl remain located in the Holmdel Radio Laboratory of BTL.

However, there were many more than twelve people involved in work on the transistor. The members of the original Shockley & Morgan group were bringing the point-contact transistor into a more satisfactory state and did not come up with any major breakthrough. Shockley, however, was always ahead of the group with his way of thinking. He started to work with new strong individualities. Morgan Sparks and Gordon Teal moved the projects ahead. Mechanical Engineer J. B. Little conducted the initial single-crystal growth experiments. W. L. Bond designed all micromanipulators used for transistor assembly. Robert M. Mikulyak did crystal growing and transistor assembly. J. R. Haynes, R. J. Kircher, F. S. Goucher, W. R. Sittner, G. L Pearson, and Miss A. D. Mills performed numerous measurements. This is just a fraction of the names I encountered when doing my research. The transistor was the result of a well-conducted research effort which was led by strong individualities whose egos could not be stopped by setbacks.

It has been proven by history that creativity lies in the individual mind and not in the team. The team is the nemesis of progress and change. The team is a convenient hideaway for the mediocre and the weak. Ambition is what drives the good scientist and artist to try great things, and try again when they screw up. "The team is stronger and more capable than the one" is the mission statement of collaboration. It is also the motto of communism. Mervin J. Kelly did not call for team collaboration; instead he created in Bell Laboratories culture respecting creative individuals. Kelly, a brilliant physicist himself, understood that great engineers have almost no social skills and they cannot comply with politically correct and hypocritical rules. Such "inequalities" are well balanced by such qualities as: smartness, never satisfied with status quo, aggressive and willing to take a risk, driven to excel, and mainly, able to solve a problem.

References

[1] M. Sparks, G. K. Teal, U.S. Patent 2,631,356

[2] G. K. Teal, M. Sparks, E. Buehler, "Growth of germanium single crystals containing P-N junctions," Phys. Rev. Vol. 81 (1951), p. 637

[3] W. Shockley, "The Theory of p-n Junctions in Semiconductors and p-n Junction Transistor," Bell System Technical Journal, Vol. 28 (1949), p. 435

[4] W. Shockley, M. Sparks, G. K. Teal, "Physical Principles Involved in Transistor Action," Phys. Rev. Vol. 83 (1951), p. 151

[5] J. L. Moll, I. M. Ross, "The Dependence of Transistor Parameters on the Distribution of Base Layer Resistivity," Proc. IRE., Vol. 42 (1954), p.1761

3

Shockley Semiconductor Laboratories

Shockley is certainly one of the most creative men that I've ever known in my life, or hope to. He had a marvelous way of simplifying a problem and getting at the fundamental part of it, cutting away all of the extraneous information and getting a model simple enough to be handled mathematically or experimentally. I think it must be the same working with any really creative individual – the ideas flow so fast that it keeps you busy trying to keep up. Consequently, your rate of learning is very, very high because you are working so hard to understand what is being given to you. This is a characteristic of a good professor at an advanced level.

Robert Noyce
"Working with Shockley", January 1966

At the end of 1954, Shockley spent time at Caltech as a visiting professor while exploring possibilities for his new business. In February 1955, the Los Angeles Chamber of Commerce, where Arnold Beckman was vice president, organized a gala banquet to honor Lee De Forest and Shockley for their scientific contributions. Shockley and Beckman quickly became friends.

Six months later Shockley called Beckman and asked if he would be willing to serve on the board of Shockley's company. After a brief discussion, Beckman realized that he would serve on the board of a company with several persons who were in direct competition with Beckman Instruments Company. Shockley had no business experience and Beckman realized that Shockley's plan had no chance to succeed. Beckman, however, was an excellent businessman and saw immediately a tremendous opportunity. He invited Shockley to Newport Beach to discuss the details for few days.

Shockley envisioned a company for large-scale, high-volume production of silicon semiconductor components – transistors and electronic switches us-

FROM: WOLCOTT & ASSOCIATES
PUBLIC RELATIONS
1308 WILSHIRE BOULEVARD
LOS ANGELES 17, CALIFORNIA
DUnkirk 5-1439
(Robert B. Wolcott, Jr.)

FOR: BECKMAN INSTRUMENTS, INC.

FOR RELEASE A.M.'S TUESDAY
FEBRUARY 14 AND FOLLOWING

Beckman Instruments, Inc., yesterday announced the establishment in the Stanford community of the Shockley Semiconductor Laboratory to develop and produce transistors and other semiconductor devices in the field of advanced electronics for automation.

The news was revealed by Dr. Arnold O. Beckman, founder-president of Beckman Instruments at a luncheon for scientists, educators and the press at the Hotel St. Francis.

Headed by Dr. William Shockley, inventor of the junction transistor, as director, the nucleus of the rapidly expanding research team consists of four Ph.D's: G. Smoot Horsley, formerly of Motorola and Bell Laboratories; Leo B. Valdes, formerly of Pacific Semiconductor, Inc., and Bell Laboratories; William W. Happ, formerly of Raytheon Manufacturing Co., and Sylvania Electric Products; and R. V. Jones, who has just completed training at the University of California at Berkeley. The first three are experts in the field of semiconductors, the basic material of transistors which are revolutionizing the electronics field by replacing the vacuum tube. Jones' research work, in a different branch of physics, has direct application to some basic semiconductor problems.

Fig. 3.1. Front page of the Beckman Instruments Press Release announcing Shockley Semiconductor Laboratory

ing a solid state diffusion. Bright Beckman realized that such devices might revolutionize the Beckman Instruments Company and he convinced Shockley to abandon his idea of an independent new company and instead set up an organization under the Beckman Company umbrella. By the end of Shockley's visit the Beckman attorney drafted a letter of intent to form a subsidiary of Beckman Instruments: Shockley Semiconductor Laboratories. Shockley assured Beckman that the new company would produce large quantities of semiconductor devices within two years. With a written agreement in hand, Shockley returned to the East Coast and started searching for recruits.

In 1955 the profits of Beckman Instruments surmounted one million dollars for the first time. Beckman swiftly acquired transistor patent licenses from Western Electric and in February 14, 1956 announced in San Francisco the launch of Shockley Semiconductor Laboratories, located in Mountain View, California (Fig. 3.1.)

Gordon Moore, a lifelong and harsh critic of William Shockley, suggested that Shockley moved to the West Coast because *"unstable Nobel laureate (Bill Shockley) simply needed to be near his mother"* [1]. Nothing could be farther from the truth. When William Shockley left BTL, he approached several companies and even the Rockefellers with a reasonable business plan to launch manufacturing of the improved diffusion transistor. The business plan was asking for \$1,000,000 payable in a three year period with an initial installment of \$500,000 for the first year and \$250,000 each consecutive year. The East-cost-based Raytheon seriously considered this plan and Shockley actually started to work for the company. Although the plan was very reasonable and very modest by today's standards, Raytheon's management backed off from the plan, and Shockley separated from the company after one month.

> Another matter which will need serious consideration is whether or not we should start in Fullerton or in Palo Alto. At present, I think the decision may be postponed, but it may soon become important if we get strong reactions from future candidates such as Tanenbaum or some of the people I may try to find at Bell or RCA.

Fig. 3.2. Paragraph from W. Shockley letter to A. Beckman from October 6, 1955

Shockley originally suggested Fullerton as a place for the new company and since October 1955 he resided near the Beckman Facility at Fullerton. One other place considered was Pinellas County, Florida which offered very attractive business conditions including a governor's invitation. Arnold Beckman, who recently acquired the Specialized Instruments Corporation – SPINCO, located in Stanford University Business Park – saw several advantages in locating Shockley Semiconductor on the West Coast. Especially when Beckman's friend, then Dean of Engineering, Fred Terman of Stanford University, expressed the desire to bring a new business into the area.

W. Shockley was not terribly excited about this option, after a meeting at the Los Altos Chamber of Commerce. Local officials were more enthusiastic about the farming industry; they had no idea what a transistor was and offered no support for Beckman's new company. Another problem Shockley faced was obtaining qualified personnel for his company. Shockley contacted several people in Bell Labs and RCA. Some of them were willing to go to Fullerton, but none to Palo Alto. Actually Morris Tannenbaum, Morgan Sparks, W. J. Pietenpol, and R. L. Wallace, Jr. visited the Beckman facility in Fullerton on September 30, 1955. They considered Fullerton but rejected Palo Alto. At that time the east-coast-based RCA and Bell Laboratories were the best research organizations in the nation, and very likely in the world. They were well equipped, surrounded by the best universities, and located in an already civilized part of the country. In contrast, Santa Clara County had, at that time, no direct long distance phone, and San Antonio Road, where Shockley located his facility had a traffic light at the intersection with two

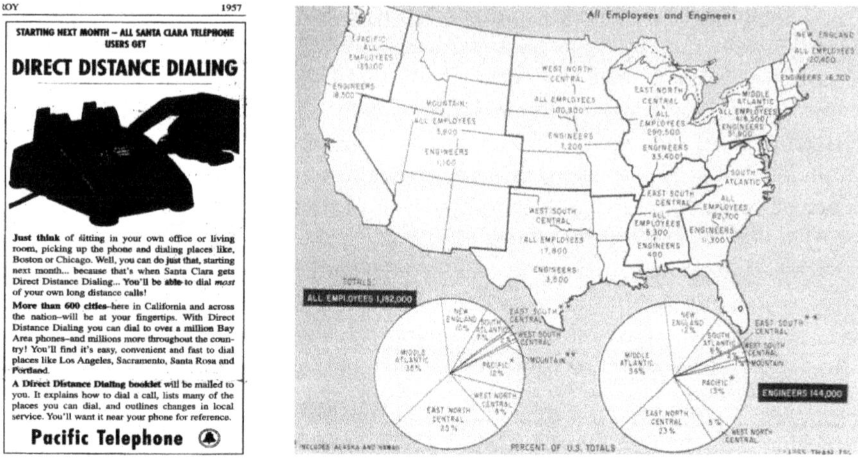

Fig. 3.3. Pacific Telephone's 1957 announcement of direct dialing service for Santa Clara County; and the chart of U.S. Engineering manpower of the electronics industry as of January 1, 1960

lane 101 highway. The only port connecting Santa Clara County with the rest of the world was a train and the San Francisco airport. In 1959, 1,182,000 employees of the electronics industry, including 144,000 engineers turned out products worth \$12 billion. The bulk of the industry was concentrated in the East Central and East Coast. Only 13% of the engineers employed in the industry were working in the Pacific region (Fig. 3.3.)

In February 1956, Shockley rented a modest, wood construction building[1], at 391 San Antonio Road in Mountain View with approximately 5000 square feet, and with four other engineers launched Shockley Semiconductor Laboratories, a division of Beckman Instruments, Inc. The four others were: Leopoldo B. Valdes, a former Bell Labs, General Electric and Pacific Semiconductor employee; G. Smoot Horsley, who met Shockley during his brief stay with Bell Laboratories and who worked for four months at Motorola; William W. Happ formerly with Raytheon an Sylvania; and R. V. Jones, a fresh Berkeley graduate. The team hired Carolyn S. Himsworth as a secretary. Professor John G. Linvill of Stanford University had signed as a consultant of the new company on February 29, 1956.

In a letter to Robert Sproull of Cornell University, Shockley wrote: *"my group just leased a 5000 square foot building on the southern edge of Palo Alto and we are busy gathering together laboratory equipment and additional personnel. ...Do you have in mind any good experimental or theoretical Ph.D's?"*

[1] A few years later Kurt Hubner, one of the engineers who worked on the development of Four Layer Diodes, asked Shockley why he settled for such an "unattractive' place. Shockley responded with a kind of nostalgia *"you do not know how hard I worked to get here."*

Because nobody knew where Mountain View was, Shockley referred to this location as a "southern edge of Palo Alto." In letters dated October 24, 1955 Shockley personally contacted all of his friends at universities and asked for good students of Physics or Chemistry, but it was not easy to get the right people to move to the West coast. The advertisement Shockley Laboratories placed into Chemical and Engineering News in February and March 1956 yielded about twenty five replies, one of them from Robert N. Noyce.

Chemists and Metallurgists Needed for Semiconductor Research and Development[1]

BY WILLIAM SHOCKLEY[2]

RECEIVED JANUARY 15, 1956

Semiconductor chemistry and metallurgy are expanding fields. Examples of the creation of new science in the field of solid solutions in silicon and germanium are easily found[3] and this science is being effectively applied to transistor technology.[4] There will be many opportunities for professional advancement in this field as semiconductor electronics production will grow rapidly[5, 6, 7] with transistor production increasing by a factor of 100 to 1,000 in the next 5 to 10 years.[8]

New semiconductor devices and manufacturing methods will be developed by the Shockley Semiconductor Laboratory of Beckman Instruments, Inc.[9] The Laboratory will also afford opportunities for basic research and publication to its scientific personnel[10] and will be located on the Stanford University Estate a short drive south from San Francisco. One or two chemists and metallurgists are needed now to participate in the formative phase of the activity. Experience in semiconductor work is desired, but it is less important than outstanding technical ability and personal maturity.[11] All inquiries will be promptly answered and held in confidence if so desired.

Additional Note – Openings for electronic engineers exist in the Shockley Semiconductor Laboratory and in Beckman Divisions in Fullerton, South Pasadena, Newport Beach, and Richmond, California. Send resume to Fullerton address.

(1) This work is supported by Beckman Instruments, Inc.
(2) Formerly with Bell Telephone Laboratories, Inc.
(3) Bull. Am. Phys. Soc. V. 30, No. 7, Nov. 25, 1955, pg. 24, 31, 33, and V. 30, No. 8, Dec. 28, 1955, pg. 22.
(4) For example, seventeen papers pertaining to such applications were presented on October 13, 1955 at the Semiconductor Symposium of the Electronics Division at the Pittsburgh meeting of the Electrochemical Society.
(5) Aviation Week, Jul. 25, 1955, pg. 50.
(6) Ibid. Aug. 1, 1955, pg. 36.
(7) Business Week, Sept. 17, 1955, pg. 108.
(8) Unpublished operations research by the author.
(9) Business Week, May 22, 1954, pg. 178; Factory Management and Maintenance, Apr. 1955, pg. B-2.
(10) Private communication from Dr. A. O. Beckman.
(11) A quantitative analysis by the author of the economics of employing outstanding contributors has been reported in Science, Nov. 4, 1955, pg. 879; and Chemical Week, Nov. 28, 1955, pg. 148.

BECKMAN INSTRUMENTS, INC.
FULLERTON 10, CALIFORNIA

Shockley called the Director of Research at Philco, W. E. Bradley, where Noyce was working. Bradley provided a good reference and added "his brother was teaching at Berkeley and Noyce wants go to West."

Fig. 3.4. Dean Knapic, William Shockley and Smoot Horsley in February 1956

One graduate, Jay Last, who worked with Beckman instrumentation at MIT, came up with some ideas on how to improve the reliability of instruments. Jay was referred to Beckman and subsequently to Shockley. Shockley asked Last to meet him at the Cosmos Club in Washington and wrote in his diary *"Last will let me know by Dec. 15, 1955."*

In April 1, 1956 the Shockley Laboratories consisted of four Ph.D's in Physics, one in Electrical Engineering and a M. Sc. in Electrical Engineering, a secretary and a general handyman. Shockley was reporting to Arnold Beckman:

> *"... we have recently formalized some organizational structure up here, my senior assistant is Dean D. Knapic and my junior assistant is Smoot Horsley, Robert Noyce, who has recently joined us from Philco, is acting as Horsley's advisor on technical and scientific problems, and Mrs. Bolender, our office manager , is working under Horsley on purchasing and accounting, and my secretary, Mrs. Himsworth is handling recruiting correspondence for Horsley and myself ... "*

W. Shockley's official title was Director, and there were no business or marketing personnel.

On March 13, 1956 Shockley placed a new ad into the New York Herald Tribune and The New York Times:

> *Shockley Semiconductor Laboratory of Beckman Instruments, Inc. has openings for qualified electronics engineer with research and development experience in semiconductor device design and/or application, computer circuitry or related fields. Some of the many benefits of joining our firm include (1) the opportunity to grow with present small organization (2) located in California near Stanford Univer-*

sity (3) where you will work in close contact with scientists of varied disciplines. For interview call MURRAYHILL 2–7430."

A list of almost 300 names who responded to this ad were after initial contacts labeled with status "Dead" or "Limbo." About 90 names had a status "We have the ball" or "He has the ball."

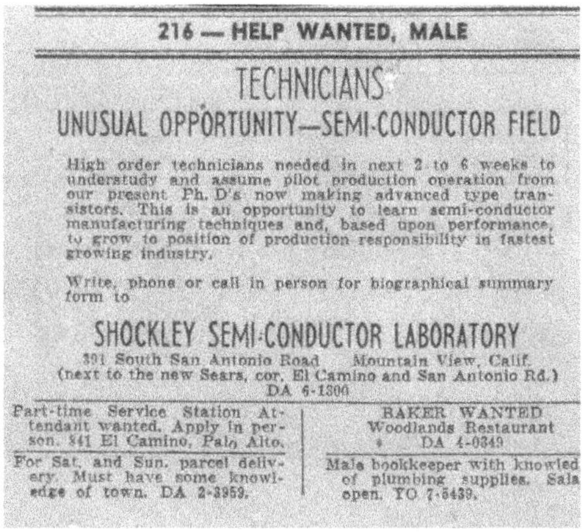

216 — HELP WANTED, MALE

TECHNICIANS

UNUSUAL OPPORTUNITY—SEMI-CONDUCTOR FIELD

High order technicians needed in next 2 to 6 weeks to understudy and assume pilot production operation from our present Ph. D's now making advanced type transistors. This is an opportunity to learn semi-conductor manufacturing techniques and, based upon performance, to grow to position of production responsibility in fastest growing industry.

Write, phone or call in person for biographical summary form to

SHOCKLEY SEMI-CONDUCTOR LABORATORY

391 South San Antonio Road Mountain View, Calif.
(next to the new Sears, cor. El Camino and San Antonio Rd.)
DA 6-1300

Part-time Service Station Attendant wanted. Apply in person. 841 El Camino, Palo Alto.

For Sat. and Sun. parcel delivery. Must have some knowledge of town. DA 2-3959.

BAKER WANTED
Woodlands Restaurant
DA 4-0349

Male bookkeeper with knowled of plumbing supplies. Sala open. YO 7-5439.

Fig. 3.5. Shockley Semi-Conductor Laboratory ad in the Palo Alto Times

Shockley very quickly realized that recruiting qualified people to the "southern edge of Palo Alto" was the main problem he was facing. Shockley wrote in an inter-office memo to W. W. Wright of Beckman Instrument:

"Suitable candidates for this job are extremely difficult to find according my experience at Bell Labs. The difficulty is generally this: If the man has adequate technical competence he does not have the patience to try to communicate with the experimentalist who cannot follow his theoretical reasoning or appreciate the elegance of his points which he may have developed. On other hand if he is a man of lesser capacity, he may be quite incapable of dealing with the difficult problems which arise in actual experimental situations."

After four months of head hunting, at the beginning of March 1956, the entire staff of Shockley Semiconductor Laboratory was: Barclay, Bolender, Breen, Grunwald, Happ, Horsley, Himsworth, Jones, Knapic, Pretzer, Roberts, Valdes, Wagner, and Zinn.

C. Sheldon Roberts got an excellent recommendation ("the brightest student we have") and he had an excellent reputation in Dow Chemical. He joined Shockley Semiconductor on February 8, 1956. Robert Noyce was interviewed on March 3, 1956 in Fullerton, and joined the company in April 1, 1956.

David Allison, Gordon Moore, and 13 others met Shockley after a meeting of the American Physical Society on Sunday, February 5, 1956 where Shockley organized a headhunt. Dr. Spencer who provided a reference for Gordon Moore said to Shockley: "*good ability to get along, does not cause trouble.*" Moore indicated to Shockley that his current salary at Johns Hopkins Lab was $7500, his offer from Lockheed was $9000, and he told Shockley that he will be in California during February 16 through 23, 1956, and would like to see if Shockley could do better then Lockheed. Gordon Moore joined Shockley Semiconductor in April, 1956 after the group at the Applied Physics Laboratory of Johns Hopkins University was disbanded. Gordon Moore received a B. S. degree in chemistry from UC Berkeley and a Ph.D. in chemistry in 1954 from Cal Tech. He said that before his job at Shockley Semiconductor, he had no idea what semiconductors were. He attributed his interest in chemistry to the Russians. While still in grade school he knew how to use nitroglycerine and was experimenting with rockets. At that time, he joked: I maintained a "solid lead on The Russian." The Russians ruined Gordon's illusion in 1957.

Initials	Name		Salary	Age	Starting Date
HB	Breen		312.00	52	3-19-56
RG	Grunewald	(48 hrs)	585.00	50	1-16-56
CH	Himsworth		375.00	35	3-14-56
AP	Pretzer	(48 hrs)	475.00	36	2-29-56
GS	Stout		277.00	21	4-25-56
ET	Troyer		425.00	31	6-15-56
TZ	Zinn		329.00	45	1-24-56
			2,778.00		
Monthly:					
DA	Allison		575.00	30	4-23-56
JB	Blank		835.00	30	4-11-56
LB	Bolender		500.00	57	2-13-56
WG	Gadbury		500.00	30	5-21-56
WH	Happ		825.00	36	11-15-55
SH	Horsley		950.00	40	12-15-55
VJ	Jones		750.00	27	12-15-55
EK	Kleiner		1000.00	33	6-15-56
DK	Knapic		1450.00	38	4-1-56
JL	Last		675.00	26	5-21-56
GM	Moore		750.00	27	4-14-56
RN	Noyce		1000.00	29	4-1-56
SR	Roberts		1000.00	29	3-13-56
WS	Shockley		2500.00	46	10- 1-55
LV	Valdes		900.00	28	10-10-55
RW	Wagner		500.00	38	4-10-56
			14,710.00		
	Total		$ 17,488.00		

Fig. 3.6. Payroll of Shockley Semiconductor Laboratory as June 1, 1956

"*Good insight, original, had been very effective and good to be around*" was Linus Pauling's characterization of Jean Hoerni. Hoerni was actually considering returning to Europe. Shockley noted in his diary, after an interview with Hoerni, that Hoerni was the only one who asked what type of work he would be doing and did not ask questions about salary. Hoerni joined Shockley Semiconductor in July, 1956.

With E. Kleiner's and E. G. Troyer's acceptance of job offers, and outstanding job offers of C. E. Benjamin and V. H. Grinich, during that month Shockley was able to establish Production and Research groups in his organization. The Production group was headed by D. D. Knapic with members R. C. Wagner, D. F. Allison, W. Gadbury, R. Grunewald, A. Pretzer, and G. Stout. The Research group was headed by G. S. Horsley and included R. Noyce, L. Valdes, W. W. Happ, R. V. Jones, S. Roberts, G. Moore, and J. Last. The administrative staff was comprised of L. Bolender, C. Himsworth, T. Zinn, and H. Breen. In addition, H. Packer, C. T. Sah, D. Fok, J. K. Clifton, J. S. Astin, and L. C. Nofrey were considering job offers.

When Jack A. Morton of Bell Labs provided Shockley with silicon crystals, Shockley Semiconductor could finally start operations. The scope of the research work in 1956, can best be seen from the released Technical Memorandums:

- 9/21/56 W. Shockley – Current voltage relationship in an ideal junction transistor with floating base in the presence of light
- 9.21/56 S. C. Roberts – Metallographic study of silicon crystal surfaces after abrasion
- 9/21/56 W. Shockley – Localized radiation damage as a means of studying vacancies and interstitials
- 9/21/56 S. C. Roberts – Report on BTL visit
- 10/10/56 J. A. Hoerni – Solution to diffusion problems
- 10/1/56 W. Shockley – N-skin speculation
- 10/4/56 J. A. Hoerni – Diffusion of concentration profiles for aluminum diffusion and phosphorus N-skin
- 10/4/56 W. W. Happ – Correlation of layer thickness and surface conductance
- 10/9/56 W. W. Happ – Signal flow graphs
- 10/15/56 R. N. Noyce – Carrier generation and recombination in the space charge region of p-n junction
- 10/17/56 W. W. Happ/J. Hoerni – Interpretation of layer thickness and surface conductance in NPN structures obtained by diffusion
- 11/6/56 C. T. Sah – Avalanche multiplication in the space charge region of Si p-n junctions with carrier generation centers
- 12/4/56 C. T. Sah – Investigation of the Si p-n junction current voltage characteristics

The major task, however, was to set up manufacturing and research equipment and instrumentation. The majority of facility equipment needed to be internally developed. In 1956, no equipment for processing semiconductors was available on the market. Although Shockley Labs had full access to design documentation for equipment developed at Bell Labs, the young and inexperienced Shockley engineers needed advice and guidance when developing and using new tools. Such work requires focus and day-by-day supervision.

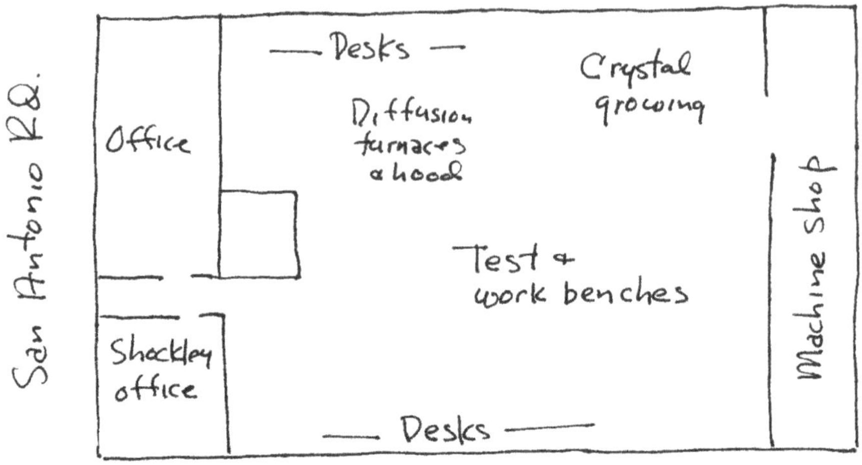

Fig. 3.7. Shockley Laboratories at San Antonio Road as remembered by Jay Last (in upper right corner was crystal pulling equipment operated by Shelton Roberts; hood and diffusion furnaces were located on the tables in the center of the space.)

Timing was not fortunate for Shockley, and a major distraction and derangement happened: John Bardeen, Walter Brattain and William Shockley won the Nobel Prize on November 1, 1956.

Shockley certainly deserved the Nobel Prize, but the timing could hardly have been worse for him. Shockley was flooded with letters and mail, countless invitations; every journal asked him to be on their Editorial Board; the news media asked him for interviews etc. With his trip to Stockholm and Europe, Shockley had no time to work and could pay very little or no attention to his company.

To start up a new company, especially a company which is using a new technology, a number of different people with different skills are needed to contribute to the development of new tools. The problem is that the exact nature of the interaction could not be predicted or well planned. It is up to the leader and management to coordinate new situations and define daily adjustments as they are required. The major requirements for effective interaction are good communication and the ability of individuals to help and be flexible with problems their colleagues may have. In order to complete some projects, other projects needed to be rejected. The choice of good criteria must apply; however, Shockley as director of Shockley Semiconductor Laboratories was not clear on his goals.

Arnold Beckman, Shockley's financier and a brilliant man himself, equally failed as an entrepreneur. Beckman's personal style of management rather depended on the sensing of opportunities and backing individuals while avoiding meetings and managerial confrontations. Such an approach was no longer adequate for a corporation involved in the start-up of a new and revolutionary technology. After gifted individuals like John Bishop, who quit Beckman Instrument in 1959, left the company because he knew what was wrong but could not change it, Beckman frankly admitted that *"certain inadequacies in the internal organization and controls contributed to the company deficit."*

When Beckman met Shockley in 1955, he realized that Shockley had no idea how to run a business; however, he did not provide enough guidance to successfully launch a new start-up operation. In Beckman Instruments' Annual Report for 1956 on page 13, Beckman stated, *"an outstanding example during the past year was the establishment of the Shockley Semiconductor Laboratory. Dr. W. Shockley, internationally renowned inventor of the junction transistor, joined the company to carry on basic and applied research in the exciting new field of semiconductors."* Shockley was assuming that he could continue with the basic research in the same way as he did in Bell Laboratories. On the other hand, Beckman felt that a bright scientist would have no problem to run a business. The problem was that Shockley was never exposed to a business environment and had no ability to compromise. When Shockley designed, for example, a micromanipulator, he could not use a version which was already developed and would do the job, he needed to design the world's best micromanipulator. Shockley did not find anything that could not be improved. Beckman had little compunction about reminding Shockley what the job really was.

Shockley's autism was unavoidable and was a reason for many of Shockley's social confrontations. Shockley came from a long, aristocratic American family, directly descending from John Alden and Priscilla Mullins from the Mayflower in 1620 on his father's side. Shockley's father William Hillman Shockley was born on September 18, 1855 in New Bedford, MA. He held a Master's Degree from MIT (1875) in Mining Engineering. As a botanist he discovered a new genus, Hacastocleis Shockley. He studied music and spoke German, French, mandarin Chinese, Russian, and Italian. Between 1896 to 1913 he worked on mining projects in China, Mongolia, Manchuria, India, Russia, Australia, Korea, Peru, Sudan and England. In 1908 W. H. Shockley married Cora May Wheeler, born in 1879 in Moberly, Missouri. Cora's mother Sally Jane divorced architect Nicholas H. Wheeler and re-married S. K. Bradford, mining surveyor. Cora was adopted by her stepfather and changed her name to May Bradford. She graduated from Stanford University in 1902 with a degree in Art and Mathematics and become the first woman in the U.S. to be appointed a deputy mineral surveyor in both Nevada and California by the U.S. Surveyor General. May Bradford Shockley was a very successful painter, exhibiting her work in San Francisco, Los Angeles, Pasadena, Santa Barbara, New York and Washington. Mrs. Herbert Hoover sponsored her exhibit in the National Arts Club in 1922; her paintings were in the Lincoln bedroom at the White House during the Hoover administration. Their only child, William Bradford, was born in London on February 13, 1910.

But at the end of 1956 nobody was concerned that all was not right with the new venture. All of Shockley Semiconductor Laboratory's staff gathered at Rickey's restaurant on El Camino Road to celebrate the Nobel Prize. For a while, the enthusiasm and euphoria of all the employees was unabashed. Their boss, Bill Shockley, head of start-up Shockley Semiconductor, had just won the 1956 Nobel Prize in physics for his part in inventing the transistor. The future seemed to bubble up in the glasses held by these young men and women toasting their boss.

Shockley, 46, could be very difficult, overbearing and suspicious. He cut people down. He brought them up short. That day, however, the drinks were on him. "I never adjourned to start drinking champagne at nine o'clock in the morning on any other occasion in my life!" recalled Gordon Moore.

In his Nobel lecture, Shockley states:

> *"Frequently, I have been asked if an experiment I have planned is pure or applied science; to me it is more important to know if the experiment will yield new and probably enduring knowledge about nature."*

The Nobel prize only boosted Shockley's scientific attitude and it was only a question of time until his business "experiment" would fail.

In January 1957, Beckman opened a new Spinco (Specialized Instruments Corporation) facility in the Stanford Research Park, where some of the production machinery (crystal growing equipment for example) of Shockley Semiconductor was located. The first indication of trouble with Shockley's team started in early in January 1957. R. V. Jones had some ideas about molten zone refining and had a disagreement with Grinich and Hoerni. On January 16, Jones complained to Gordon Moore about *"mental stagnation."* The next day Jones resigned.

Horsley suggested Shockley talk to Noyce. Noyce was very factual and characterized the situation as *"lab having family troubles"* and if these troubles were allowed to proceed three persons will be leaving. One of them was Hoerni. Shockley felt that "it would be catastrophic if Roberts and Hoerni left." Hoerni was literally begging Shockley to allow him to be more involved in the experimental work. Shockley wanted to keep him as a theoretician. Roberts, who re-designed and significantly improved Bell Laboratories' crystal growing machine, and who also designed an improved version of the micromanipulator, was unhappy with Shockley's "passion for mechanical engineering" and arguing about every small detail.

Noyce pointed out that a lot of work was done behind Shockley's back because some individuals felt they could not tell Shockley. Those who had tried were afraid that they would be fired. Noyce agreed to work out the problems and act as a mediator. Shockley concluded the discussion with Noyce by stating *"I believe it can be still made to go successfully. I would prefer to try to do so."*

In January 1957, Shockley Semiconductor Laboratory's expenses were approaching $ 1 million. The total R&D budget of the Beckman Instrument

Fig. 3.8. Celebration at Rickey's, November 1956. Standing (from left to right) Gadbury, Horsley, Peterson, Lee, Noyce, De Bernardi, Asemissen, Last. Sitting: Clifton, Moore, Roberts, Jones, Shockley. Turned back: Wagner

Company reached almost 14% of sales revenue and after a public announcement of these results the Beckman Instrument's stock dropped by 5% in March 1957.

In 1956 the R&D expenditure of Beckman Instruments reached 10.6% of sales, and 13.6% in the first quarter of 1957. During a Conference between all divisions, Beckman Scientific, Helipot, Berkeley Process Division, Spinco, Shockley Semiconductor, and Liston-Backer plant, Arnold Beckman ordered the total R&D budget be reduced to 8.6% of sales.

Jack Bishop was delegated to meet with Shockley to discuss the new situation. Shockley was now surrounded with hostile forces not only from Beckman's discontent from the top, but with growing difficulties from the staff under him.

Bishop, Beckman and Shockley decided to focus mainly on two projects: the main project was the four-layer diode and the internal project was the Junction Field Effect Transistor. The four-layer diode was in 1957 a very revolutionary device which might enable many new electronic applications. The project was supported by military contracts (NONR 2934(00) Navy, AF339616)-6707 Wright-Patterson, and DA36-039 SC 85239 Signal Corps) and could not be canceled. The JFET request originated internally from the Beckman instrumentation group for chopper application in DC amplifiers.

The four-layer diode group was managed by Horsley, and Allison was in charge of device development, Hoerni did the diffusion work, evaluation,

Fig. 3.9. Standing (from left to right): Blank, Kleiner. Back of table: Knapic, Emmy Shockley, Horsley. Sitting on left side: Sah, Allison, Shaffer, Pretzer, Troyer, Happ. Sitting on right side: Ms. Bolander, Kreick, Grinich, Brown, Ms. Breen, Hoerni

and control. Last was responsible for all measurement and data interpretation. Roberts did soldering and dicing, Knapic did the mounting of the device to the fixture, and Sah did the study on capacitance effects. The field effect transistor project was directed by Shockley, Kleiner and Happ, with technicians Farwell, Parker, Pleibel and Tavares responsible for fabrication and evaluation. Grinich with Paterson worked on circuit application; Noyce, Blank and De Bernardi worked on diffusion; Harry Sello with Brown worked on wax masking and etching. Fok with technicians Lewis, Smith and Wagner were developing gallium diffusion. Gordon Moore and Blank, with technicians Ford, Grunewald, Clifton and Madden, grew crystals and worked on some diffusion furnace hardware.

Gordon Moore in an interview for Scientific American [3] said:

> *"We fiddled around, trying to make some devices. Then Shockley changed direction. When I first went to work for him, he was thinking of making a transistor. But then he decided he wanted to make a rather obscure device called a four layer diode."*

Once again Gordon Moore was not well informed. Shockley did not change the direction of the company; there was strong interest in four layer diodes. Professor Jim Gibbons, who worked part time as the first Shockley application engineer, worked with MIT on core computer memory switched by a four layer diode. Philco Corporation ordered diodes for detonator firing applications, where they needed to switch high energy in milliseconds. The market potential for the four-layer diode was claimed to be immensely large. Beckman Instruments' Annual Report from 1957 stated:

> *"Shockley Semiconductor Laboratory have developed and are manufacturing a new four-layer diode that weights less than a quarter of an ounce. A transistor-like device can do the work of five related electronics components ten times its size and functions under high temperatures and gravitational forces."*

Contrary to common myth, Shockley Semiconductor Laboratory never worked on a mesa transistor similar to the device which was developed in Bell Laboratories. From the early days of Shockley Semiconductor Laboratory, the four-layer diode was the main focus of all activities. Shockley and Beckman licensed Bell Laboratories technology which they were not directly using, the license only helped to maintain unrestricted flow of information between both places. Shockley Laboratory was using silicon crystals prepared by Bell Labs, blueprints for some manufacturing equipment and had detailed information about every single technological step to produce silicon semiconductor devices.

There were two key processing steps which were known, (except for Bell Lab engineers) only to Shockley Semiconductor employees. The first was the idea of using photoresist for patterning of semiconductor structures. When Shockley was still in Bell Labs, he hired Jules Andrus, a professional commercial artist as one of his technicians. Ray Warner, who was carpooling with Andrus, remembers that Andrus commented one day in 1954: *"I won't be with you fellows in the carpool tomorrow. The boss has some crazy idea about using photoresist on germanium, so he is sending me to Rochester (Eastman Kodak) to pick up a bottle of the stuff and some instructions on how to use it."*

Morgan Sparks sent a letter to W. Shockley with a note that Bell Labs group was attempting to develop resist patterning of transistors (Fig. 3.10). Shortly after Shockley left BTL for Caltech, Andrus' patent was issued in 1964 without Shockley's name. It is not clear if Shockley simply "gave" the invention to Andrus, or if Bell Labs' Patent Department processed the patent application and removed Shockley's name as a result of Shockley's departure [4].

The second, equally important invention was the idea of using silicon oxide to protect the semiconductor surface during diffusion. The very early experiments performed in Bell Labs by C. J. Frosch indicated that a silicon surface processed in the vapor-solid-state-diffusion systems at temperatures above 900°C with non-oxidizing gases such as H_2 or He was seriously pitted. In addition, in some situations, a significant evaporation of the silicon surface occurred.

BELL TELEPHONE LABORATORIES
INCORPORATED
MURRAY HILL LABORATORY
MURRAY HILL, NEW JERSEY
SUMMIT 6-6000

May 14, 1954

PROFESSOR WILLIAM SHOCKLEY
The Athenaeum
551 South Hill Avenue
Pasadena 5, California

Dear Bill:

I am enclosing a clipping from last Sunday's New York Times which will be of interest to you. I have forgot-

---

Your talk with Jules Andrus spurred development there, and we are now attempting to make some transistors from ~~induced~~ diffused layers using a photoengraving technique.

Best wishes,

Morgan

1152-MS-JE

Fig. 3.10. M. Sparks letter to W. Shockley with information about J. Andrus effort at Bell Labs

C. J. Frosch and L. Derick, assisted by C. S. Fuller, M. Tanenbaum, J. Moll and F. M. Smith, found by accident that oxidizing gasses such as O_2 or CO_2 or non-oxidizing gases containing sufficient concentration of oxygen or water vapor will not change the surface of silicon. Frosch and Derick discovered that the silicon oxide surface layer provides a selective mask at high temperatures against diffusion of some donors and acceptors into silicon. In a Bell Labs memo titled "Masking by SiO_2 Film" Frosch and Derick described in detail the masking effectiveness of the oxide layer and its dependence on the time, temperature, carrier gas composition, impurity type and pre-oxidation conditions.

Very likely, because Bell Laboratories researchers did not consider photoresist as a viable technology, Shockley did not follow his own suggestion and take advantage of Andrus work. Shockley rather adopted a wax masking technique developed by Miller in the Western Electric Allentown facility. As everybody knows today, this was the wrong decision.

The Shockley group had full access to Bell Laboratories' know-how. When something was not clear they could just call for clarification. The ambitious group was very interested in Bell's diffused transistor developed by Miller in the Allentown facility, and the new masking method using resist and an oxide as a protective layer. Shockley routed all Bell memoranda around to his staff. The routing list attached to Miller's memo (Fig. 3.12) describing all details of the new diffusion transistor is initialed by Noyce, Allison, Hoerni, Last,

Formulae for Etching Intricate Patterns in Oxide Layers on Silicon - [redacted]

[redacted]

October 16, 1956

MEMORANDUM FOR FILE

ABSTRACT

A new etch has been formulated which will remove a film that is oxidized on the surface of silicon crystals.

The photo-engraving technique is used to secure a desired pattern. By etching the exposed portion of the oxide film that is not protected by the photo-resist, a pattern of oxide is left on the surface of the crystal under the solid portion of the photo-resist. This oxide pattern can now be used as a diffusion mask and also as an acid resist mask where removal of silicon is desired, using the special silicon etch formulas specified in this paper.

INTRODUCTION

An effective technique to coat silicon surfaces with a thin SiO_2 (oxide film) was devised by C. J. Frosch and L. Derick.(1) They found this oxide film was useful as a mask for diffusion purposes. The photo engraving technique was tried to photographically reproduce a desired pattern but unfortunately the emulsion (photo-resist) did not adhere to the oxide film long enough for the pattern to be etched through the oxide film because of the strong concentration of Hydrofluoric (HF) acid that was being used as an etch.

ACKNOWLEDGMENTS

The authors wish to acknowledge the help rendered by C. J. Frosch and L. Derick for supplying us with oxide coated silicon crystals used in our experiments, also E. Berry for oriented and polished silicon crystal blanks.

(1) [redacted] C. J. FROSCH and L. DERICK MASKING BY SiO_2 FILM

(2) [redacted] J. ANDRUS - CONTROLLABLE CHEMICAL ETCH

(3) [redacted] J. ANDRUS and W. L. BOND PHOTOENGRAVING IN TRANSISTOR FABRICATION

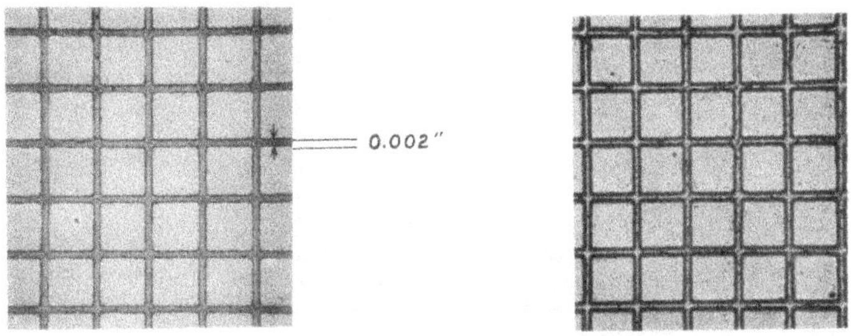

a. Dark lines are portion covered with photo-resist.

b. Lines have oxide film, squares are portion that were etched.

Fig. 3.11. Title and abstract of Bell Telephone Laboratories Memo describing the use of oxide as a diffusion mask dated October 16, 1956

and others. If there were any remaining questions they would be answered by Bell Labs staff. Shockley engineers, with access to Bell Labs' information about the latest technology, learned in a matter of months all they needed to produce the best conventional transistor at that time.

BELL TELEPHONE LABORATORIES
INCORPORATED
MURRAY HILL LABORATORY
MURRAY HILL, NEW JERSEY
CRESTVIEW 3-6000

Reprint 838

November 9, 1956

DR. W. SHOCKLEY
Shockley Semiconductor Laboratory
of Beckman Instruments, Inc.
391 South San Antonio Road
Mountain View, California

Dear Bill:

I am enclosing a copy of the memorandum by Bob Miller which I mentioned in our telephone conversation. This work was undertaken to clear up some uncertainties about the solubility and diffusion constants of aluminum in silicon, and has not been followed up as a possible device process.

We have not experienced your trouble with phosphorus contamination from quartz tubes so far as we know. A couple of years ago Bruce Hannay had analyses made on all samples of quartz which we could lay our hands on. These mass spectrographic analyses didn't turn up phosphorus in any samples except some natural Brazilian quartz. He did find a number of other elements including lithium, which was our main concern at that time.

We have done very little work with graphite as a possible environment for diffusion. Friedolf Smits uses a small graphite tube as a holder for gallium since gallium wets everything else we know of. This tube was vacuum baked at about 1800°C before use, and has proven quite satisfactory.

BELL TELEPHONE LABORATORIES
INCORPORATED
MURRAY HILL LABORATORY
MURRAY HILL, NEW JERSEY
CRESTVIEW 3-6000

July 3, 1957

DR. WILLIAM SHOCKLEY
Shockley Semiconductor Laboratories
of Beckman Instruments Inc.
391 S. San Antonio Rd.
Mountainview, California

Dear Bill,

This afternoon I received a telephone call placed by Bob Noyce (I did not get to talk to Noyce, but to one of your other men whose name escaped me), asking for information on the design of our PNIP silicon power transistor.

Since talking this over with Jack Morton, we feel that it would be desirable to send this information directly to you for routing to your people.

Attached is a copy of a memorandum by J. T. Nelson, J. E. Iwersen and F. Keywell which gives the information in which Noyce is interested. We expect that this will be published in more or less this form about the end of the year.

Please let me know if you need additional information which is not included in the memo.

Sincerely,

W. J. Pietenpol

W. J. PIETENPOL
Director of Development-
Solid State Devices

MH-252-WJP-JEG

Att.
Paper

Fig. 3.12. Front page of letters mailed to W. Shockley by M. Sparks and W. J. Pietenpol

The conventional transistor was no challenge to Shockley. He wanted a different transistor. Early in 1957 Shockley and Noyce envisioned a junction field effect transistor for application as a low level switch. Their work later became U.S. Patent 2,967,985. The junction field effect transistor (JFET) is operated by increasing and decreasing the width of the space-charge region in the channel of the transistor. As the thickness of the space-charge region increases, current flow from the source to the drain through the channel decreases. To achieve a reasonable frequency response, the length of the channel should be in the order of ten microns. To manufacture such small devices in 1957 was unrealistic.

The typical structure of a junction FET as used at Shockley Semiconductor Laboratory in 1957 is shown in Fig. 3.13. Then most of the problems with this device development were related to the diffusion and contacting the diffusion regions. Poor junctions and even worse contacts were reasons for a large reverse leakage current.

The channel length was 15–20 μm wide, junction depth was 4–5 μm and the grooves were defined by Apezion wax and etching in 90% HNO_3 and 1–10% HF. The etching rate for a 1% HF solution was 0.8 μm/min @ 25 °C. Plated contacts on the top side were evaporated through a shadowing mask.

On May 15, 1957 Noyce tried make a transistor by evaporating aluminum across the entire structure and etching grooves through 10–20 μm thick wax mask grooves. He found that he could not obtain a uniform depth of grooves by etching through the aluminum.

Noyce experimented with any possible metal to form a reliable contact. An example of the experiment showing the dependence of reverse current

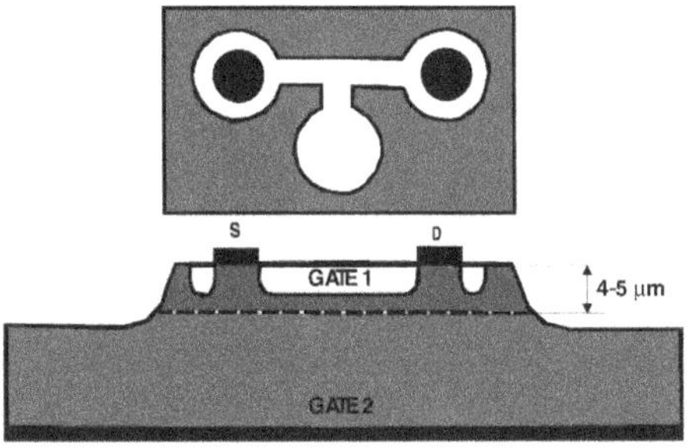

Fig. 3.13. Structure of Junction Field Effect Transistor designed in Shockley Semiconductor Laboratory in early 1957

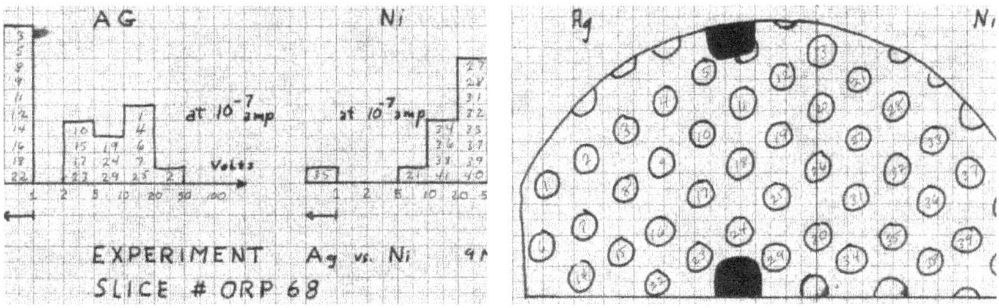

Fig. 3.14. Dependence of reverse current on material used to form a contact of JFET device

on material used for contacts is shown in Fig. 3.14. Harry Sello made a major improvement to the device pattern definition when he found that the wax evaporated from the heated container at 300–320°C exhibited the best adhesion to the silicon surface. Hoerni suggested the use of Gallium as a dopant to form a more abrupt junction. All these changes improved the device behavior, but over all the device performance remained very erratic.

The junction FET was planned to be used in the all-transistor DC amplifier in Beckman Instruments/Data Control system unit (Model WN). W. F. Gunning of Data Control division and Noyce worked really hard, but sometime even hard work was not enough. The Luck Factor is often needed in such a situation.

The four-layer diode was a very versatile circuit element which was considered to have an extensive use in electronic circuits. The four-layer diode was a two-terminal device which could stay in either of two states – an "open" or low conductance state corresponding to 10 to 100 MΩ and a "closed" or high conductance state corresponding to 3 to 30 Ω.

The diode is switched from one state to the other by controlling the voltage and current through it. If the voltage exceeds the "breakdown" voltage V_b, the device will change from open to closed, provided sufficient current is available to hold it in the closed state. The necessary current is called the

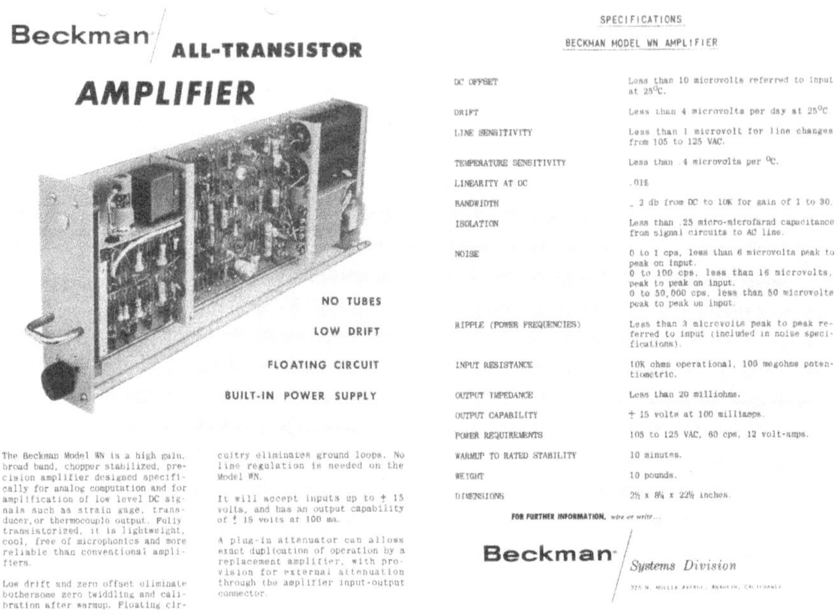

Fig. 3.15. Beckman Amplifier which should be using junction FET as a chopper

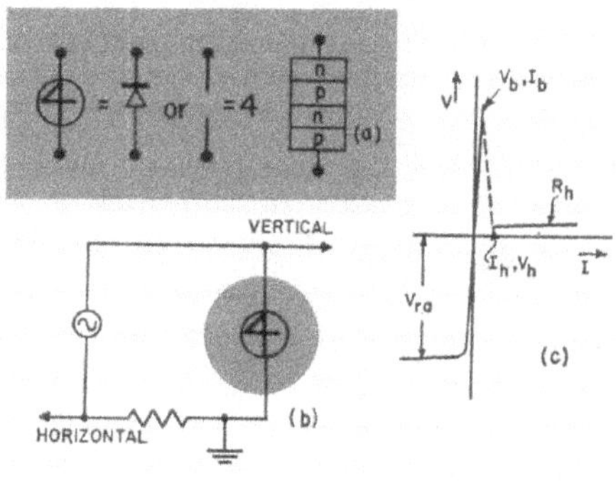

Fig. 3.16. AC test circuit and typical scope display during test of four-layer diode

holding current I_h. If the current falls below I_h, the diode will switch back to open.

The processing of the first version of the four layer diode consisted of the following steps:

Starting Material: Silicon 0.5 Ω-cm, 42 micron thick

1. Cleaning (Initial Oxidation)
   Oxidation 1300°C/10–15 minutes
   HF Oxide strip

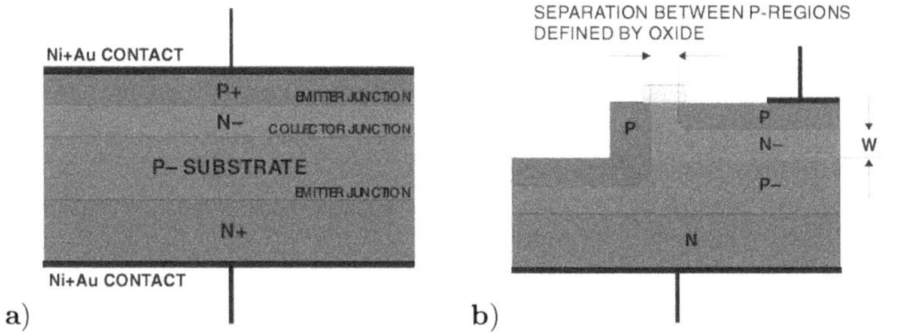

Fig. 3.17. Circular structure of four layer diode which was produced since the second half of 1958 (**a**), and proposed partial planar structure of Four Layer Diode (**b**)

2. N-Predeposition (P_2O_5)/30 min
   Wafer temperature 800°C
   P_2O_5 Temperature 210°C
   Layer resistance 30 Ω/sq
3. Wax mask
4. Glass and oxide etch
5. Wash wax (HF/20 min)
6. Drive-in 1300°C/90 min
7. Wax mask (Etch step)
8. HNO_3 /HF etch (80 sec)
9. Glass and oxide etch (form oxide strip)
10. P-Diffusion (B_2O_3)
    Wafer temperature 1200°C/30 min
    Layer resistance $< 1\Omega$/sq
11. Glass and oxide etch
12. Wax mask
13. Contacts
    Nickel plate

The thickness of each layer was approximately 8 µm. Hoerni worked mainly on the diffusion and samples evaluation; Harry Sello developed the wax masking; Sheldon Roberts and Noyce worked on the ohmic contacts. The first samples were completed in the wood structure of Shockley Semiconductor Laboratory with a wood ceiling, where space was shared between the machine shop, diffusion furnaces, dust and dirt. By a miracle, some of the diodes worked. It became very clear that production could only be successful in a reasonable clean environment. For this reason Beckman Instruments management set up the first clean laboratory space in a new Spinco Division building in the Stanford industrial Park.

The processing of the Junction FET and the four layer diode in today's environment is trivial, and most college students have no problem building such devices. In 1957, when diffusion was a very erratic process step, and working with wax masks was an experience as fulfilling as a spiritual affair. The working device was more than an accomplishment. Shockley and his new team completed both the junction FET and the Four-Layer Diode in 1960. Unfortunately for Shockley, a new semiconductor industry was already working on integrated circuits, and Shockley, for the first time in his life, was not leading this revolution.

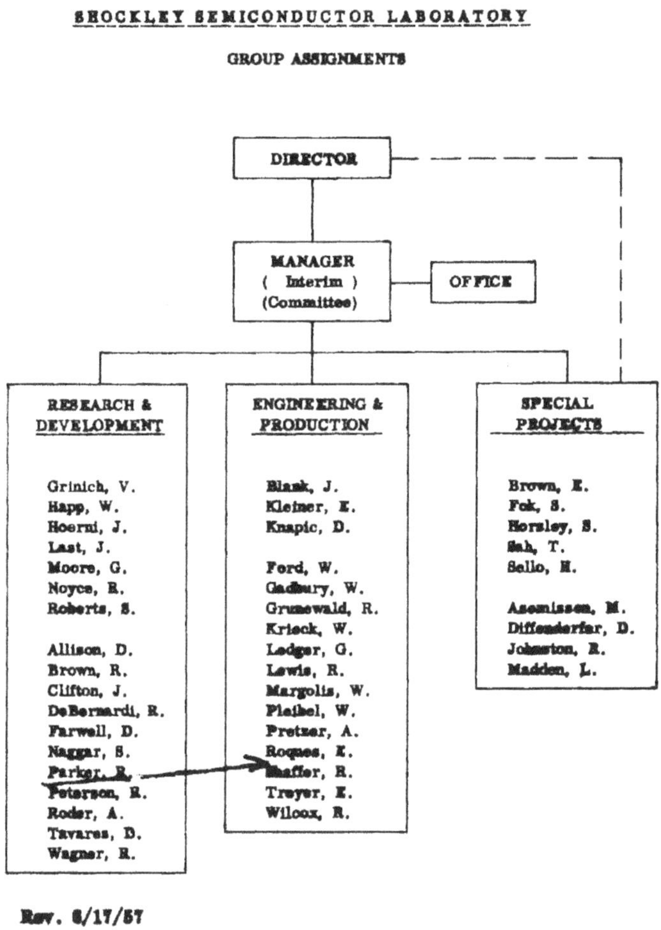

Fig. 3.18. Shockley Semiconductor organization chart as of June 17, 1957

In the environment of "try-learn-try-again" most technical people such as Hoerni, Roberts, Allison, and Noyce in Shockley's group learned semiconductor manufacturing know-how with extreme speed. During April 1957, Hoerni, and Last switched to open-tube diffusion, similar to the Bell Labs. The closed-tube effort was stopped. In Shockley's Semiconductor Report dated April

4, 1957, Hoerni and Last referred to the Bell Labs memo recommending to switch to oxygen instead of nitrogen as a carrier gas during the boron diffusion. Hoerni suggested switching from B_2O_3 to BCl_3 as a boron source.

Hoerni was reporting to Horsley and Shockley:

> *"one of the principal points discussed was: are we making optimum progress towards production or would we get to production more rapidly by starting on a program of making Chinese copy of present method used by BTL. It was decided that the program was quite close to fruition so we would continue on the present course. However, to prevent the possibility of pursuing an elusive solution for another week or so it was decided that deadline should be establish and if the present program did not have evidence of a predictable, the effort should be shifted to the present BTL method."*

Shockley, Sah and Noyce explained the recombination effect in the space charge region and published their study, which today is considered as a major contribution to the understanding of semiconductor devices. In other work, Shockley and Last described the behavior of defects in semiconductors, work which became a building block to future understanding of diffusion in semiconductors.

The one who benefited the most from these remarkable experience was the group of young engineers who were well equipped for their future enterprise. With "hands on" experience they had an advantage to do experimental work of their choice. The group of Shockley's young scientists, with the strong ego and desire of prestige, rewards and motivation to prove themselves, started to run into more frequent conflicts with Shockley.

In March 29, 1957 Shockley wrote down a note in his notebook: "*... surprised to find Kleiner had gone to East. On Tuesday March 26, 1957 gone to LA for metal show. I thought he will be gone 3 days. Actually he will be gone altogether 9 working days. Production chart is 4 weeks out of date.*" Kleiner left to go East to explore the possibility of the financing a new company.

Controversial Dean Knapic started to push his proposal to manufacture and sell the silicon crystals. William W. Happ was most often a target of criticism of young scientists. Happ was really incompetent and Shockley would have avoided a lot of troubles if he had removed Happ from any decision making.

Shockley's troubles started to accumulate. Shockley suffered from insomnia. Strange phone calls ringing late in the night or early morning in his residence only worsened Shockley's very poor social skills.

Shockley's suspicions pointed to Shelton Roberts. The bizarre situation culminated when Shockley's secretary cut her finger on a small bit of metal sticking out from the door to her office. Shockley accused Roberts of sabotage. Roberts examined the piece of metal under a microscope and found out that it was a pushpin that had lost its cap!

This "Pin Affair" entered into history: Gordon Moore became instrumental in forming a group of dissidents and acted as their spokesperson. The group congregated at the Black Forest restaurant in Los Altos and Gordon Moore called Dr. Beckman.

Beckman agreed to meet with Moore and his compatriots, in confidence, over dinner. Moore examined the situation at Laboratories and presented an ultimatum to Beckman: *"Unless some kind of solution could be found, the group will resign together."* Moore's requirements put Beckman into an unsolvable situation. On one hand Shockley, Nobel Prize winner and recognized achiever, on the other hand group of young scientists who one year ago knew nothing about the transistor. Beckman had no other choice and said to the group *"I am not going to fire Shockley at this stage of the game."* He did not want to abandon the group either. Beckman decided to break the threat of resignation to Shockley in person and expected that Shockley would handle the situation himself. Beckman set up dinner with Shockley and asked him to change his managerial approach, otherwise a significant part of the group would leave the company. Shockley was stunned by the news, but he was not able to learn and benefit from Beckman's advice.

On June 5, 1957 Robert Noyce met Shockley and proposed the changes requested by the "Group." Noyce suggested that they work as a team on transistors equivalent to Texas Instruments. The next day, on June 6, Shockley called Noyce again and asked if *"does he have any new suggestion for getting together?"* It was already too late to negotiate any settlement. During Kleiner's trip to the East at the end of March 1957, Kleiner's father established a contact with a New York based investment firm, Hayden Stone & Co. A Kleiner senior passed the desire of the group to establish a new company to Alfred Coyle, then manager at Hayden Stone, and Art Rock, a fresh MBA from Harvard who worked for Coyle. Coyle was interested in diversifying Hayden Stones' investments and he took Rock's advice that a spin-off from the Nobel Prize winner Shockley is almost a certain success.

In mid June of 1957, Beckman put Joseph Lewis into a newly created position between Shockley and the company staff. Shockley responded to Beckman with a plan *"if decision is for me to go ahead in saddle:"*

1. Company policy will be controlled by Hanafin-Noyce-Knapic-Moore (if tie.)[2] When they cannot decide they bring case to me. I cannot overrule but decisions and proposals are made matter of record.
2. Crystal sales program – proposal by Knapic must be judged by Beckman (not WS)
3. Beckman must re-establish authority
4. WS security and job satisfaction firm up before September 3
5. Beckman (not WS) decide about major objections against W. W. Happ

[2] A reason why Shockley considered Gordon Moore into the management team was Arnold Beckman's demand to work with the rebellious group leader.

A few weeks later, Arnold Beckman promoted Maurice Hanafin as a General Manager of Shockley Semiconductor Laboratories. All these changes had no effect because the "Group" already had a decision. They were notified that Bud Coyle had a young assistant Arthur Rock, and they are flying to San Francisco to meet Shockley dissidents at the Clift Hotel.

Robert Noyce found himself in a peculiar situation. Noyce was well respected by Shockley and managed to deal with Shockley's complicated personality. He was the only one from the group with previous experience of germanium transistor technology from Philco and he did not believe that Gordon Moore's group would be able to find and obtain funding.

The night before meeting with financiers, Shelton Roberts negotiated with Noyce and asked him to desert Shockley. After a night of wrestling between Shelton Roberts and Robert Noyce, Noyce joined the group and got the privileged passenger seat in Robert's van on the way to San Francisco. At the end, the eight Shockley employees and two bankers agreed to start their own company. It was now up to Coyle and Rock to find a sponsor. They scheduled second round of talks for the next morning. Meanwhile the dissidents continued their work at Shockley, preparing themselves to build silicon mesa transistors with diffusion technology they learned from Shockley.

Arthur Rock contacted about 40 companies where a semiconductor operation might fit. Every one of them turned the offer down. In the beginning of August 1957, Coyle and Rock accidentally ran into Sherman Fairchild who directed them to John Carter, the chairman of Fairchild Camera and Instrument Corporation. Carter considered the offer worth considerating and dispatched Richard Hodgson to meet with the seven applicants in San Francisco. Hodgson made a deal. As soon as all legal papers were drafted, the group would resign and join a new organization. Prokhurst & Co., a New York based legal firm, which was hired to do the legal work, informed the group in the middle of September 1957 that the legal papers were ready for signatures.

Suddenly, brilliant Shockley, who was called a *"marvelous intuitive problem solver"* and a *"tremendous generator of ideas"* by Robert Noyce, turned out that he was *"hard as hell to work with"*, as his style was *"oppressive"* and he *"didn't have trust and faith in other individuals."* R. N. Noyce, E. Kleiner, C. S. Roberts, J. Blank, V. H. Grinich, J. A. Hoerni, J. T. Last, G. E. Moore submitted their resignation to William Shockley on September 18, 1957.

Shockley wrote in his notebook "Group resigns." As a few times before and many times more in the future, Emmy Shockley was the only ally who firmly stood on Shockley's side.

Three months later Dean Knapic formed Knapic Electro Physics with other principal shareholders Gerthard R. Fisher, S. Marshall Kempner, and Benjamin Swing.

Several of Shockley's former employees publicized that Shockley had an exceptional ability to find good and talented people. This claim is, of course,

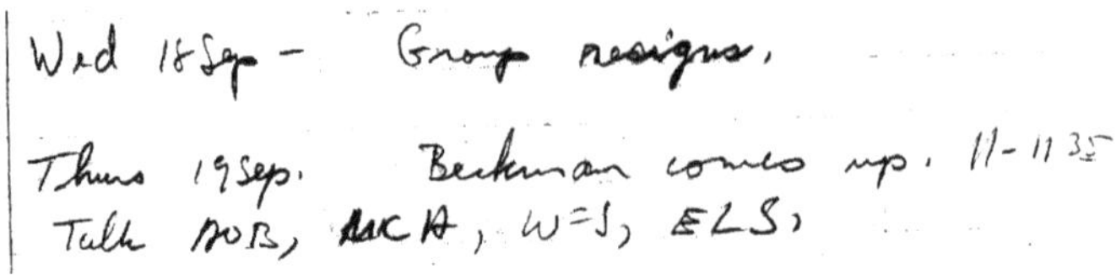

Fig. 3.19. Note in W. Shockley notebook dated September 18, 1957

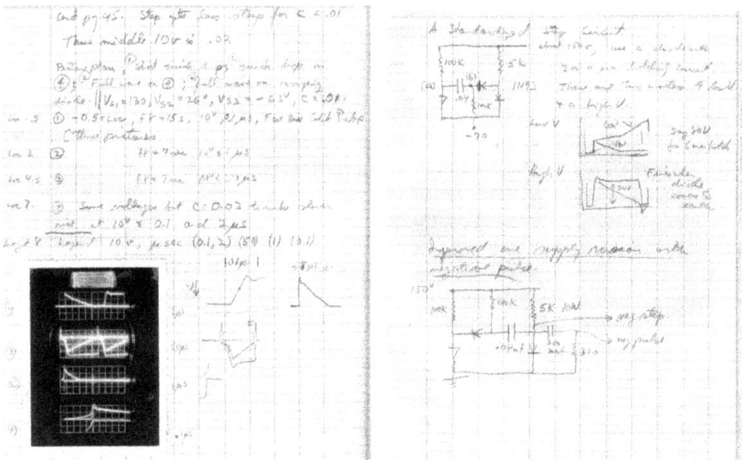

Fig. 3.20. W. Shockley Notebook # 28 last entry covering work on four layer diode before "group resign." Next entry is dated by January 29, 1958 with note "Witnessed, discussed and understood, J. Gibbons"

because Shockley selected them. History clearly proved that this myth was not true as he hired several employees of questionable integrity. For example, Dean Knapic was recommended to Shockley by Knapic's supervisor in Western Electric. Knapic falsified a record about his education in Yugoslavia and claimed that he was the Navy commander involved in highly specialized operations of U.S. Navy in the South Pacific during 1942–1946. When Knapic in 1958, stole Shockley's crystal growth technology and established Knapic Electro Physics, Inc., Shockley found out that Knapic was a con artist and never was a commander nor had a Ph.D.

On September 20, 1957 Shockley stated that the resignation of the group of defectors – reportedly referred as the "Traitorous Eight" had no real effect on the Shockley Laboratory. But he knew that this was not true.

Shockley never understood the reasoning of the group of "Eight Traitors" although he was the type of person who never gives up easily; he also never fully recovered from this set back.

Because Fairchild Semiconductor did not participate in Bell Labs "Diffusion Conferences" and did not pay a license fee for the transistor technology,

[redacted]

Beckman *Instruments, Inc.*

to Dr. Shockley

INTER-OFFICE MEMORANDUM

TO: See "Copies to:

DATE May 28, 1959

FROM: L. N. Duryea

COPIES TO: R. Erickson
W. W. Wright
M. C. Hanafin
R. J. Steinmeyer

SUBJECT: Fairchild Investigation

Tuesday evening, shortly after eight o'clock, May 26, 1959, I met with Dr. Ewart, (Ed) Baldwin, vice president and general manager of the Rheem Semi-Conductor operation, and Dr. Hank Marchart, his director of engineering.

Baldwin told me the following: From February 7, 1958 through March 5, 1959 he was vice president and general manager of Fairchild Semi-Conductor Corporation, whose president was Erwin Hale, an officer of Fairchild Controls Company, a subsidiary of Fairchild Camera Co. The actual strings were pulled by Richard Hodgson, the executive vice president of Fairchild Camera.

Fairchild Camera executed an agreement September 17, 1957 with the eight individuals formerly employed by the Shockley Semi-Conductor Division of Beckman Instruments, Inc. The contract recited that the eight were a group of scientists engaged in the research and manufacture of semi-conductor devices

The agreement had been negotiated over a six month period. The agree ment provided that Fairchild would provide 1.38 Million Dollars over an 18 month period to a corporation to be formed by the eight individuals in exchange for a

---

Other companies, specifically Pacific Semi-Conductor, Hughes, Western Electric and now Motorola Texas Instrument, have these products. Hughes and Pacific Semi-Conductor developed theirs on government contracts and Western Electric produced its product only for its own use. The government contracts were with the Signal Corps and were R & D and first runcontracts. The R & D information was, of course, published and there are other articles, namely Kramer in Germany and Early of Western Electric, concerning these techniques.

He suspected that the group had copies of their notebooks at home, as when problems came up they often stated, "I will check that at home", and the next day had the answer. They were careful not to employ Shockley personnel and did not discuss their prior activities at Shockley although the history of the formation of their group and the agreement with Fairchild was disclosed to Baldwin by Noyce. He felt the only thing that the Fairchild people themselves developed was the tooling. The only Shockley employee that they later hired with real ability was Allison.

All manufacturers except Fairchild and Shockley use a different way of diffusion, indicating that the particular process employed by Fairchild was acquired from us.

Fig. 3.21. The letter from Beckman Instruments, Inc. legal counsel L. Duryea to M. C. Hanafin

Arnold Beckman later asked his legal counsel Leslie Duryea to investigate possible legal action against Fairchild Semiconductor. The Beckman Company's legal department wanted to file suit and stop Fairchild Semiconductor from using information acquired during their association with Shockley. (Fig. 3.21) Duryea argued that the group had access at Shockley semiconductor Laboratory to all the information from BTL "Diffusion Conferences", and they were aware of the latest work on oxide masking and resist "photoengraving." Duryea was referring to a memo dated October 16, 1956 described in great detail the latest BTL development work on the oxide surface protection (Fig. 3.11.) Shockley attached a routing list to the memo and Robert Noyce, Gordon Moore, Jean Hoerni, and Jay Last initialed the routing list.

Duryea also dwelled on an announcement that Mr. Hale published in Electronic News on October 21, 1957 stating that the new group's first product, is a high speed mesa silicon transistor for computer applications. He said that the device was already in the engineering stage and production would get under way in *"a number of months."*

Duryea put together a solid case. Some of the young engineers abandoned not only Shockley Semiconductor Laboratory but also their girlfriends. One of them during deposition testifies that she knew about copying of Shockley "cook books."

Shockley, however, was not interested in any legal dispute and refused to participate in the Beckman Instruments legal process. No legal action was initiated. Eventually, Shockley's other employees joined the "Eight" at Fairchild. Between them, for example, David Allison and "Tom" C. T. Sah,

Frustrated Beckman sold the Shockley subsidiary to the Clevite Transistor Company in 1960, ending his formal association with the semiconductor industry. Shockley Semiconductor Laboratory changed name to Shockley Transistor, Unit of Clevite Transistor. Shockley placed in Physics Today series of very unusual recruiting advertisements (Fig. 3.22) and hired mostly young engineers from Europe.

Everybody who worked with Shockley described their personal experiences and portrayed all the negatives of Shockley's personality. When a mob is lynching, it is easy to add more fuel to the fire. Former Shockley chemist and later manager of Fairchild S. G. S. division in Italy, Harry Sello, who considered himself a good swimmer described Shockley's negative personality in this statement: "*He* (Shockley) *invited me to go swimming with him one lunch time at Stanford. Just, you know personal invitation. Hey come on; let's go take a swim. He had heard earlier that I liked to do a little swimming or did a lot then. He said wonderful pool. Let's go over there, take some time and relax. And we both dove into the pool about the same time and he said I'll race you. And he beat me; he was pretty fast. And we kept swimming a little bit. And after about four or five lengths I sort of hauled myself out gasping at one end of the pool. I wasn't in quite good shape. Shockley stayed in there and completed over thirty lengths. And he had to show me how much better he was at even swimming than I was. And that was a mark."* Shockley was a bad person, very simply, because he was a much better swimmer then Harry Sello. That's how simple and low some people could be in order to win their argument.

Can any person be only negative? When Shockley Semiconductor Laboratory was acquired by Clevite Transistor Product, president of Clevite, William G. Laffer was concerned with many thefts of know-how from Shockley Semiconductors, Shockley wrote on June 13, 1963 this letter:

Dear Bill,
Clevite, Cleveland

After our conversation of Thursday a week ago, I find myself much concerned about several questions you raised involving moral character of research and development personnel.

When Kurt Hubner came to my office and told me of his decision to leave (to return to Switzerland to a company supported I believe in part by government and in part by watch industry) I stressed his obligation to stay with us to achieve the technical objective of Rome four-layer diode contract. It is my understanding from A. Goetzberger that now we have never been so well ahead of schedule on a device contract as we are on this one.

I have the same reactions to the allegation that Hubner has boasted of leaving with a collection of all our technical information. I can imagine his boasting that he had learned a lot and greatly increased his ability while with us. In fact, I would be disappointed if he did not so boast.

I have also checked that he at Biesele's instructions returned all copies and technical memoranda including those he had authored himself.

I have talked to Maurice Hanafin whose integrity and sound objective judgment of people have my highest respect, and he feels also that Hubner would not perform an ethically questionable act and that it would be so completely out of character as to be inconceivable to us that he boasted of improperly taking with him proprietary information.

I believe he could have found better paying job in U.S.A. than with us. Finally, he says, and I believe him, that his wife insisted on a return to Switzerland. This involved another conflicting set of loyalties.

In brief, I think Hubner is a fine, able, effective individual. Given opportunity, I would hire him again and I hope to have an enduring good relationship with him.

With the hope that these comments may have value in Clevite decision-making.

Sincerely yours,
WS

DIFFUSION DOWN DISLOCATIONS

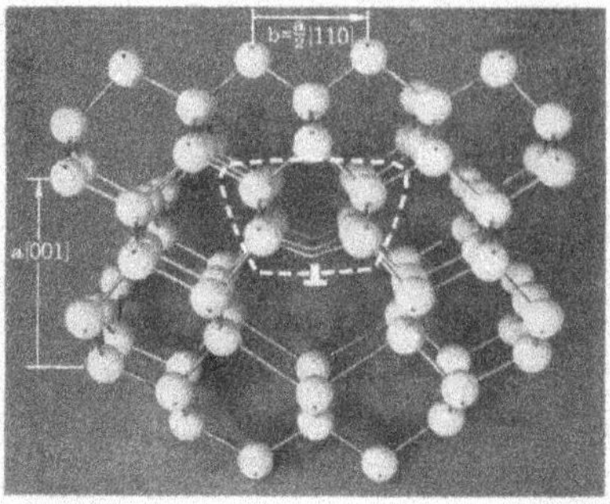

What is the core of a dislocation in silicon like? Does it have unpaired electrons in dangling bonds, or do the electrons pair up as in the model we constructed to illustrate Hornstra's[1] theory? Impurities diffuse down dislocations faster than through the bulk. Is the cause an extra concentration near the dislocation-core of impurities or vacancies or both? Are dislocations good or bad for semiconductor device development? Can they be used as guides for diffusion to produce new and useful structures?

We are trying to get answers to these questions by studying diffusion down small-angle grain boundaries in silicon composed of spaced edge dislocations. We have found a simpler way[2] of measuring the constants describing diffusion on the boundary by making p-n junctions through phosphorus diffusion into p-type silicon. The p-n junction is a

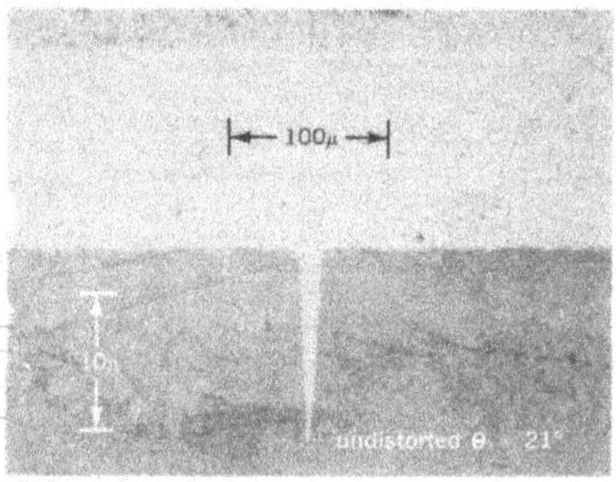

1. J. Hornstra, J. Phys. Chem. Solids 5, 132 (1958).
2. K. Hubner and W. Shockley in *Structure and Properties of Thin Films*, (Wiley, New York, 1959), p. 302; Bull. Am. Phys. Soc. Ser. II, 4, 409 (1959).
   H. Queisser, K. Hubner and W. Shockley (to be submitted to Phys. Rev.).
3. A. H. Cottrell, *Dislocations and Plastic Flow in Crystals*, (Clarendon, Oxford, 1953), p. 56.

(Publications of reference 2 describe results obtained under contract with the Air Force Cambridge Research Laboratory.)

phosphorus isoconcentration line that can be revealed by staining. The velocity v and the angle θ of the advancing spike are measured and the accompanying graph is made. The value of the slope of the straight line can be combined with Cottrell's theory[3] of atmospheres around dislocations to reach the conclusion that most of the grain boundary diffusion current flows down the enclosed two atomic columns of maximum compression, the diffusion current density being 400,000 larger there than in the bulk.

We expect to achieve bigger current concentrations by using atoms like gallium or bismuth which have larger misfit factors in silicon than phosphorus has. We also expect to make transistors of various sorts using diffusion structures built on grain boundaries.

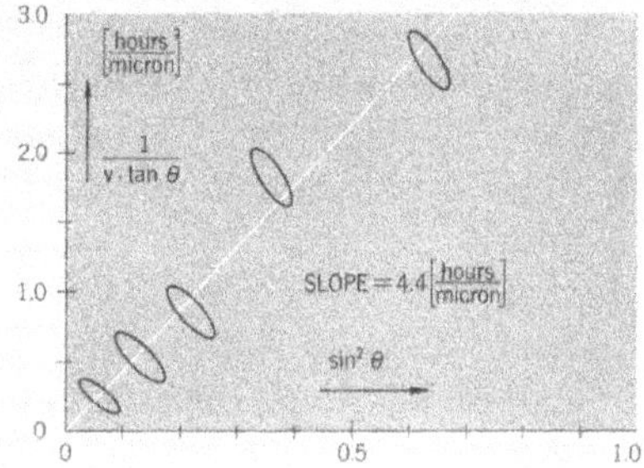

If increasing the fund of knowledge of atomic phenomena in silicon and at the same time helping get new devices to our customers challenges you, please let us know. We hope you have worked on semiconductors, but more important is your ability, general level of experience, and attitude.

For employment opportunities where this grain boundary work is being done, address your inquiries to:

Dr. William Shockley
Shockley Transistor, Unit of Clevite Transistor
Stanford Industrial Park, Palo Alto, California

For other Clevite opportunities in the semiconductor field, write Placement Director at:

Clevite Transistor, Waltham 54, Massachusetts

Clevite Research Center
540 East 105th Street, Cleveland 8, Ohio

Intermetall G. m. b. H.
Hans-Bunte-Strasse 19
Freiburg/Breisgau, West Germany

Shockley TRANSISTOR
UNIT OF CLEVITE TRANSISTOR
STANFORD INDUSTRIAL PARK, PALO ALTO, CALIFORNIA

CLEVITE CORPORATION

Fig. 3.22. One of the series of recruiting advertisement Shockley was using during 1960

When Victor Jones resigned from Shockley Semiconductor Laboratory, Shockley wrote a warm recommending letter to Prof. R. Smoluchovski. When Jay Last wanted to go to work in Europe, Shockley wrote support letter to the Beckman Company in Germany. When a student's megaphone broke down during protests at Stanford, while students called Shockley a Nazi, Shockley looked at the megaphone, fixed some wiring and returned it working back to the student. These are examples of the other side of Shockley's personality.

Young Shockley was hasty and a speeding driver. Jaguar Motors in Coventry, England, still keeps Shockley's letter dated January 28, 1958 which states

> *"My XK-120 has frequently been a pleasure to drive. This is chief reason, from my point of view, for owing such a car. However, the pleasure of driving it has been more than offset by the difficulties of maintaining it in satisfactory condition."*

Shockley had a serious automobile accident on July 23, 1961 and caused Shockley to give up sport and some other activities. The older Shockley, however, did not slow down much.

Shockley accepted a professorship at Stanford University in the summer of 1963. Stanford University News Service released this Press Release on August 5, 1963:

> *Nobel laureate William B. Shockley, co-inventor of the transistor, has been named the first recipient of the Alexander M. Poniatoff Professor of Engineering Science at Stanford University, effective September 1, 1958.*
>
> *His selection was announced today by University provost Frederick E. Terman.*
>
> *A lecturer in electrical engineering at Stanford since 1956, Dr. Shockley will continue as consultant to the Shockley Laboratory of Clevite Transistor, headquartered in Stanford Industrial park. In his new position Dr. Shockley will serve as professor at large in engineering and applied science. His new professorship is named in honor of Alexander M. Poniatoff, founder and board chairman of Ampex Corporation. The endowed chair will permit Dr. Shockley to teach one graduate level seminar regularly through the academic year and personally supervise doctoral level research work by students on campus. Dr. Shockley is one of the five Nobelist's now teaching at Stanford. The others are Prof. Felix Bloch and Robert Hofstadter, both of the Physics Department, and Prof. Joshua Lederberg and Arthur Kornberg, both of the Medical school.*

During his tenure at Stanford and later in his life Shockley became the object of great controversy. At the time when the nation was going through desegregation Shockley picked upon the work of his big idol – Alexander G. Bell, who was, beside the telephone, also interested in eugenics and served as the honorary chair of World Eugenics Conference in 1921. Shockley's interest in eugenics was prompted by an unusual news story. In 1963 *"the proprietor of a delicatessen in San Francisco was blinded by a hired acid thrower, who was one of 17 illegitimate children of an improvident, irresponsible woman with an IQ of 55 who could remember the names of only nine of her children."*

Shockley advocated more research on racial questions, which are largely excluded from the grants by big foundations. He was not exactly positive about what needed to be done, but he thought that the Academy of Science may be an appropriate institution to conduct such a study and find the data needed for accurate conclusions.

Shockley presented a proposal to the Academy in spring of 1966 and asked for a study of the causes of supposed black-white intelligence differences. The Academy of Sciences declined to pursue a such controversial subject. Shockley was lobbying his personal friends. On May 4, 1966 Harvey Brooks of Harvard wrote to Shockley:

> *"I am afraid that in the present climate it is very probably the kiss of death as far as anybody listening you objectively is concerned. If the statement about mean white and non-white IQ's did not have such touchy implications, it would probably have remained unnoticed."*

Frederick Seitz in a letter to W. Shockley from July 22, 1966 wrote:

> *"I can think of few problems more sticky than trying to decide further what can be done about them. The last thing in the world I would do is to suggest that you cease and desist in your pursuit. It may not be unfair, however, to point out that some of the great minds of scientifically advanced countries have speculated on the same issues since the days of The Origin of Species."*

Shockley responded to Seitz on September 13, 1966 *"Brook's appraisal for my effort to urge that we "do the research, find facts and discuss them widely" would lead to attaching the label "racist" to any study."*

On December 14, 1966 in a letter to Seitz, Shockley quoted a letter from Dr. Villard dated November 28, 1966 to Seitz:

> *"The subject is obviously an unpopular one. It makes me proud to belong to an Academy which includes men who are willing to seek the truth, no matter how unpalatable it may happen to be at the time. The great pitfall of democracy is the tendency of leaders whose power is derived from the majority vote, to tell their constituents not what the public ought to be hearing, but rather what they happen to want to hear.*
>
> *It seems to me that if a democratic society is to survive at all, it must include in its ranks men who have the courage of their convictions, who are not afraid to speak out, and who are not afraid to think of the 'unthinkable'.*

Many argued that Shockley had no formal education in genetics and therefore was not qualified to express his opinion. Such an argument is a typical example how a hypocritical and politically correct society deals with problems which it is not able to approach.

It is interesting to just list some of the areas in which Shockley worked: theory of vacuum tubes, solid state physics, semiconductor amplifier, electron microscope, nature of metallic state, band theory of solids, order in alloys, space charges and many others. His work resulted in more than ninety U.S. patents. Shockley actually presented the result of his genetic study in a more sophisticated manner than people with formal training in genetics.

Shockley and his friend James Fisk were assigned by Bell Laboratories in early 1940 to examine the potential of fission as an energy source. One day Shockley come up an with idea about the chain reaction of U-235. In two months, Shockley, without any formal education in nuclear engineering, with J. Fisk designed one of the world's first nuclear reactors. Shockley's and Fisk's work was classified by the government in a matter of hours and later all results of their work was transferred to the Manhattan Project.

Later during WWII, Shockley under the guidance of P. Morse applied the science of operations research, then largely ignored in the U.S., Shockley, without any formal education in operation research, with his effort in Anti-Submarine Warfare Operations Group saved thousand of lives.

Shockley quickly earned a reputation for scientific brilliance. He could look at a problem and solve it faster than others would even understand where the problem is. And he solved problems in ways they never imagined. It does not matter in which scientific field Shockley worked, he always excelled. To argue that Shockley could not learn about genetics is more than nonsense. He was not so much angry as frustrated that he has been misunderstood in his view on genetics and a proposal of this "thinking exercise."

Today, many people do not know what Shockley stood for, still many just "know" that he was "a fascist and racist." Perhaps here we may exactly clarify Shockley's statements about Negroes. In a court transcript [Civil Action File No. C81-14313 Shockley vs. Cox Enterprise, Inc. and R. Witherspoon] from September 1, 1981 Shockley under oath said:

> *"My position is not that all Negroes are inferior to all whites; instead I do believe that many Negroes are superior to many whites. In fact my statistical studies show that American Negroes achieve almost every eminent distinction that whites achieve and are ten times more successful per capita in winning Olympic gold medals. However, the probability for distinction depends upon mental powers on a capita basis and that is between ten to one hundred times smaller, and it is this probability that I fear is falling as a result of high ghetto birth rates."*

Phenomenal scientists come frequently with unusual ideas. Linus Pauling, for example, wanted to tattoo a warning on the skin of everybody who possesses a genetic susceptibility to sickle cell anemia. Pauling however, contrary to Shockley managed to maintain his reputation because he did not target any specific ethnic group and he gave up on a certain group of people, who as Pauling said, *"are taking my time to think."*

Shockley's reputation was destroyed because Shockley asked questions that no one wanted to ask, much less answer. Nobody really proved that Shockley was wrong with his eugenics opinions; however, also nobody joined him in support. He gave to his detractors easy ammunition against him and it was not too difficult for an incoming "politically correct" society to minimize his contribution to mankind.

What, then, is so different so untouchable about this war hero, transistor inventor, and Stanford University professor William B. Shockley? The answer is painfully clear. He has trespassed on a taboo subject, offending one of the most deeply held prejudices of modern American society where too many liberals and conservatives who may talk about intellectual freedom and claim it but do not extend it to anybody who seems to threaten their own system of values and doctrines.

Shockley compiled the mail he was receiving during his "Thinking Exercise" crusade. The box labeled "Friends and Cowards" contains in the folder "Friends" about twenty letters, while the folder "Cowards" is about 1.8″ thick. From one attachment to the letter in Friend folder we can learn that the Shockley phenomenon is nothing new in history.

Dr. Galileo Galilei, Director
Grand Ducal Institute for Advanced Study
Florence, Tuscany *Rome, November 15, 1615*

Dear Professor Galilei:

Your letter urging a crash program of research into the question of the earth's rotation has been studied with great care by the experts of our Academy of Sciences. I am sorry to have to say that their recommendation is negative.

Even with today's expanding astronomical knowledge it is still impossible to sort out all the complex hierarchies of angles that may influence the motions of the crystalline spheres, and to expect swift results from new research is wholly unrealistic.

We question the urgency of a program to calculate the orbits of comets. In the first place, if the orbits are at all complex, the results of such research are almost certain to be inconclusive. In the second place, it is not clear that major social decision depend on such information. It is contrary to all evidence that social problems such as flood, drought, and plague are not caused by comets.

Despite the great number of observations that have been made through your telescope, it is still not clear whether the phenomena alleged to have been seen are real or merely hallucinations induced by the devil. For example, there is no scientific basis for a statement that there are or that there are not satellites of Jupiter.

We feel that you were hasty in criticizing your colleagues when they refused to look through your telescope. To shy away from seeking the truth of one thing; to refrain from collecting still more data that would be uncertain meaning but would invite misuse is another.

With best wishes for a productive career in significant scientific research, I remain,

Yours very sincerely,
R. F. R. Bellaramine, S. J.,
Grand Inquisitor

When Shockley had overcome his shock of the Fairchild exodus, he hired a new team, mostly of young Europeans. Swiss physicist Kurt Hubner came from General Electric in Virginia, Adolf Goetzberger from Siemens in Munich. A fresh Ph.D. from Göettingen was Hans Queisser, later followed by Hans Stark and Roland Haitz from Germany and Hugo Fellner from Austria. Development and sales of four-layer diodes continued, the junction field effect transistor, however, never got into production stage. The majority of R&D projects were government sponsored. Defects and reliability of silicon devices were main research topics.

Shockley Semiconductor Laboratories were renamed Shockley Transistor Corporation when A. Beckman sold the Shockley subsidiary to Cleveland, Ohio based Clevite Corporation. The Shockley unit become part of Clevite's Transistor Division in Waltham, Mass. Clevite was bought by ITT in 1965, and the ITT closed Shockley Palo Alto plant in 1969.

References

[1] G. Moore, K. Davis, SIEPR Discussion Paper No. 00-45, July 15, 2001
[2] G. Moore, The birth of Microprocessor, Scientific American, September 22, 1997
[3] G. Moore, Solid State Physicist, interview for Time Magazine
[4] J. Andrus and W. L. Bond of BTL disclosed photoresist patterning of transistor ("Photoengraving in Transistor Manufacturing", News abstracts of Electrochemical Soc. Semiconductor Symposium, Washington, D.C. 1957)

4

Fairchild Semiconductor Corporation – Subsidiary of Fairchild Camera and Instrument Company

"You do things in this world to make it a more enjoyable place to live. Make it more comfortable, healthier. You do not do things just to make more money."

Sherman Fairchild, 1965

The morning after resigning from Shockley Semiconductor Laboratory, on Thursday, September 19, 1957, the eight former Shockley scientist and engineers became by law the so called "The California Group." The group signed a sixteen page legal document prepared by Parkhurst & Co., a New York legal firm. Parkhurst & Co., was representing Hayden & Stone.

AGREEMENT dated September 19, 1957 among FAIRCHILD CONTROLS CORPORATION, New York (herein called FAIRCHILD CONTROLS), FAIRCHILD CAMERA AND INSTRUMENT CORPORATION, a Delaware corporation (being called FAIRCHILD CAMERA); PARKHURST & Co with its principal place of business New York, N.Y. (herein called Parkhurst); JULIUS BLANK, VICTOR H. GRINICH, JEAN A. HOERNI, EUGENE KLEINER, JAY T. LAST, GORDON E. MOORE, C. SHELDON ROBERTS, and ROBERT N. NOYCE (said eight person being collectively called the CALIFORNIA GROUP.)

Fairchild Control is presently engaged in development and manufacturing activities in the electronic field and desires to expand such activities to include semi-conductors (such as transistors.) Fairchild Controls believe that its entrance into semi-conductors field will be involvement its present activities in the electronics field and any semiconductor products developed and manufactured as result thereof can be economically marketed by Fairchild Controls' present sales organization to its existing customers.

ARTICLE ONE[1]
Organization and Activities of the New Corporation

ARTICLE TWO
Voting Trust Agreement

ARTICLE THREE
Management of the New Corporation

ARTICLE FOUR
Financing the New Corporation

ARTICLE FIVE
Options of Fairchild Controls to Purchase Stock of the New Corporation

ARTICLE SIX
Purchase of Notes of the New Corporation by Stockholders

ARTICLE SEVEN
Arbitration

ARTICE EIGHT
Miscellaneous

ARTICLE NINE
Governing Law

EXHIBIT B
SEMI-CONDUCTOR PROJECT
ESTIMATION CASH EXPENDITURES
FIRST EIGHTEEN MONTHS OF OPERATIONS

[1] Only the titles of the Articles, not the contents, are included here

Equipment, including both capitalized items and other equipment chargeable to expense		*\$386,600*
Salaries of the group to include a total of 30 persons of which 22 will be technicians and junior engineers.		
Salaries per annum of primary group:		
Robert N. Noyce	*\$15,600*	
Eugene Kleiner	*\$14,700*	
C. Sheldon Roberts	*\$14,700*	
Julius Blank	*\$13,800*	
Victor H. Grinich	*\$13,800*	
Jean A. Hoerni	*\$13,800*	
Jay T. Last	*\$13,800*	
Gordon E. Moore	*\$13,800*	*\$171,000*
Salaries of all others including a General Manager		*\$279,000*
General Expenses – all other costs including but not limited to, rent, utilities, sales expenses, administration expenses, materials and labor for samples or prototypes, payroll fringe benefits, travel expenses; purchased services and any other items ordinarily considered as general overhead, administrative and sales expense		*\$436,000*
Direct labor and material for the production of items to be built for inventory and sales to customers		*\$116,000*
		\$1,388,600

Note:
The above are gross expenditures. Money received from sales will not be considered as an increase in the amount available for gross expenditures.

Each of "The Eight" owned 100 shares of Fairchild Semiconductor stock, Hayden Stone owned 225 shares, and 300 shares were held in reserve for future key managers. Fairchild Camera and Instrument agreed to loan \$ 1.38 million over a period of eighteen months with the following conditions: each member of the group will put in \$500 and Fairchild had an option to buy the company from the group.

The new company Fairchild Semiconductor Corporation was established in a 17,000 square feet building at 844 Charleston Street in Palo Alto. Dr. Robert N. Noyce was put into the position of director of research and development; Richard Hodgson, executive vice-president of Fairchild Camera and

Mr. Robert Noyce
11645 Lundy Lane
Los Altos, California

Dear Bob,

Attached is a check for $3,000. as advance to you to cover necessary expenditures in setting up the semi-conductor operation until such time as the Corporation is formally organized.

Please keep a record of the expenditures so that Fairchild Camera can eventually be reimbursed by the Fairchild Semi-conductor Corporation.

We have a sizeable quantity of surplus shop equipment which you might be able to use, so let me know your requirements.

The group will be interested to know that it appears possible to extend Fairchild's program on group life and medical insurance to your operation. I'll let you have the details later.

Best regards.

Sincerely,

RH:mtb
enc. ck #9-5003

Richard Hodgson

Fig. 4.1. Hodgson's letter to Robert Noyce dated October 5, 1957

Instrument Corporation, had been named chairman of the new semiconductor division but remained in his office in New York. H. E. Hale, vice-president of Fairchild Controls Corporation and general manager of the Components division, became president of the new group.

In reality, until the end of November 1957, the Fairchild Semiconductor ran its operation out of Vic Grinich's garage. When they moved on December 8, 1957 into a new space, their equipment was basically tables and chairs. The first employee hired by the "Eight" was Murray Siegel. By December 1, 1957 Fairchild Semiconductor had 15 employees. Employee number fifteen was Thomas H. Bay who joined Fairchild as a Marketing manager.

On November 6, 1957 the Fairchild "Eight" formed two transistor design teams: one under former Paramount Pictures consultant and Shockley Semiconductor Laboratory chemist, Gordon E. Moore, to develop NPN's, and one under Jean Hoerni, to develop PNP's.

Jean Hoerni and David Allison (also an ex–Shockley employee) developed the diffusion processes in diffusion furnaces they designed with Cecil "Art" Lasch, Jr. The group extended the Bell Telephone Lsboratories research work on diffusion at Shockley Labs and then refined the process at Fairchild to enable the production of commercial manufactured transistors.

Transistor technology used by transistor companies on the East coast was relatively well developed in 1957. However, the mesa structure and silicon diffusion technology was still a novelty. By the end of 1957 only Bell Telephone Labs and Shockley Labs had some experience with this device. From all other transistor manufacturers, only Motorola hired a Western Electric engineer to start their Mesa Transistor Program in 1958.

The Fairchild "Eight" made two important decisions regarding the mesa transistor: they adopted the same approach as Shockley Laboratories with wax masking and patterning structure. Both base and emitter were formed by diffusion. However, very early on, the Fairchild group switched to photoresist. Although the photoresist patterning of semiconductor structures was developed at Bell Laboratories, Western Electric engineers considered this process non-manufacturable, and they preferred the metal shadow masks. There was no photoresist for semiconductor masking operations. All the work at Bell Labs and Fairchild was done with photoresist developed by Eastman Kodak for patterning of printed circuits.

The "Eight" were greatly motivated. They disagreed with Nobel Prize winner and wanted to prove that they were right. They had a remarkable relationship at the beginning. They knew all the tasks necessary to make transistors; they divided them up and worked together. At the start each of the "Eight" did his part without a whole lot of formal supervision and they worked mostly with their hands. They did not even have a copy machine. They depended completely on their own resourcefulness and they were not worried about the Wall Street rating. In addition, Good Luck was with them.

Fairchild Camera and Instruments Corporation learned lessons from Beckman's failure, and they were well aware of risking more than a million dollars on an idea that was not proven, with engineers who had no managerial experience. John Carter was especially nervous that he had nobody old enough back in the West coast and he placed an advertisement in the December 1957 The Wall Street Journal (Fig. 4.2), seeking a senior person who knew the semiconductor business. One of the applicants was Dr. Ewart M. Baldwin. He was a product engineering manager at Hughes Semiconductor. Hughes, at that time, was a well established company and was the world's biggest producer of silicon diodes with a lot of military business.

Ewart (Ed) Baldwin impressed Richard Hodgson and he joined Fairchild on February 7, 1958. Baldwin was originally offered the same deal as the members of the "Traitorous Eight." Baldwin would pay $500 and own the same number of share as the others. Baldwin was not satisfied with this proposal and after lengthy negotiations no resolution was established. Baldwin never acquired any equity in Fairchild Semiconductor.

The future success of Fairchild Semiconductor is partially due to Baldwin's early accomplishments. In his first talk, to about 35 Fairchild employees in mid-February 1958, he emphasized quality control, lifetime testing and manufacturing discipline. These were very new subjects to The California Group who had a very hazy idea about manufacturing. Baldwin was able to convince Fairchild Camera Corporation to build a new 60,000 square feet

Fig. 4.2. Fairchild Camera and Instrument Advertisement in The Wall Street Journal (December 1957)

manufacturing facility at the end of 1958, at a time when no transistors were yet delivered to customers. Ed Baldwin changed an original goal of $60,000 in sales in eighteen months with the 30 employees set up by the "Traitorous Eight," to $500,000 at the end of the year, and $5,000,000 by the end of second year. All goals were met.

It was not very difficult for Sherman Fairchild, who was IBM's largest shareholder and a member of IBM's board committee, to get an order for a hundred transistors at a hefty price; one hundred fifty dollars per piece. In December 1957 Tom Bay and Bob Noyce visited IBM's primary military facility in Oswego. IBM was involved in a government contract to replace an analog computer in the B52 bomber, and they were looking for core memory drivers which could switch 150 mA and 40 V and could work at 85°C. Noyce had never made such a device but he said *"we can do it."* In February 1958 Fairchild received a purchase order for one hundred transistors at the $150.00 price per piece

Sherman Mills Fairchild was a man of many talents – and even more interests. He was a real pioneer in many respects – loved beautiful girls, and would be seen at all the fancy nightclubs in New York, but never got married; he always managed to step aside just at the right time. Fairchild was founder of Fairchild Camera & Instrument Corporation, and of Fairchild Engine & Airplane Corporation and became a pioneer in the fields of photography, aviation, and sound engineering. Fairchild was not only an indefatigable investor in those fields, but his inquiring mind also led to patents on such disparate products as engraving mechanisms and carriers for fruit picking (the latter his last patent at age 73). In his lifetime he was granted thirty patents. He designed his own indoor tennis court. He was a gourmet cook. He was a voracious reader and subscribed regularly to 150 trade and technical journals. He was a partner in a music publishing firm. Fairchild's curiosity and inventive mind was obvious early and they were encouraged by his father, George Fairchild who was a remarkable man himself. The elder Fairchild started out as a printer's devil, later financed the manufacture of one of the earliest rotary newspaper printing presses in the U.S., and was one of the founders – and first chairman – of International Business Machines (IBM). George Fairchild also served for twelve years in the U.S. House of Representatives.

One hundred and ten days after completing the blueprints for the Fairchild Lab, the gas plumbing (Oxygen, Hydrogen, Nitrogen, Argon, and compressed air) of Fairchild Lab in Palo Alto was completed on January 31, 1958 by 34 Fairchild employees (15 professionals, 5 skilled, and 14 semi-skilled employees.)

The first seven bench diffusion furnaces used in the corner of Fairchild's Palo Alto facility were very similar to furnaces used by Shockley Semiconductor. Cecil A. (Art) Lasch, Jr., a technician who worked at Fairchild, did most of the work on the diffusion furnace hardware.

The quartz tube of Fairchild's diffusion furnace had an inner diameter of 2″ and a quartz liner between the processing tube and heating element. A very simple "variac" adjustable variable transformer was used to control the temperature of the single zone. The tube was equipped with a network of

Fig. 4.3. Sherman M. Fairchild (1896–1971)

quartz pipes, quartz valves, and flowmeters. All systems resembled research laboratory equipment rather than industrial manufacturing equipment.

The wafers were pushed into and pulled out of the tubes manually. Cooling of wafers usually took place in a cool zone within the diffusion tube, which allowed the wafers to cool to a temperature suitable for handling. The programmed heating and cooling was not used because of the difficulty in maintaining good temperature profiles.

The used gases were just emitted into the atmosphere by vents placed just above diffusion tube exhausts. There was no "clean room etiquette", and the chain smoker Robert Noyce frequently smoked in the Fairchild manufacturing facility.

Jay Last, who had a background in optics from the time of his studies at MIT, paired with Robert Noyce and designed contact masking equipment that aligned masks to the wafer flat. Fairchild 0.5″ to 1″ diameter wafers had quite a long "flat edge" (Fig 4.4) so aligning the mask to the end point of the flat and to the wafer crown was quite satisfactory.

The structure and dimension of the Fairchild mesa transistor was identical to the Bell Laboratories transistor 2N560. Shockley Laboratory was provided with all details of new Bell Labs devices and "The California Group" had full access to this information.

The geometry of the 2N560 and the Fairchild transistor was huge. The diameter of the mesa ring was approximately 750 μm. Alignment error was not better than 100–150 μm.

Three masks were necessary to complete the transistors:

1) Emitter region
2) Base and emitter contacts
3) Metal

The wafers were placed with their front side (resist) on the emulsion side of the masking plate, and exposed for 60 sec by mercury lamp. The exposed wafers were developed and backed.

Fig. 4.4. Fairchild 1″ diameter wafer

Process Flow: MESA NPN 2N696

1) 200 μm thick N-type Si substrate (1–1.4 Ω-cm)
2) Substrate etch and polish to thickness 120 μm
3) Surface oxidation (wafer flat on quartz boat)
   - 1200°C/16 hours dry O_2 , ~ 10000 Å
4) Base Deposition
   - Source Ga_2O_3 is inserted into furnace during surface oxidation, nitrogen flush 1200°C/5 minutes, 500 cm^3/min Nitrogen
5) Base Drive-in ($X_j \sim 3.5$ μm)
   - 1200°C/30 minutes, 90 cm^3/min Nitrogen, 10 cm^3/min Hydrogen
   - Flash – 1200°C/3 minutes, 500 cm^3/min Nitrogen
   - Base Deposition and Drive-in are performed in the same tube by switching gasses
6) MASK – Emitter
   - Wet etch oxide, emitter region and back side (17°C HF, NH_4F solution)
   - Strip resist (acetone)
7) Emitter Deposition
   - Wafers flat on quartz boat
   - P_2O_5 Source temperature 200°C, Wafer zone temperature 1000°C, 200 cm^3/min Hydrogen, deposition time 60 minutes
8) Evaporation of Nickel on back side
9) Emitter Diffusion ($X_j \sim 2.6$ μm)
   - Wafers flat on quartz boat
   - Wafer zone temperature 1100°C, 200 cm^3/min Oxygen, diffusion time 45 minutes

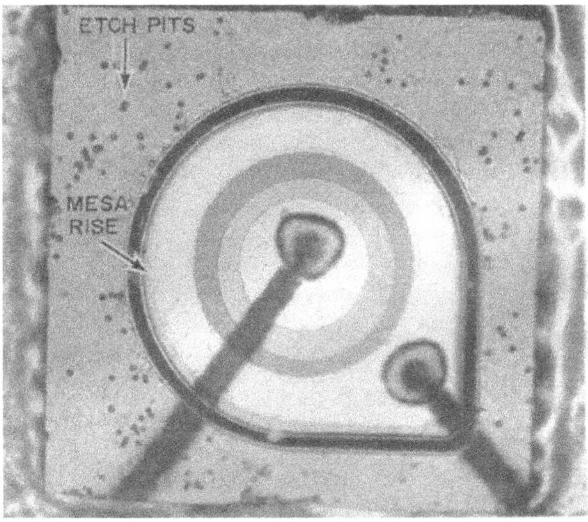

Fig. 4.5. Fairchild Mesa 2N696 transistor

10) MASK – Base and Emitter Contacts
11) Vacuum Evaporation of Aluminum
12) MASK – Metal
13) Metal etch
14) Metal alloy (~ 600°C/ 5min, Argon, rapid cool)
15) Backgrinding to thickness 60 μm
16) Nickel backside plating
17) Wax deposition on front side through a glass screen covering mesa region
18) Mesa etch (~ 15 μm)
19) Wafer dicing
20) Thermocompression wire bonding

Process Flow: MESA PNP 2N1131

1) 200 μm thick P-type Si substrate (0.7–1.3 Ω–cm)
2) Substrate etch and polish to thickness 120 μm
3) Base Deposition
   - Source Sb_2O_3 at 650°C, wafers zone temperature 1120°C, 250 cm^3/min Nitrogen, deposition time 25 min
4) Base Drive-in ($X_j \sim 6$ μm)
   - 1200°C/15½ hours, 250 cm^3/min Oxygen Thickness of oxide grown during drive-in ~ 1 μm
4) MASK – Emitter
5) Wet etch oxide (oxide grown during base diffusion)
6) Emitter diffusion
   - Wafers are placed vertically into the quartz boat
   - Wafer zone temperature 1130°C, 400 cm^3/min Nitrogen, 3 cm^3/min Oxygen, flush time 5 minutes
   - Wafers are inserted into processing tube
   - Additional 15 cm^3/min Hydrogen (thin oxide)
   - BCl_3 is added to the gasses (3 minutes)
   - 400 cm^3/min Oxygen (3 minutes)
   - Wafers are removed from the processing tube
7) Front side of wafer is covered by wax and glass plate
8) Back side of wafer cleaned in HF
9) Wafer backside sand paper scratch
10) Evaporation of Nickel on the back side
11) Nickel diffusion
    - Wafers flat on quartz boat
    - Wafer zone temperature 1130°C, 400 cm^3/min Oxygen, diffusion time 11 minutes
    - Slow cool to 200°C
12) MASK – Base and Emitter Contacts
13) Vacuum Evaporation of Aluminum and Phosphorus

14) MASK – Metal
15) Metal etch
16) Metal alloy (~ 600°C/ 5min, Argon, rapid cool)
17) Back grinding to thickness 60 µm
18) Nickel backside plating
19) Apiezon wax deposition on front side through a glass screen covering mesa region
20) Mesa etch (~ 15 µm)
21) Wafer dicing
22) Thermocompression wire bonding

THE ONLY MANUFACTURER OF SILICON MESA TRANSISTORS OFFERING A YEAR AND A HALF OF PRODUCTION EXPERIENCE, A WHOLE FAMILY* OF PRODUCTS IN VOLUME PRODUCTION WITH ASSURED DELIVERIES ON SCHEDULE AND THE ULTIMATE IN QUALITY OF WORKMANSHIP.

*GENERAL PURPOSE types suitable for switching RF and DC applications over a wide current range.

2N696 & 2N697

*PNP COMPLEMENT to the 2N696 and 2N697

2N1131 & 2N1132

*HIGH VOLTAGE type particularly suited to video amplifiers and RF oscillators.

2N699

*LOW STORAGE types optimized for high current saturated switching circuitry.

2N1252 & 2N1253

*HIGH SPEED LOGIC transistor suitable for saturated switching circuitry without sacrificing speed.

2N706

AVAILABLE IN QUANTITIES OF 1-999 FROM DISTRIBUTOR STOCKS OR DIRECT FROM THE FACTORY FOR ORDERS OF 1,000 OR MORE. COMPLETE SPECIFICATIONS FROM EITHER SOURCE.

FAIRCHILD SEMICONDUCTOR CORPORATION

Fig. 4.6. List of mesa silicon transistors which were offered by Fairchild in November 1959

There was a competition for six weeks between Gordon Moore and Jean Hoerni. Both transistor projects were successful, but the NPN transistor was easier to manufacture, mainly because of the contact to the base. To form an Aluminum ohmic contact to the moderately doped Antimony base of a PNP transistor is much more difficult than in the case of Boron' doped base of an NPN transistor. The yield of the NPN transistors was higher and G. Moore, as head of group, had the power to choose the NPN transistor as the first Fairchild transistor. Volatile, often sarcastic, Swiss born, and not necessarily always nice, scientist Jean A. Hoerni, was clearly unhappy that his PNP transistor did not prevail. Significant differences between Moore's and Hoerni's personalities only increased tension between Hoerni and his boss.

Many years later, Gordon Moore wrote that "*Hoerni, our theoretician, sat at his desk and thought.*" [2]. This is a very unfair statement. Jean Hoerni became a very good experimentalist and he did with David Allison the most work on the boron diffusion, which was at that time very difficult.

The first 100 Fairchild transistors were mailed to IBM in July 1958. A tentative Specification for the 2N696 was issued in August 1958 and an advertising campaign started in September 1958. The Fairchild silicon mesa transistors significantly improved several transistor parameters and there was no competitor who could offer similar products.

In 1959, there were 26 companies manufacturing transistors and sales of transistors reached approximately $0.8 Billion. The transistor manufactures were:

Advanced Research Associates, Kensington, MD
Amperex Electric Corporation, Hicksville, NY
Bendix Aviation Corporation, Long Beach, NJ
Bogue Electric Company, Paterson, NJ
CBS-Hytron, Lowell, MA
Clevite Transistor Products, Waltham, MA
Delco Radio, Kokomo, IN
General Electric Company, Syracuse, NY
General Transistor Corporation, Jamaica, NY
Hughes Aircraft Company, Los Angeles, CA
Industro Transistor Corporation, Long Island City, NY
Minneapolis-Honeywell Regulator Company, Minneapolis, MN
Motorola, Phoenix, AZ
Pacific Semiconductor, Culver City, CA
Philco Corporation, Lansdale, PA
Radio Corporation of America, Somerville, NJ
Raytheon Manufacturing Company, Newton, MA
Rheem Semiconductor Corporation, Mountain View, CA
Silicon Transistor Corporation, Carle Place, NY
Sprague Electric Company, North Adams, MA
Sylvania Electric Products, Woburn, MA
Texas Instruments Incorporated, Dallas, TX
Transitron Electronic Corporation, Wakefield, MA
Tung-Sol Electric, Newark, NJ
Western Electric Co., New York, NY
Westinghouse Electric Corporation, Youngwood, PA

Approximately 2.4 million transistors were exported to the United States from Japan compared to 10,620 units in 1958.

In November 1959 there was no product on the market that could compete with portfolio of eight Fairchild transistors. (Fig. 4.6.) Despite that the yield of mesa transistors was very low, Fairchild sales reached $65,000.00 in July and August 1958 because of the transistors' high selling price.

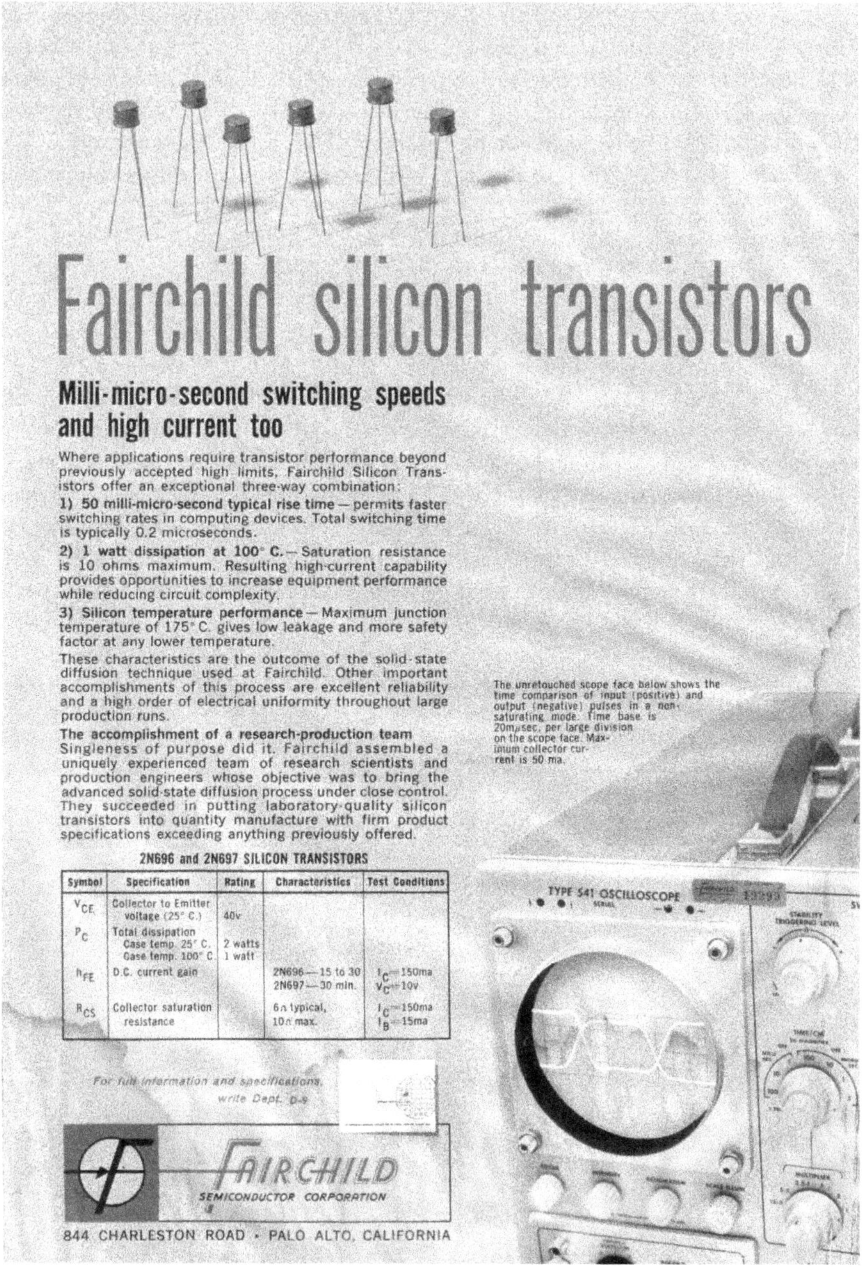

Fig. 4.7. Fairchild Semiconductor advertisement for 2N696 and 2N697 released in September 1958

In March 1959 Ed Baldwin abruptly resigned as a general manager of Fairchild Semiconductor and with several other defectors from Fairchild established Rheem Semiconductor Corporation in Mountain View. At the end of 1959 a Bell Labs group headed by Ian Ross developed an epitaxial transistor and published a paper describing the process at the Solid State Circuit Conference in June 1960 and during 1960. Several companies, TI, Motorola, Sylvania, and Transitron jumped on the process without hesitation. They saw an epitaxial transistor as the answer to Philco Corporation's micro alloy (MAT) and micro alloy diffused transistors (MADT) which were the fastest devices on the market.

Ed Baldwin hired a young ex-Bell Labs and ex-Shockley Laboratories engineer Leopoldo B. Valdes as his manager of Research & Development.

Valdes worked at Bell Labs on resistivity and lifetime measurements and he had a lot of experience with epitaxial films and mesa transistors. Baldwin's business plan was based on an assumption that Rheem could produce a better mesa silicon transistor and be the first on the market with an epitaxial device. Rheem produced a better 2N697 device; the problem was that Lippincott, Ralls and Hendricson, Fairchild's legal counsel, filled suit against Rheem. The court marshals raided Rheem premises and found the Fairchild's proprietary material which Ed and his group took from Fairchild.

Fig. 4.8. L. B. Valdes (1958)

The attorney for Rheem Semiconductor Corporation had only one argument which he could use to defend the actions of the not well executed startup: he told the judge that Dr. E. M. Baldwin did nothing different than the "Traitorous Eight" who took Shockley Laboratories' "cook books." The litigation was quietly settled out of court. Fairchild Semiconductor did not want to generate bad publicity about the "Traitorous Eight," and Rheem agreed that they would not produce mesa devices based on Fairchild proprietary technology.

At the time of litigation in 1960, the Rheem operation had about 500 employees and reported a profit for the first nine months of $71,573 after tax provision on sales of $748,435. This was the only period of time when Rheem made money. The settlement doomed Rheem and in October 1961 Rheem Semiconductor Corporation was acquired by Raytheon Company, Waltham, MA.

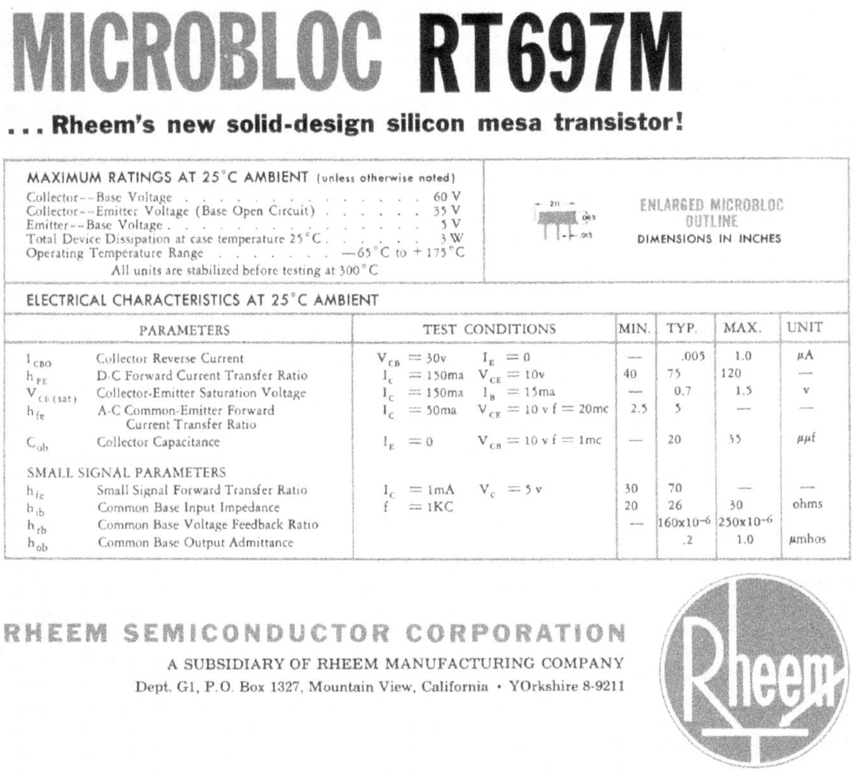

MICROBLOC RT697M

... Rheem's new solid-design silicon mesa transistor!

MAXIMUM RATINGS AT 25°C AMBIENT (unless otherwise noted)

Collector--Base Voltage	60 V
Collector--Emitter Voltage (Base Open Circuit)	35 V
Emitter--Base Voltage	5 V
Total Device Dissipation at case temperature 25°C	3 W
Operating Temperature Range	—65°C to +175°C

All units are stabilized before testing at 300°C

ENLARGED MICROBLOC OUTLINE
DIMENSIONS IN INCHES
.211
.063
.013

ELECTRICAL CHARACTERISTICS AT 25°C AMBIENT

PARAMETERS		TEST CONDITIONS		MIN.	TYP.	MAX.	UNIT
I_{CBO}	Collector Reverse Current	V_{CB} = 30v	I_E = 0	—	.005	1.0	μA
h_{FE}	D-C Forward Current Transfer Ratio	I_C = 150ma	V_{CE} = 10v	40	75	120	—
$V_{CE(sat)}$	Collector-Emitter Saturation Voltage	I_C = 150ma	I_B = 15ma	—	0.7	1.5	v
h_{fe}	A-C Common-Emitter Forward Current Transfer Ratio	I_C = 50ma	V_{CE} = 10 v f = 20mc	2.5	5	—	—
C_{ob}	Collector Capacitance	I_E = 0	V_{CB} = 10 v f = 1mc	—	20	35	μμf
SMALL SIGNAL PARAMETERS							
h_{fe}	Small Signal Forward Transfer Ratio	I_C = 1mA	V_C = 5 v	30	70	—	—
h_{ib}	Common Base Input Impedance	f = 1KC		20	26	30	ohms
h_{rb}	Common Base Voltage Feedback Ratio			—	160x10^{-6}	250x10^{-6}	
h_{ob}	Common Base Output Admittance				.2	1.0	μmhos

RHEEM SEMICONDUCTOR CORPORATION
A SUBSIDIARY OF RHEEM MANUFACTURING COMPANY
Dept. G1, P.O. Box 1327, Mountain View, California • YOrkshire 8-9211

Rheem

Fig. 4.9. Rheem Semiconductor Corporation 2N697M released in August 1960

Good luck struck the Fairchild "Eights" (after Shockley aborted the court action and after the settlement with Baldwin) for third time: Fairchild Semiconductor received an important contract. Fairchild was awarded a contract in excess of $1,381,105 by the Boeing Co., in Seattle, for the manufacturing of transistor to be used in the Air Force Minuteman program. Autonetics, a division of North American Aviation, Inc., associate prime contractor for the inertial guidance and flight control systems of the missile, had recently accepted Fairchild's highly reliable and stable planar transistors for use in the Minuteman.

This was in addition to a half-million dollar contract which Fairchild also received from the Boeing Co., making a total of 8 million dollars in contracts and purchase orders which the company had given to Fairchild in 1961. The new contract called for Fairchild Semiconductors to provide transistors which have a proven high reliability. Most of the transistors provided were silicon planar devices.

The transistors would be used in the manufacture of the ground support equipment of the missile. The Boeing Co. had the responsibility for overall assembly of the Minuteman in addition to developing and manufacturing the ground support equipment.

The Minuteman project was one of the most important projects for the electronic components industry because the project served as a catalyst for improved reliability of electronics assemblies. Motivating force behind this program was a missile concept which had an initial objective for its airborne guidance and control system, a Mean-Time-Between-Failures of 7,000 hours. The Minuteman missile eliminated a costly warm-up of electronics systems by keeping the electronics "hot." The entire system was operating for the complete period of readiness. The target data was set into the missile at the time of its installation in the silo and operation of its guidance system was continuously monitored. When the supplier is selected, he must agree to perform tasks specified by the Autonetics Engineering Department with Device Failure Analysis.

The 30 kg and 0.05 m^3 Minuteman computer was consuming 250 W of electric power, mainly for hundreds of Fairchild mesa silicon transistors used in a computer designed with discrete Transistor-Diode Logic.

Shortly after the contract was signed, Autonetics discovered that the Fairchild double-diffused mesa transistors had a reliability problem: an un-passivated base-collector junction was very susceptible to electric breakdown. Fairchild was facing serious troubles, but their lucky star was still shining, and one of the "Traitorous Eight", Jean A. Hoerni, made the star even shinier.

In many ways Jean A. Hoerni had personality similar to Shockley's. Born to an upper middle class family in Switzerland, he had a double degree in physics, one from the University of Geneva and a second from the University of Cambridge. After studies in Cambridge he was at the California Institute of Technology in Pasadena, where he did postdoctoral work in Linus Pauling's Department. He applied for a position in Bell Laboratories at the end of 1955 and ran into William Shockley. At the Shockley Laboratories, Jean

Fig. 4.10. The guidance computer of the MINUTEMAN I intercontinental ballistic missile

Hoerni was hired as a theoretical physicist and worked for the first months on diffusion theory, and then started laboratory work on diffusion. The general concept of boron diffusion from boron trichloride was known from the work done at Bell Labs. Transition from concept to practical utilization, however, is often long and tedious. Hoerni became a very skillful experimentalist. The learning curve was not easy and cheap. As with every good experimentalist, Hoerni destroyed and broke things. One of Hoerni's boron tubes exploded during experimental work he was conducting and created a crater in the wall and left shattered quartz debris everywhere. But the persistent Hoerni, stimulated by rejection of his PNP transistor, was determined to design the best transistor ever.

The idea to passivate the semiconductor surface was known since the first junction transistor. For example, in June 2, 1952 Sidney G. Ellis and Jacques I. Pantchechnikoff (known better under his later name Jacques Pankove) of RCA, filed a patent application 2,796,562 "Semi-conductive Device and Method of Fabricating Same." The patent description states: "The characteristics of semiconductor devices are very strongly dependent upon conditions existing at the surface of the semiconductor material. It is known, for example, that water vapor lowers the impedance of contacts to junctions and results in an appreciable loss in gain. A protective material such, for example, as silicon monoxide, silicon dioxide, or magnesium fluoride may be used as a protective material." Dietrich A. Jenny of RCA filed a similar patent application describing a method for "protecting the surface of a semi-conductive device having a P-N junction" in April of 1953. A similar application was filed one year later in December 1953 by Horace E. Haring of BTL providing "a surface coating of a material which will furnish oxygen for adsorption of the germanium surface." (Fig. 4.11.)

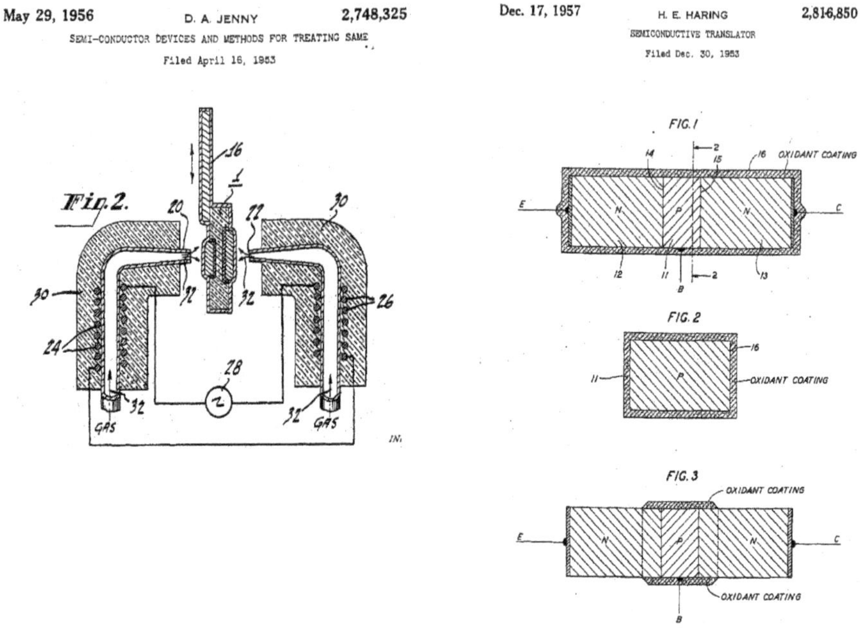

Fig. 4.11. Patent applications of D. A. Jenny (RCA) and H. E. Haring (BTL) suggesting surface protection of semiconductor devices

The RCA engineers C. W. Mueller and N. H. Ditrick also presented a paper in June 22 1955, at the IRE-AIEE Conference on Semiconductor Research in Philadelphia with the title "Uniform Planar Alloy Junction for Germanium Transistor." In this work the authors proposed a new technique for forming planar parallel junctions with uniformity of junction depth of 0.5 microns over 90 per cent of the junction diameter (Fig. 4.12.)

Hoerni was aware of the work done by C. Frosch and L. Derick who maintained a close working relationship with Shockley even after Shockley left Bell Labs. Shockley was copied with each memo or experimental result BTL produced. Hoerni also attended a meeting of the Electrochemical Society in 1958, where Mohamed "John" Atalla presented a paper about passivation of PN junctions by oxide. Atalla's presentation was based on the experimental results described in 1957 BTL memos. All this information served as a precursor of planar devices.

Hoerni, when asked how he got a "planar idea", answered that the idea struck when he took a morning shower and was thinking about the Attala device. Indeed, K. E. Daburlos and H. J. Patterson of Bell Laboratories continued on the work of C. Frosch and L. Derick, and developed a process similar to Hoerni's about the same time. They evaluated the process in Allentown, but they give up on the future development because of manufacturing difficulties.

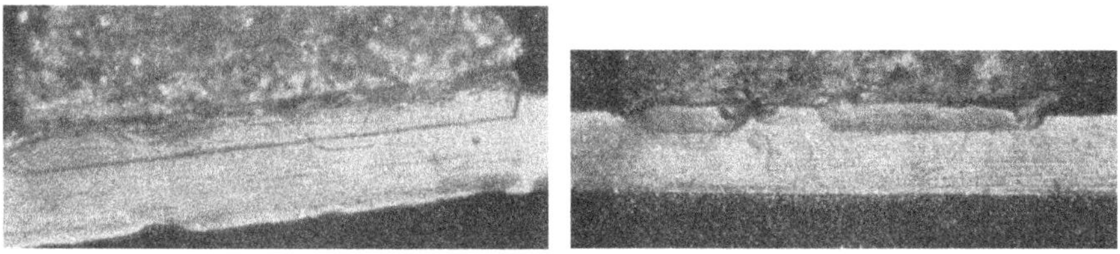

Fig. 4.12. The Planar PN Junction Alloyed Diode developed by RCA in 1955

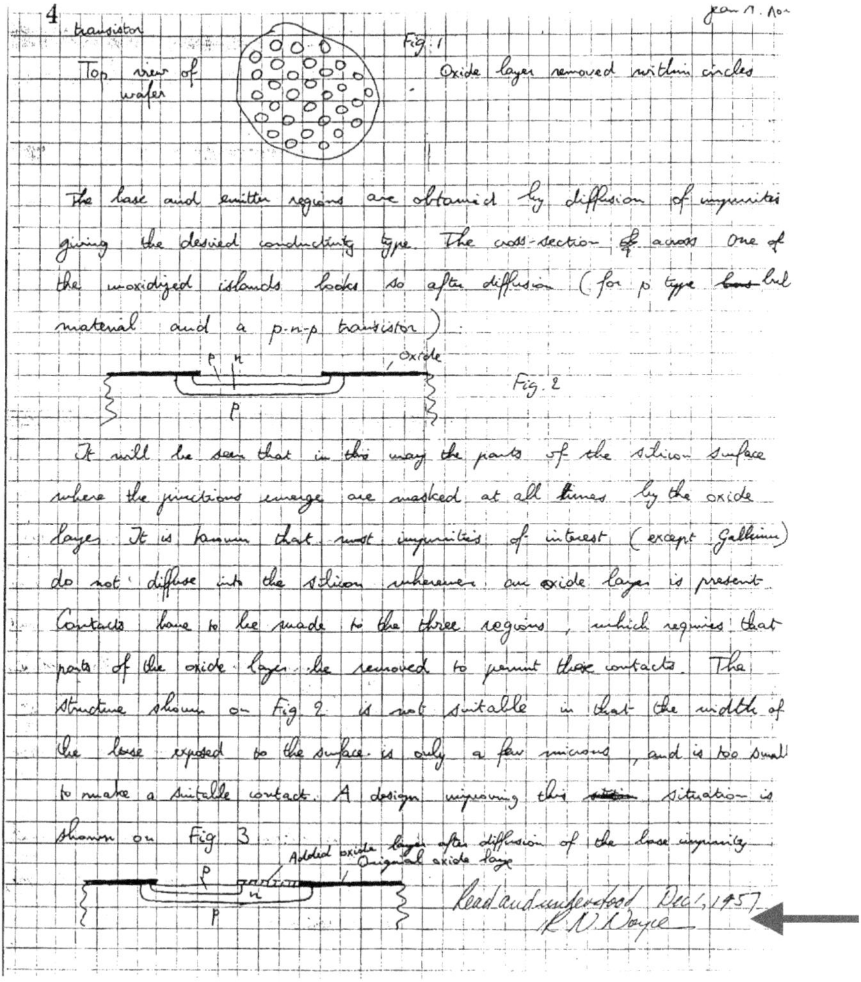

4 transistor — Jean A. Hoerni

Top view of wafer — Fig. 1 — Oxide layer removed within circles

The base and emitter regions are obtained by diffusion of impurities giving the desired conductivity type. The cross-section ~~of~~ across one of the unoxidized islands looks so after diffusion (for p type ~~base~~ bulk material and a p-n-p transistor):

It will be seen that in this way the parts of the silicon surface where the junctions emerge are masked at all times by the oxide layer. It is known that most impurities of interest (except Gallium) do not diffuse into the silicon wherever an oxide layer is present. Contacts have to be made to the three regions, which requires that parts of the oxide layer be removed to permit these contacts. The structure shown on Fig 2 is not suitable in that the width of the base exposed to the surface is only a few microns, and is too small to make a suitable contact. A design improving this ~~situ~~ situation is shown on Fig 3

Read and understood Dec 1, 1957
R. N. Noyce

Fig. 4.13. Page 4 from Jean Hoerni Fairchild's notebook describing the "Method of protecting exposed p-n junctions at the surface of silicon transistors by oxide masking technique." (December 1, 1957). The entry is also sign by Robert. N. Noyce

Hoerni wrote down a brief description of his ideas on page 3 and 4 of his Fairchild notebook and titled it as "Method of protecting exposed p-n junctions at the surface of silicon transistors by oxide masking technique." Notes are dated December 1, 1957. The entry was witnessed and signed by Robert N. Noyce on the same date.

The next month, January 1958, Fairchild's founders were finishing the facility and were busy with development work on the device which already had a proven record and which they needed to deliver: diffused mesa transistor. There was no time to explore the potential of Hoerni's unproven ideas. The interest in Hoerni's idea occurred when the massive reliability problems of Fairchild's mesa devices began.

At the end of April 1958 Jean Hoerni re-started work on his old idea of "protecting the exposed P-N junctions" and finished the work in an extremely short period of time, when he worked alone, as was his custom. Hoerni ran most of his experiments at night, without any research budget. He later boasted, "*it worked basically the first time.*"

Fig. 4.14. Jean A. Hoerni and his furnace gas distribution system in 1959

Hoerni, driven by rejection of his mesa transistor by Moore a few months earlier, kept his work away from Gordon Moore. Jay Last made the extra mask Hoerni needed for his new device. Jay, who probably knew Hoerni best because they were close buddies, described Hoerni as *"most productive when he was discontented, and when he was upset over something."* Fortunately for semiconductor business, Hoerni was nearly always discontented or upset over something.

Hoerni's, silicon planar device was demonstrated on Wednesday, March 4, 1959, exactly one week after Ed Baldwin left Fairchild. At that time, almost nobody recognized that Hoerni's planar process was the major change in transistor structure, which enabled the high volume production of reliable commercially manufactured transistors.

The main processing steps of Hoerni's Planar Process are shown in Fig. 4.15 for an NPN transistor, and they may be summarized as:

1) N-type starting silicon material
2) Base masking oxide
3) MASK – Base
4) Base diffusion and oxidation
5) MASK – Emitter
6) Emitter diffusion
7) MASK – Base and Emitter contacts
8) Vacuum evaporation of Aluminum
9) MASK – Metal

Hoerni found that the problems associated with the approach pursued by Bell Laboratories were mainly due to pinholes in oxide film. He developed a more sophisticated oxidation process. The significant advantage of the planar process is oxide passivation of the semiconductor substrate, which improved the electrical parameters such as reverse leakage current, breakdown voltage, noise figure, and low current h_{FE}.

The planar process was, after the invention of the junction transistor, the most important invention of microelectronics. The planar process removed the main disadvantage of the mesa transistor – an exposed collector-base junction, which is particularly vulnerable to contamination during contacting and assembling.

Producing such a device in "primitive" conditions, as part of Hoerni's moonlighting, was very difficult. The new device had all kind of problems and yield was not higher then 2–5%. As can be seen in Fig. 4.16 almost 50% of devices had mechanical defects (mostly scratches.)

Every engineer who has done something new knows that there is always the same scenario. Every new idea starts with problems. The outcome depends on the type of problem. If the problem does not need money or some-

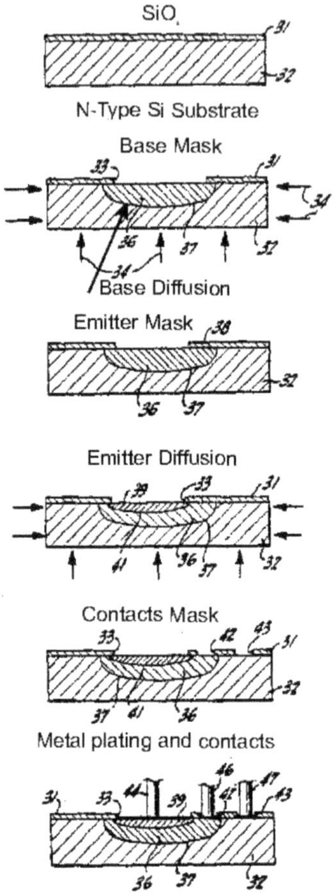

Fig. 4.15. The main processing steps of the planar transistor (Hoerni's Patent # 3,064,167 filed to the U.S. Patent Office on May 1, 1959.)

thing that a good "Capricorn"[2] engineer can provide, the new idea succeeds (and management often takes credit for this effort). If the new idea needs money or resources which the engineer is not able to provide, the project is often killed. An idea is considered as bad, and the engineer is blamed for failure, because the management knows from the beginning that this was a bad idea.

[2] The way to succeed in spite of artificially created burdens is to have some combination of the following character traits: persistence, tenaciousness, uncompromising, stubbornness and confidence. Most managers who are obsessed with being politically correct would interpret these traits as arrogance. I do believe they are essential to innovation. Observation suggests that stubbornness is to be one of the main features of engineers described in this book.

Gordon Moore wrote in [3]:

> *"Fairchild engineers developed the planar transistor, a structure wherein the sensitive portions of the transistor were protected by an adherent glassy layer of silicon dioxide."*

First of all, the "Fairchild engineers" was only Jean Hoerni, and Gordon Moore was one of the few who opposed Hoerni's idea. The reason why the world got the planar transistor is that Fairchild salesman Bob Graham convinced Autonetics that the planar device was superior to the mesa transistor. The timing for Hoerni could not have been better. Noyce and Moore, who emerged as Fairchild managers, were under pressure from Autonetics. The reliability of the oxide passivated and planar devices was unprecedented, and Autonetics actively forced the company to improve the new process and put it into production. Autonetics also made a very crucial decision – they were willing to purchase future planar transistors. Hoerni, determined to succeed, moved his office from the laboratory in Palo Alto to the manufacturing facility in Mountain View.

If Hoerni followed the "politically correct" way, developments of transistor would maybe take a different route. The hardheaded Hoerni, motivated by rejection of his ideas by Gordon Moore, had only one alternative to succeed – he must create an engineering masterpiece. And he did!

In the history of transistors and integrated circuits, there are only a few cases when well-planned and managed projects resulted in success. The transistor, Diffusion Technology, and the Planar Process were not the result of co-ordinated and supervised effort. The occurrence of such a constellation of facts and imperatives are now interpreted as results of exceptional Fairchild business management. This is not the case, for the planar transistor. Who then was responsible for such great achievements?

The planar transistor was the product of a well-educated man with mixed social skills, who could be very charming, but also garrulous, and not always friendly. Hoerni was the only one from the "Traitorous Eight" who stood on the level of William B. Shockley. And in the same way as Shockley, Hoerni got no recognition for his planar process.

The planar transistor was the result of personal individuality and personal individuality only. The planar transistor was the product of a scientist who was characterized as a moody and not a very polite and friendly person. Does this make the significance of the planar device less significant?

Hoerni knew that the support, which he suddenly got from Noyce and Moore, was mainly due to pressure from Autonetics. Hoerni had no major word in decisions where the young Fairchild Semiconductor was heading. Although his title was the Head of the Physics Department, he was the only employee in the department. There was not even a technician.

It took another year before Fairchild Semiconductor first introduced the diffused planar transistor during the Institute of Radio Engineers show in New York in March 1960. In August 1960 the planar version of the 2N696

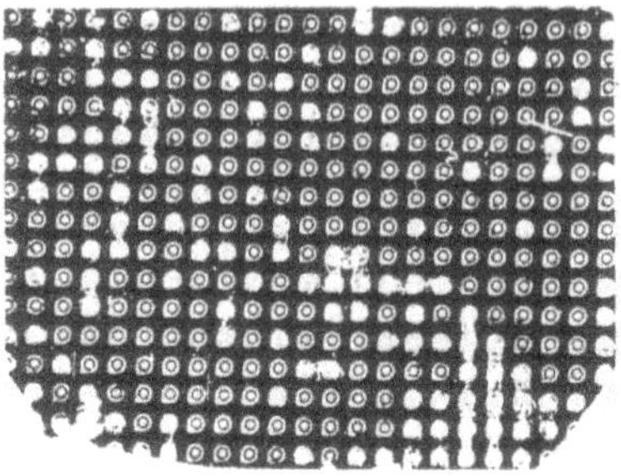

Fig. 4.16. The yield of first planar transistors was 2–5%

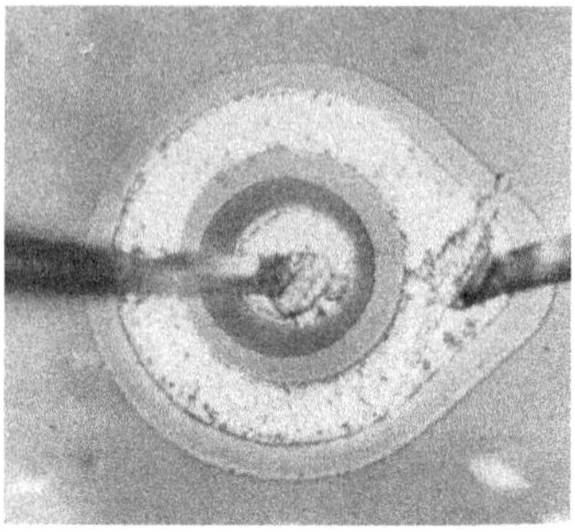

Fig. 4.17. Emitter and Base Contacts of Jean A. Hoerni's Planar Transistor. Outside diameter is ~ 754 μm. The thickness of passivating oxide was about 1.5 μm

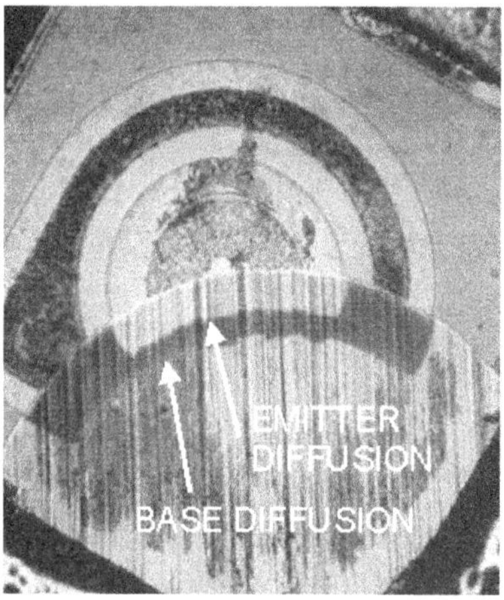

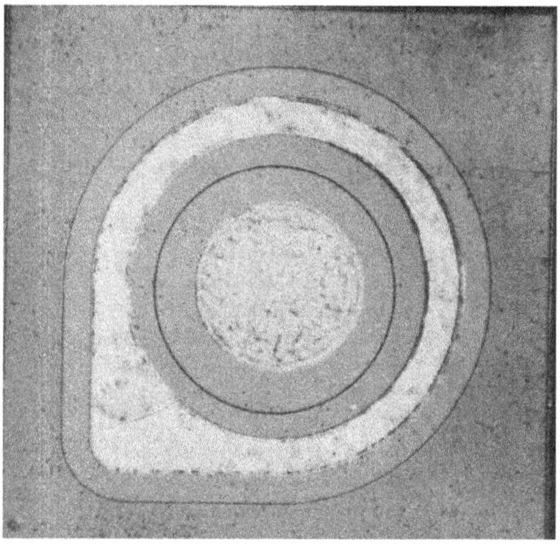

Fig. 4.18. "Decorated" Emitter and Base Diffusion of Fairchild Planar Transistor. The photographs of emitter and base diffusion layer showed in picture is from the time when the author did reverse engineering work on Fairchild transistors

was ready for sampling. In October 1960, Robert Noyce announced that *"now Fairchild intends to convert all of its transistors to the new construction as quickly as possible."*

There was no transistor on the market which could compete with Hoerni's device. Hoerni's transistor made all other transistor technologies obsolete. The 1000 hour test confirmed that the planar transistor has less than one nA of leakage at $V_C = 60$ V compared to typically 10 nA at $V_C = 30$ V for mesa transistor (Fig. 4.19). The collector-base leakage current as a function of temperature is shown in Fig. 4.20.

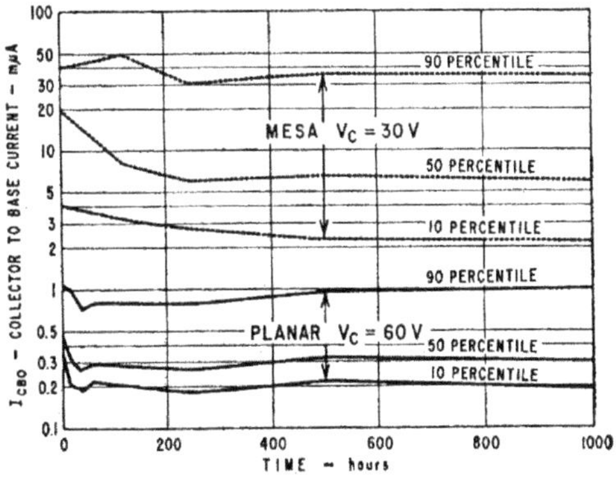

Fig. 4.19. Room-temperature leakage current after 1000 hours life test at 300°C for planar (2N1613) and mesa transistors

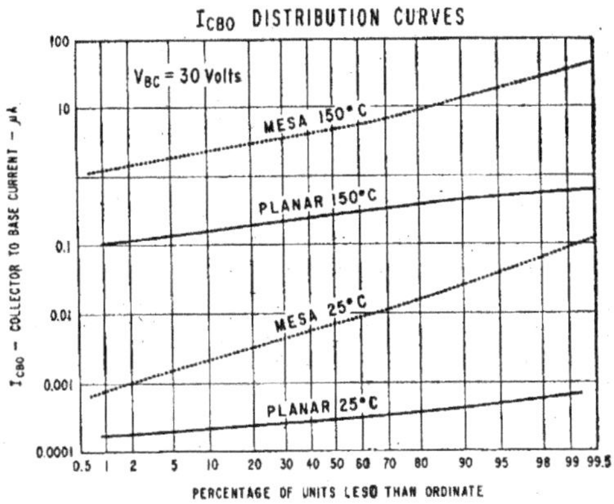

Fig. 4.20. Collector-base leakage current for planar and mesa transistors

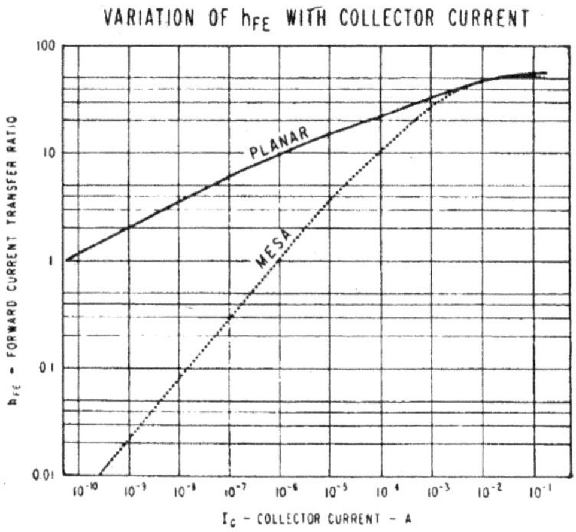

Fig. 4.21. Variation of beta with collector current for planar and mesa transistors

Fig. 4.22. Whisman Road, Mountain View, California

During his work on planar transistors, Hoerni envisioned two additional devices. In his Monthly Report to Gordon Moore dated July 1, 1960, Hoerni described transistor logic using N-MOS transistors. The brief description of his idea is shown in Fig. 4.23. Hoerni did his work at the same time when Dawon Kahng of Bell Labs filled his patent application # 13,688 and # 32,801 in March and May 1960 for MOS transistors. The second Hoerni idea was a planar diode manufactured a similar way as the transistor. In mid-1960 Fairchild management had no idea what an MOS device is, and did not pay any attention to Hoerni's device. Hoerni, pre-occupied with problems of planar technology, had no time to go back and continue work on his unipolar transistor. As a result of Hoerni's effort, Fairchild Semiconductor opened a "diode facility" in San Rafael, and at that time the more practical diode got into production.

By the end of 1960, Fairchild Semiconductor employed 1550 employees. Thanks to Ed Baldwin, Fairchild had a state-of-the-art new facility completed in August 1959 and expanded in 1960, with total area of 108,000 sq. ft. at 545 Whisman Road in Mountain View (Fig. 4.22.) The facility in San Rafael had about 55,000 sq. ft. and the facility in Palo Alto had 56,000 sq. ft. The major problem was to find a qualified labor force. At that time Fairchild Semiconductor was hiring almost indiscriminately (Fig. 4.24.)

The company was growing significantly, and as always, the growth brings problems. Transfer from the development facility in Palo Alto to manufacturing facilities was slow and difficult, and management needed to solve many problems which may be tougher then technical ones. For example, in December 1960, Eugene Kleiner asked Gordon Moore to solve a problem when the staff in Palo Alto was accusing people from Mountain View of taking Palo Alto's supplies of paper and pencils.

When Charles E. Sporck arrived as a new Production Manager in October 1959 he noticed that Fairchild had *"no structured manufacturing organization."* The Fairchild Research and Development was in the same situation. At the end of 1959 Gordon Moore ran the Chemistry Section with Bernard Rabinovitch (surface characterization), Worden Waring (packaging), Paul Ignacz (electrochemical) and Bernard Yurash (analytical services). In the Physics

F.S.C. CONFIDENTIAL

FAIRCHILD SEMICONDUCTOR CORPORATION
Research and Development Department

PROGRESS REPORT - Physics Section July 1, 1960

MOS DIODES

MOS diodes with high surface states density were made by doping n-type oxidized wafers with manganese and heat treating at 750°C. The capacitance ratio between 0 and 10 volts was 40, with most of the drop taking place between 5 and 7 volts.

UNIPOLAR TRANSISTOR LOGIC

The introduction of surface states on p-type oxidized wafers results in n-type inversion layers. Channels of various shapes can be "printed" on a wafer by suitable removal of the oxide (and inversion layer) on some areas of the wafer. Assuming that control of channel conductance by means of a metal electrode on top of the oxide is feasible, it is then possible to design integrated micrologic

PROGRESS REPORT - Physics Section July 1, 1960
Page 2

elements similar to the ones using junction transistors by replacing the basic inverter unit,

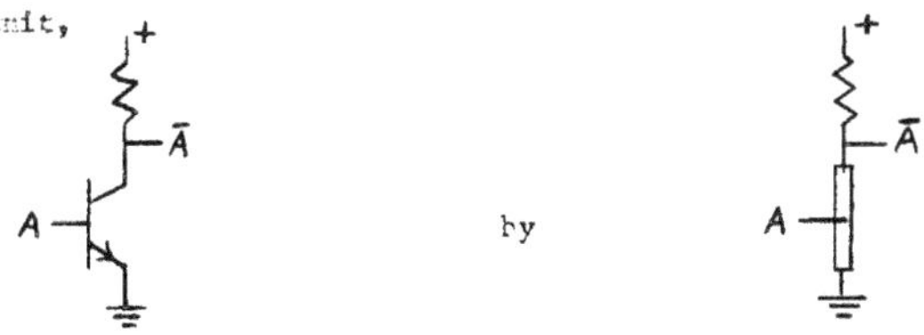

where the second symbol refers to the unipolar unit with n-type channel. Work is in progress to make single unipolar units by the method described here.

Fig. 4.23. Hoerni's report describing N-MOS transistors in July 1, 1960

Section were C. T. Sah, D. A. Tremere (tunnel diode), B. D. James and Fred Schulenberger (Aluminum Alloying), Otto Leistiko, A. P. Halle (vacuum deposition of SiO (not SiO_2!)), Tom Burke and Phillip S. Flint (mesa transistor), O. V. Hatcher (low capacitance diode), C. A. Lasch, E. W. O'Keefe (diffusion and furnace operation), Sheldon Roberts and L. Lynn (material research.)

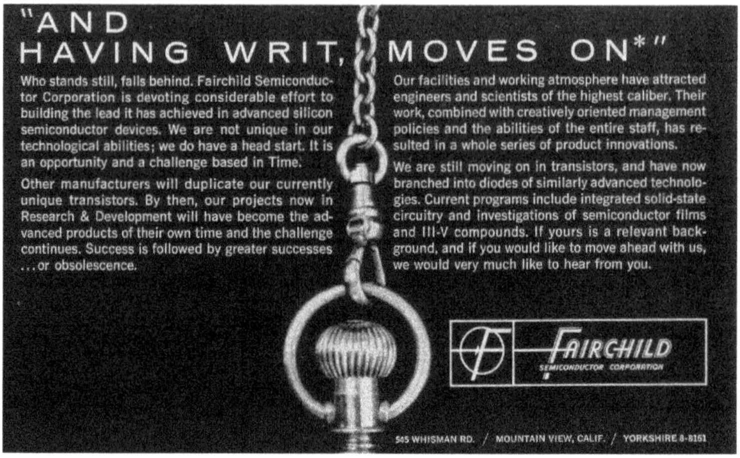

Fig. 4.24. Fairchild recruiting advertisement published in the March 1960 issue of Electronics Magazine

The members of the transistor Development Sections were M. Weissenstern, S. Levine, Garry Parker, R. Brown, P. James, R. Craig, Dave Allison and B. Bently.

Victor H. Grinich was running the Engineering Department and the Device and Reliability Evaluation Department. After Ed Baldwin's defection, Bob Noyce became the General Manager and Vice-President. Julius Blank was running Fairchild Facility, and Eugene Kleiner was running business operations in the Mountain View Facility. Jay Last and Jean Hoerni were in the company's shadow. Jay was working on the parametric diode and integrated circuits and Jean was preoccupied with production of planar devices in Mountain View.

Fig. 4.25. Transistor assembly line in new Fairchild facility (1960 – still no hairnet!)

Fig. 4.26. Advertisement for Fairchild's first planar transistor released in April 1960

Until October 1959 the R&D facility in Palo Alto was used as a production facility for some devices. On October 11, 1959, Gordon Moore sent a special memo to Bob Noyce with the message: *"October saw the last transistor production line moved out of the R&D building."* This date originated problems haunting Fairchild Semiconductor Corporation for many years.

For the time being the Fairchild parts were superior to all other manufacturers and the market demand skyrocketed. With Fairchild sales increased by more than 80% and booking more than 90% (Fig. 4.27) Fairchild management was preoccupied with manufacturing problems and paid very little attention to future products.

Fairchild Camera Sales, Net Soar in Quarter, Half Year

Special to Electronic News

SYOSSET, L. I., N. Y. — Fairchild Camera & Instrument Corp. chalked up a record June quarter and first half, with results sharply ahead of the like 1959 periods.

Net profits in both periods were more than double those of a year ago, while sales increased 83 per per cent in the quarter and 62 per cent in the half year.

In the three months ended June 30, net profit was $882,000, or 85 cents a share, as against $410,000, or 40 cents a share in the second quarter of last year. Net sales and machine rentals in the period rose to $16,447,000 from $9,006,000.

For the first half of this year, net earnings were $1,682,000, or $1.62 a share, compared with $795,000, or 77 cents a share in the like six months a year ago. The half-year net sales and machine rentals totaled $30,285,000, against $18,741,000.

John Carter, president, also pointed out that the earnings for the first six months of 1960 were more than 22 per cent above earnings for the first nine months of 1959. These figures, it was stated, do not include earnings or tax benefits from the recent acquisition of the Allen B. Du Mont Laboratories. Du Mont was merged into Fairchild on July 5, last.

The company said new orders booked in the first half totaled $38,994,000, up 91 per cent from the $20,444,000 in the like 1959 period. As of June 30, 1960, backlog was $28,532,000, compared with $19,823,000 at the Dec. 31, 1959 yearend.

Fig. 4.27. 1960 Fairchild Camera Sales (Electronics News, September 1960)

Jack Kilby's Texas Instrument patent demonstrated that multiple transistors, resistors and capacitors could be formed on the same piece of silicon substrate. However, there ware only a very few electronics circuits which could be assembled from transistors with all collectors common. The majority of electronic circuits required an electrical isolation of the individual electronic components. Therefore the concept of appropriate device isolation was one of the principal problems of the integrated circuit. The second problem, equally important for integrated circuits, was the interconnection of individual isolated components. Although none of these issues were addressed in the Texas Instrument invention, the company launched a large publicity campaign promoting the Kilby patent. Texas Instruments ran this campaign for more than three years without delivering any substance or device which was successful on the market. At the time when Fairchild was already shipping planar devices, Texas Instruments was still "bothering" with "molecular electronics" (Fig. 4.28). Phil Ferguson, who joined Fairchild from Texas Instruments remembers that Texas Instrument's General Manager Mark Shepard commented about planar technology *"It is not significant."*

In response to the TI campaign Robert Noyce asked at WESCON Conference, August 18–21, 1959 Jay Last to start a development program with the goal to create Fairchild integrated circuits. Except for this encounter, Noyce paid very little attention to Jay's work. The management of the Fairchild Semiconductor, a division of Fairchild Camera and Instrument Corporation was preoccupied with exploiting the new exciting product – the planar transistor, including a new plant in San Rafael making diodes based on Hoerni's planar principle. The development of the first planar integrated circuit was certainly not the mainstream activity at the company.

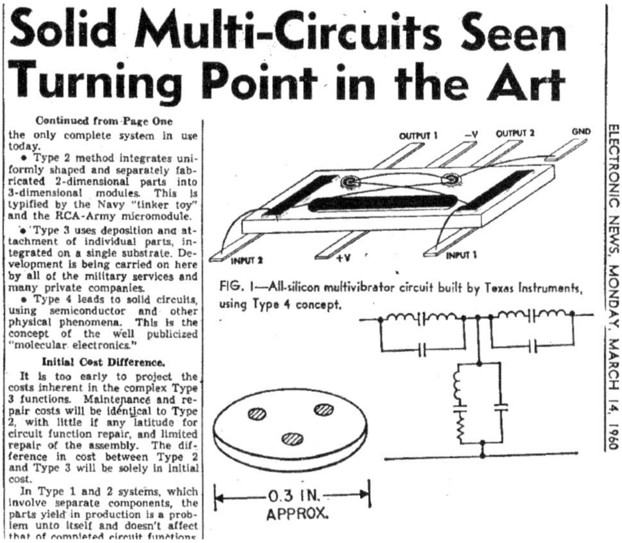

Solid Multi-Circuits Seen Turning Point in the Art

Continued from Page One

the only complete system in use today.

• Type 2 method integrates uniformly shaped and separately fabricated 2-dimensional parts into 3-dimensional modules. This is typified by the Navy "tinker toy" and the RCA-Army micromodule.

• Type 3 uses deposition and attachment of individual parts, integrated on a single substrate. Development is being carried on here by all of the military services and many private companies.

• Type 4 leads to solid circuits, using semiconductor and other physical phenomena. This is the concept of the well publicized "molecular electronics."

Initial Cost Difference.

It is too early to project the costs inherent in the complex Type 3 functions. Maintenance and repair costs will be identical to Type 2, with little if any latitude for circuit function repair, and limited repair of the assembly. The difference in cost between Type 2 and Type 3 will be solely in initial cost.

In Type 1 and 2 systems, which involve separate components, the parts yield in production is a problem unto itself and doesn't affect that of completed circuit functions

FIG. 1—All-silicon multivibrator circuit built by Texas Instruments, using Type 4 concept.

Fig. 4.28. Texas Instruments "molecular electronics" devices

Jay Last's original team included Lionel Kattner, Isy Haas, and Robert Norman. Norman worked in the Appliations Department and reported to Victor Grinich. The group selected the simple bistable RS (Reset/Set) Flip-Flop constructed using four NPN bipolar transistors and two resistors diffused into a single monolithic chip of silicon as the test vehicle. (Fig. 4.29). Bob Norman designed the circuit, the layout was prepared by Lionel Kattner and is shown in Fig. 4.30.

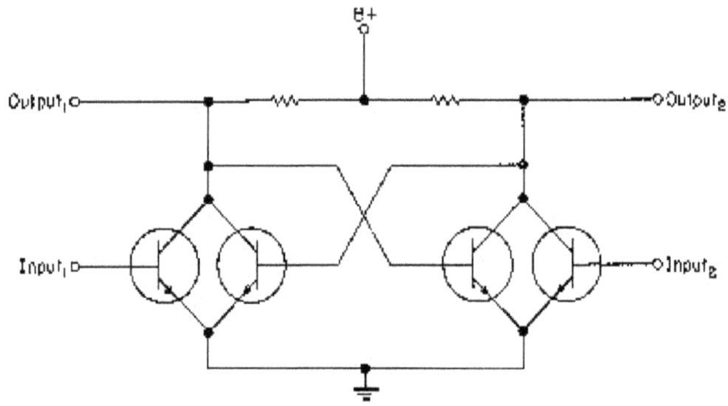

Fig. 4.29. RS Flip-Flop used as test vehicle for first Fairchild integrated circuit

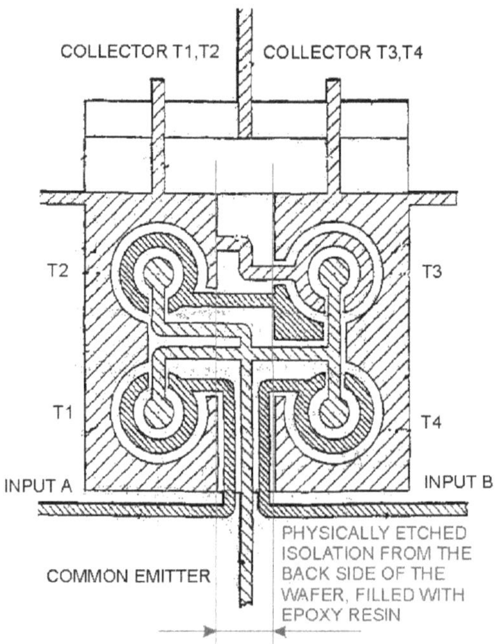

Fig. 4.30. Layout of RS Flip-Flop prepared by Lionel Kattner

The main processing steps are shown in Fig. 4.31. After metal patterning identical to Hoerni's planar devices, the front side of the wafer was coated by Apiezon wax with melting point 140°C and the whole wafer was mounted on the quartz plate. The isolation mask was applied on the back side of the wafer using a special alignment "jig" for masking at the front and the back side, designed by Jay Last. Using selective wet etch the complete wafer thickness was removed in the isolation regions. This region was called "Physical Isolation." The etched trenches were manually filled with epoxy resin. After resin curing the wafer and quartz were heated and the wafer was separated from the quartz plate, and the wax was removed from the front side of the wafer. The circuit used deposited carbon as the resistors.

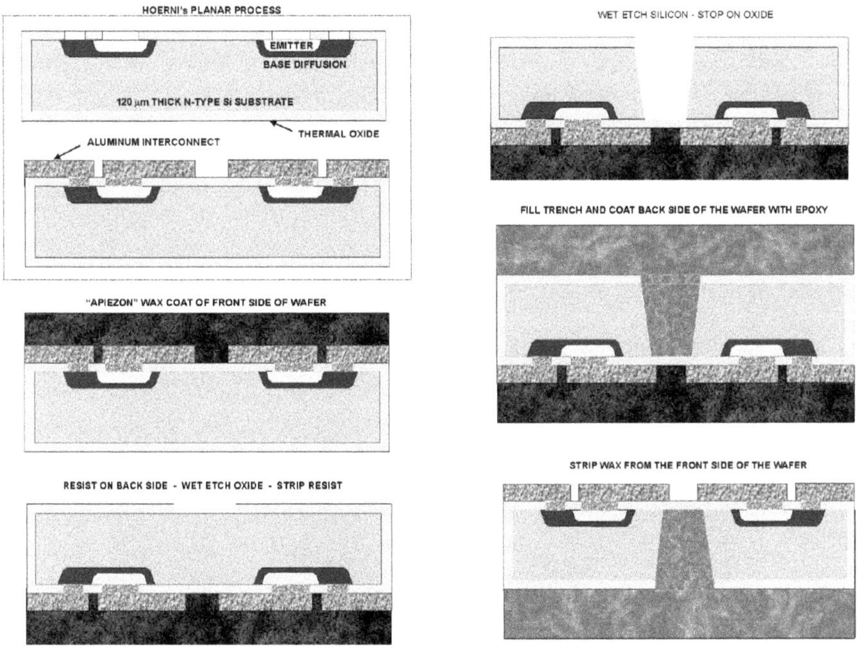

Fig. 4.31. Processing steps used for the very first planar integrated circuit (Last, Haas, Kattner, May 1960.)

Lionel Kattner produced the very first integrated circuit (Fig 4.32) in May 1960. Haas and Kattner developed the critical diffusion process. The circuit was tested by Haas, and to surprise of all, functional with the maximum operating clock was 1 megahertz and the delay 60 nanoseconds.

As always, there were many problems. An original epoxy Epocast 203 distorted or fractured the oxide film. Stycast 3020 epoxy worked much better, but still not perfect. Four units successfully passed thermal cycling at 150°C, but failed at 175°C.

In August 1960 Isy Haas replaced the Carbon resistors with diffused resistors and improved transistor geometry. Approximately 100 units passed room temperature test but failed at 70°C. In September Jim Nall design a new "topographical control and optical jigging for Micrologic elements" and S. M. Fok was testing sugar (melting point 186°C) as a mounting wax. Sugar did not work – all parts had broken oxide.

Isy Haas and Lionel Kattner tried long boron diffusion which was necessary to produce a P-type isolation region. This diffusion step required eighteen hours diffusion through a patterned oxide on the front side of the wafer. All of Fairchild's experts in boron diffusion thought that oxide would not be able to withstand the long diffusion time. However, the experiment was successful and the first devices were made by this technique in October, 1960.

The group's persistence brought a major breakthrough. In October 1960, Gordon Moore reported to Bob Noyce: "*A method was suggested to obtain electrical isolation and prototypes were made with the close cooperation of L. Kattner. The process is not yet under satisfactory control. We have made about 8 units that prove a feasibility of the concept. However, only two of those had no apparent defects. One was given to R. Norman's group. In general these units were slow (no lifetime control),* V_{CE} *were high due to high starting resistivity and there was a large spread among the DC parameters.*"

Fig. 4.32. The world's first planar integrated circuit produced by Jay Last, Lionel Kattner, and Isy Haas. Functional units were demonstrated in May 1960

Although it is difficult to contemplate the production of a circuit with this technology, the circuit with etched isolation demonstrated the feasibility of the planar integrated circuit in the form as used in modern integrated circuits.

Bob Norman, Jay Last and Isy Haas reported preliminary data at the IRE Solid State Conference on February 12, 1960 (Fig. 4.33). The team who developed the world first integrated involved: Jay Last, Bob Norman, Isy Haas, James Nall, Lionel Kattner, Gary Tripp, Robert Marlin, Chester Gunter, James Wilkerson, Jerry Lessard and Melvin Hoar.

FRIDAY, FEBRUARY 12, 1960 **Irvine Auditorium-9:00 A.M.-12:00 Noon**

SESSION VII: Microelectronic Considerations

7.4: **Solid-State Micrologic Elements**

R. NORMAN, J. LAST AND I. HAAS

Fairchild Semiconductor Corporation Palo *Alto, Calif.*

Fig. 4.33. The first presentation describing planar integrated circuit (ISSC 1960)

On November 30, 1960, Victor Grinich published a "Micrologic Application Handbook" and the group made several hundred units with approximately 5% yield. The main cause of loss of yield was the poor alignment of masks. New mask revision was expected at the end of the month.

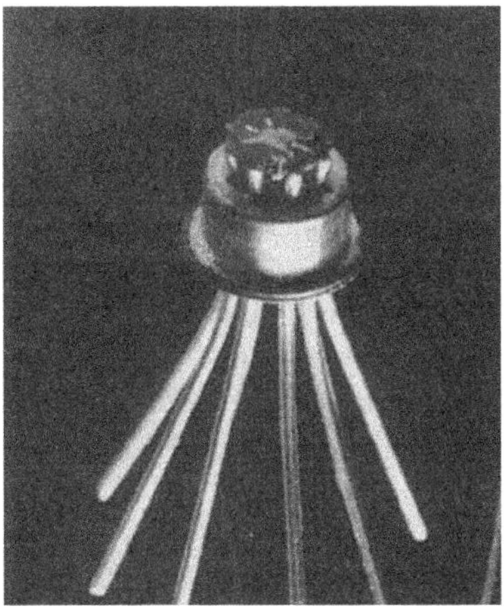

Fig. 4.34. One of the very first version of Micrologic circuits (note mounting of die to holder)

In the first experiments, a methyl borate process was used for electrical isolation. Jean Hoerni, Bob Martin and N. Gault continue work on boron diffusion using BBr_3. Occasionally both systems caused pitting of the silicon surfaces which was attributed to poor pre-deposition control. The recipe used was using pre-deposition with 0.8–1.8 Ω/sq, and drive-in at 1280°C for 22 hours. Diffusion depth of 43–48 μm was commonly obtained.

The project was reformulated, and called Phase I (Physical isolation with carbon resistors), Phase II (Physical isolation with diffused resistors), and Phase III (Electrical isolation with diffusion resistors). The key contribution to the integrated circuit was worked out by Jim Nall. Jim Nall in cooperation with Ferrand Control, Inc. improved the step and repeat control of mask alignment equipment (alignment error better than 2.5 μm, in comparison with 100 μm of the old masking approach). Also, G. Greenslitt designed a new quartz boat with holes, to make the backside pre-deposition less dependent upon the history of a boat.

The process and stability of the circuits were improving rapidly and it was clear that the project would be successful. In November 1960, Jay Last approached the Fairchild management team including Gordon Moore and Robert Noyce, and proposed a new product based on the latest successful integrated circuit research work. Among the attendees of the meeting were Fairchild VP of Marketing, Tom Bay. Tom was vigorously arguing that Fairchild could not get into the integrated circuit business because Fairchild would lose all the main valuable transistor customers. Tom Bay said: *"Jay Last has pissed away a million dollars on this integrated circuit project. I say we should shut it down."* Except for Jay and his group, no one firmly stood behind these planar integrated circuits, and no one said, "Tom, you are wrong." Not even Bob Noyce, or the future father of so called "Moore's Law", Gordon Moore. The cohesion of the "Traitorous Eight" was disrupted.

In a January 1961 Monthly Report to Robert Noyce, Gordon Moore and Victor Grinich were still "*somewhat skeptical*" about integrated circuits (Fig. 4.35.) When Jay Last wanted talk to his pal Bob Noyce, Bob could not find any time. Jay Last left for a while as visiting Ford Lecturer to MIT. Two months after the November meeting, on January 31, 1961, Jay Last, Jean Hoerni, and Sheldon Roberts resigned from Fairchild Semiconductor. Isy Haas followed Jay, Jean, and Sheldon few months later.

During 1960 to 1961 Gordon Moore and Victor Grinich had different secretaries, and it is not clear who compiled Gordon and Victor's Monthly Report to Robert Noyce with copy to Richard Hodgson. The fact is that February 8, 1961 Report contains the following paragraph:

> *"The organizational structure was perturbed rather considerably by the resignation of Jay Last and Jean Hoerni. When the pieces settle down, it seems that we will be much more nearly projectized than we have operated in the past. The administrative functions are, in general, working fairly well, the principal problem being in the personal are,*

where we suffer both from the fuzzy way in which responsibility of our personnel organization is defined and perhaps from the personnel involved. If this does not shape up shortly some moderately drastic action will be required."

INTER-DEPARTMENTAL CORRESPONDENCE

FAIRCHILD
SEMICONDUCTOR CORPORATION

TO:	R. N. Noyce	CC:	
DATE:	January 9, 1961		
SUBJECT:	Research & Development Progress Report - 1/1/61	FROM:	G. E. Moore and V. H. Grinich

Progress Report - R&D -- 1/1/61
Page 2 January 9, 1961

Micrologic:

The first several flip-flops with the new pattern completely isolated by diffusion were mounted during the first week of January. These operate at 4 mc's and early estimates suggest they will have useful fan-out over most of the temperature range. Several more runs right behind these suggest that January will produce several thousand flip-flops. This is enough that we should be able to determine finally if we have a useful device or not. We are still somewhat skeptical.

G. E. Moore/V. H. Grinich

GEM/VHG:hb/jj

Fig. 4.35. Fairchild R&D Monthly Report (January 9, 1961)

Although such characterization of reasons why Jay and Jean left Fairchild is at least questionable, the most important is that nothing or very little changed in Fairchild's managing style. Many "heavy weight" and very capable engineers were driven out of Fairchild. It is important to underline that money had no role in Last's and Hoerni's decision.

Because Texas Instruments and Westinghouse had just entered into a new pact with Autonetics, and TI and Westinghouse "Molecular Electronics" publicity was everywhere, Bob Noyce had no other choice, except that the Micrologic project must continue. The work on the first family of Micrologic elements was to continue under Lionel Kattner's direction. G. Moore wrote in a memo to R. Noyce:

"He had full responsibility for getting out this first family of integrated circuits, both from a developmental and a limited production point of view until Mountain View could be brought into production."

Fig. 4.36. The advertisement for Micrologic Flip-Flop [Electronic News, May 8, 1961]

Additional members joined Kattner: R. Anderson, R. Crippen and D. Thorn. Don Farina was transferred from the Application Engineering Section to Device Evaluation to support full time work on the Micrologic. The decision had been made to announce Micrologic elements at the March 1961 IRE Conventions. Jim Null continued with his work on masking and on January 31, 1961 Gordon Moore reported *"the masking is in the best shape we can remember."* In February, Phase III Micrologic "F" element had gone through shock, centrifuge, vibration fatigue, and variable frequency vibrations test with no failure. In April the masks for "G" and "S" element were completed.

Original Micrologic process flow (Fig. 4.39):

1. N-Si substrate polishing (80 µm $\pm$ 5 µm)
2. Oxidation (wet oxide 8000 Å)
3. MASK 1 (Isolation)
4. Wet etch oxide
5. Boron Deposition and Drive-in
6. MASK 2 (Base and P-Resistor)
7. Boron diffusion ($\sim$ 6000 Å oxide, $\sim$ 150 Ω/sq
8. MASK 3 (Emitter and Collector Contacts)
9. Phosphorus Deposition and Drive-in ($\sim$ 2 Ω/sq and $X_j \sim$ 1.4–1.6 µm)
10. Resist (front side)
11. Wet etch oxide (back side only)
12. Vacuum Evaporation of Gold on the back side ($\sim$ 400 Å)
13. Gold Diffusion ($\sim$ 1050°C/$\sim$ 15 min with fast cool)
14. MASK 4 (Contacts)
15. Evaporate Aluminum (front side, 0.01 Ω/sq)
16. MASK 5 (Metal)
17. Wet etch metal (25% solution of sodium hydroxide)
18. Metal alloying ($\sim$ 600°C/ Argon)

In parallel, the members of the Physics Section, A. N. Saxena, B. D. James, F. Schulunburg started work on "epi apparatus" in October 1960. In November "some other experiments were being tried, information was obtained as to how the Bell Telephone and the Merck people were doing their epitaxial deposition." A new reactor with RF heating instead of resistive heating was set up during February 1961.

In March 1961, L. Kattner designed an integrated epitaxial transistor (0.2 Ω-cm N-type on P-type substrate) with hope of simplifying the problem of isolation. The design of the epitaxial transistor was based on the Fairchild discrete device 2N914; however, the integrated transistors were poor – *"the major problem seemed to be in the deposited* (epi) *N type silicon layer."*

The photograph in Fig. 4.37 show the RS Flip-Flop from at the time when the original transistors with circular geometry were replaced with "stripe" geometry. Such change was possible only after Jim Nall's stepping camera was able to aligned the masks with a better accuracy, and when epi grown layers were under control.

In April, L. Kattner reported: *"considerable time has been devoted to making the necessary arrangements for effective transfer of the "F" element to Mountain View production by June 1961."* Don Farina issued a report of customer response to Micrologic after the March IRE Conventions: "*The response of people that count and of people who have been in the business a long time has been very gratifying. This is contrast to others who insist on "custom" circuitry at "any cost" because they want "design freedom." One such proud designer who wanted his pet circuit integrated, require a transistor with minimum beta of 90. Its operation even at room temperature is questionable in the writer's mind but it is his own."*

When the dust settled after the Last and Hoerni resignation, the promised more drastic measures announced by Gordon Moore in February 1961 never took place. Not unexpectedly, David F. Allison (formerly head of the Device Development Section) Mark Weisenstern and Lionel Kattner (both members of the group) and David James (formerly head of the Physics Section) resigned from Fairchild by August 1961 and formed Signetics. Fairchild announced that Victor Grinich will be acting head of the Device Section, while C. T. Sah, will be acting head of Physics Section.

This time, no "cook books" needed to be moved when "cooks" move. It took to David Allison 10 months to start his company and launch a new Signetics product, R/S Flip-Flop SE-120. The DTL 8 MHz Flip-Flop was announced at the 1962 IRE Convention and it was the only product Signetics had at that time.

Fairchild systematically reduced the prices of transistors and challenged the semiconductor industry to a price war. Some competing companies lost money before they started production. For example, in February 1960, IBM opened multi-million dollars automated assembly line to produce 1,800 NPN alloy junction transistors per hour in Poughkeepsie, NY.

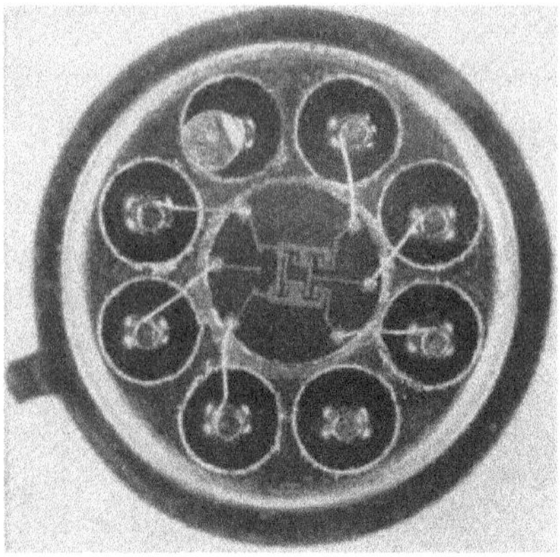

Fig. 4.37. Bonded R/S Flip Flop circuit produced with technology later used for Micrologic (mid of 1961)

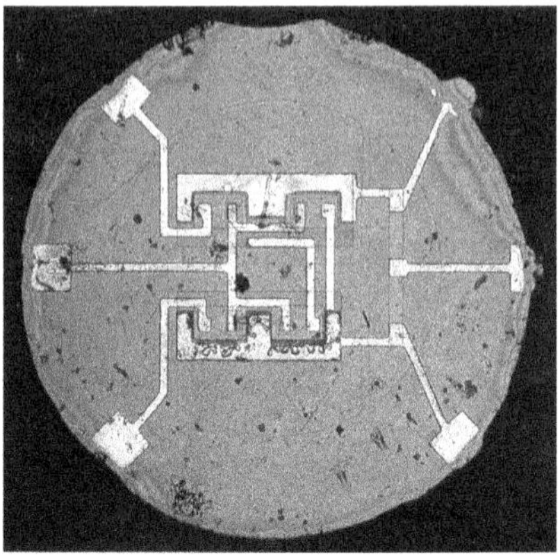

Fig. 4.38. Detail photograph of Flip-Flop from Fig. 4.37

By the end of 1960 no new design considered application of an alloy junction transistor. During the decade 1950 to 1960 semiconductor business grew from zero to over $500,000,000. Fairchild just began its romance with planar devices and nobody was worried about any "downturn."

In December 1961, Fairchild Semiconductor started construction of a new $1,500,000 Research and Development Center on a nine-acre site in Stanford Industrial Park. Gordon Moore, Fairchild's director of R&D, said the new H-shaped building will double the space of the present building, will be fully air-conditioned and mainly *"no manufacturing will be done in the new site."*

One month earlier, Device Evaluation Section and Application Section was removed from the R&D organization and became part of the marketing organization. The Device Evaluation Section was headed by Robert H. Norman, who was not getting along with Gordon Moore well. Norman wanted to file a patent application for a semiconductor memory. Gordon Moore considered *"such idea so ridiculous, that filing the application would be waste of company money."* A few years later such semiconductor memory became commodity products.

Bob Norman designed the first Micrologic Flip-Flop circuit. Additional six or seven Micrologic elements were introduced in summer 1961. All Micrologic elements were designed by Bob Norman and Bob Anderson. The family consists from the following elements:

1. F element – Flip-Flop (4 transistors, 2 resistors)
2. S element – Half-shift register (9 transistors, 5 resistors)
3. G element – Gate (3 transistors, 1 resistor)
4. B element – Buffer (3 transistors, 3 resistors)
5. H element – Half Adder (4 transistors, 3 resistors)
6. C element – Counter Adapter (6 transistors, 5 resistors)
7. R element – Shift Register (17 transistors, 9 resistors)

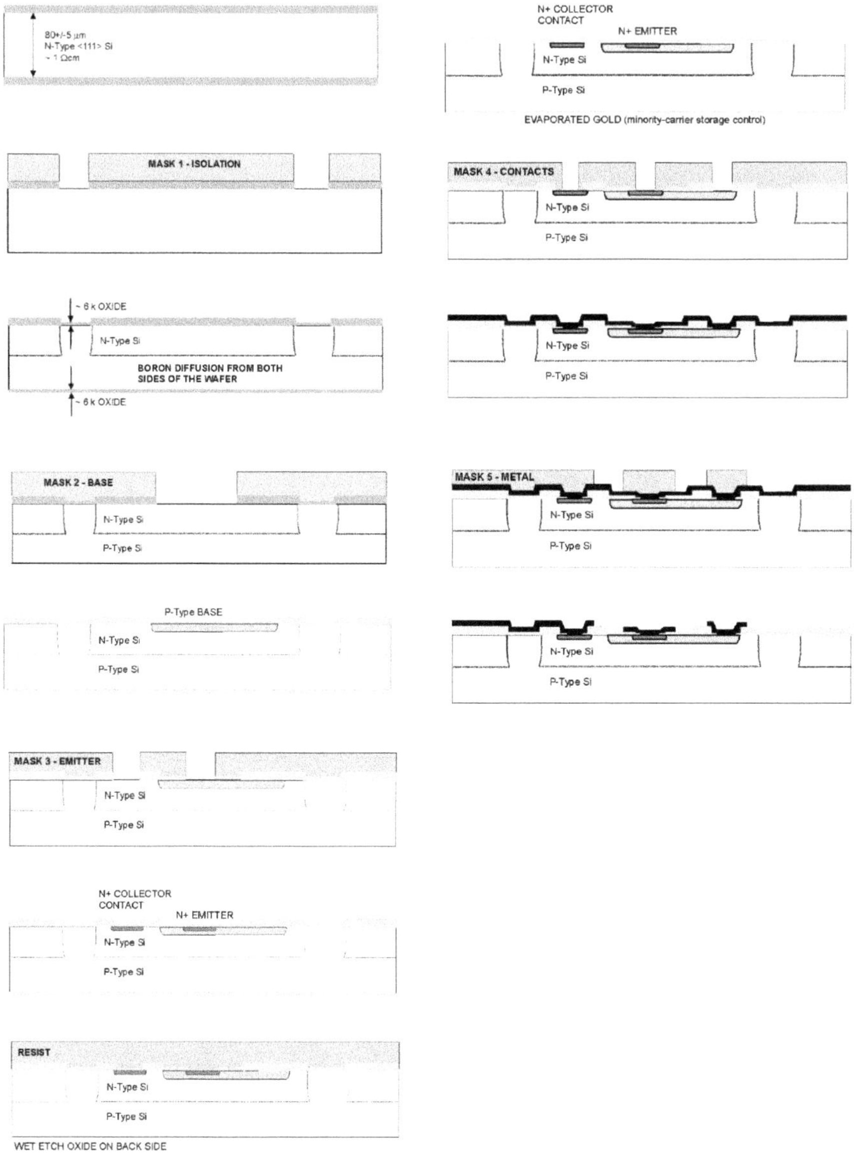

Fig. 4.39. Micrologic process flow using an N-epi layer (end of 1961)

According to Bob Norman: "*the firm has a little difficulty in convincing computer manufacturer of the advantages of Micrologic. The replaceable module has been unquestioned and the low cost, $120.00 each in small quantities upon introduction, allows the large commercial computer manufacturer a reduction in maintainability requirements.*"

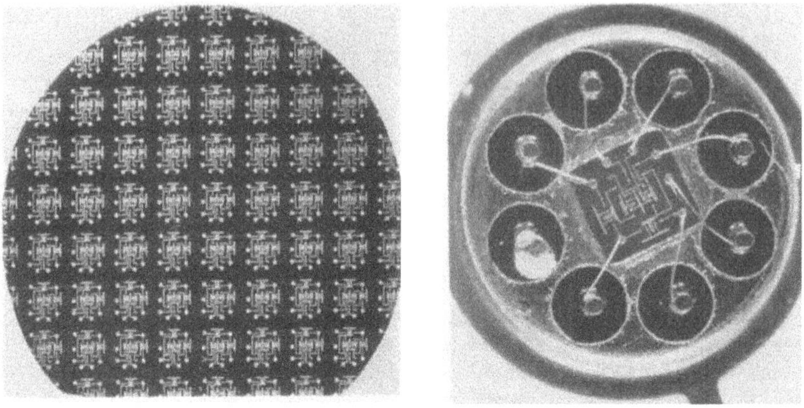

Fig. 4.40. Fairchild 1″ wafer with S Micrologic element – 1 MHz half-shift register

Eventually, Fairchild produced more sophisticated Micrologic RTL circuits. In Fig. 4.43 is shown the four input NAND/NOR gate FμL 907 which resembles the modern integrated circuits. Its layout clearly demonstrates the advances in IC technology that had been achieved. Not only did the 907 have many more components than its predecessor, laid down in a way that greatly increased the chip's performance, it also contained isolation channels and buried layers – features that were to become commonplace in ICs. The 907 consists of two logic gates, each composed of four transistors and an equal number of resistors. The transistors are the small green boxlike features with the overlapping beige bars (within the two inner boxes encircled by the dark red lines); the resistors are the horizontal green bars. A large fifth resistor lies at the bottom. The dark red lines are the visible edges of isolation channels, while the beige features are aluminum conductors. Actual size: 0.96×1.2 mm. The status of products as of summer 1965 is shown in Fig. 4.42.

RTL (Resistor-Transistor Logic) logic as used in Micrologic was a very important step in development of integrated circuit technology. But Fairchild was pushing this approach too long and after Signetics introduced their DTL products and when Texas Instruments was gaining the digital market with much more sophisticated TTL products, Fairchild technology was lagging the state-of-the art circuit development.

In May 1961 issue of the Solid State Journal Sidney L. Siegel, Vice-President Marketing for Pacific Semiconductor commented on the "state of affairs of the semiconductor industry." Siegel warned that at the moment *"the industry looks like a gold mine attracting all kind of "operators" anxious to make a fast stock profit. Many new, unproven companies have been brought on the market, and stock prices have doubled before a single product was ever manufactured. The semiconductor industry is loaded with immature managerial talent exploiting the magical romance of financial operators with semiconductor business."*

Fig. 4.41. Fairchild's very first advertisement for "F" Micrologic element released on March 3, 1961

MICROLOGIC INTEGRATED CIRCUITS	HIGH SPEED μL MICROLOGIC −55°C to +125°C	INDUSTRIAL FμL MICROLOGIC +15°C to +55°C	MILLIWATT MWμL MICROLOGIC −55°C to +125°C	TRANSISTOR/ TRANSISTOR TTμL MICROLOGIC −55°C to +125°C	DIODE/ TRANSISTOR DTμL MICROLOGIC −55°C to +125°C	COMPLEMENTARY TRANSISTOR CTμL MICROLOGIC +15°C to +55°C	COUNTING CμL MICROLOGIC 0°C to +75°C
GATES	NAND/NOR 903 3-input 907 4-input 914 Dual 2-input 915 Dual 3-input	NAND/NOR 903, 915 3-input 907 4-input 910, 911 (see MWμL) 914 Dual 2-input Epoxy	NAND/NOR 910 Dual 2-input 911 4-input	103 Dual 4-input 104 8-input	930 Dual 4-input 946 Quad 2-input 962 Triple 3-input	NOR 952 Dual 2-input Inverter AND 953 2-2-3-input 954 Dual 4-input 955 8 input	
ADDERS	904 AND/OR 2-level Half Adder	904 AND/OR 2-level Half Adder 908, 912 (see MWμL)	908 EXCLUSIVE OR 912 AND/OR 2-level Half Adder				
BUFFERS	900 Low Impedance	900 Low Impedance Epoxy 909 2-input inverting	909 2-input		932 Dual 4-input 944 Dual 4-input	956 Dual 2-input	
STORAGE ELEMENTS (Flip-Flops)	902 RS 905 GATED RS with inverter 906 GATED RS 926 J-K	923 Epoxy J-K 905 GATED RS with inverter 913 TYPE D 926 J-K	913 TYPE D		931 CLOCKED 945 CLOCKED Low Impedance 948 CLOCKED Low Impedance 950 A.C. Coupled R.S.	957 DUAL RANK	958 DECADE COUNTER CASCADING 959 BUFFER/STORAGE (coming 2nd qtr. '65) 960 DECODER/ DISPLAY DRIVER (coming 2nd qtr. '65)
OTHER	901 Counter Adapter	921 EXTENDER Dual 2-input	921 EXTENDER Dual 2-input		933 EXTENDER Dual 4-input 951 MONOSTABLE MULTIVIBRATOR		
PERFORMANCE (Logic form, speed, power, noise immunity, packaging)	RTL; 12 nsec; 15 mW/node; 300 mV; TO-5; Cerpack ¼" x ¼" flat	RTL; 12 nsec; 2-15 mW/node; 300 mV; TO-5, Metal or Epoxy	RTL; 40 nsec; 2 mW/node; 300 mV; TO-5, Cerpack ¼" x ¼" flat	TTL; 25 nsec; 22 mW/gate; 450 mV; Cerpack ¼" x ¼" flat	DTL; 25 nsec; 5 mW/gate; 1.0V; TO-5, Cerpack ¼" x ¼" flat	CTL; 5 nsec; 35 mW/gate; 600 mV; Dual In-Line Industrial type	RTL; TO-5, Dual In-Line Industrial

Fig. 4.42. Fairchild Micrologic Products Line (Summer 1965)

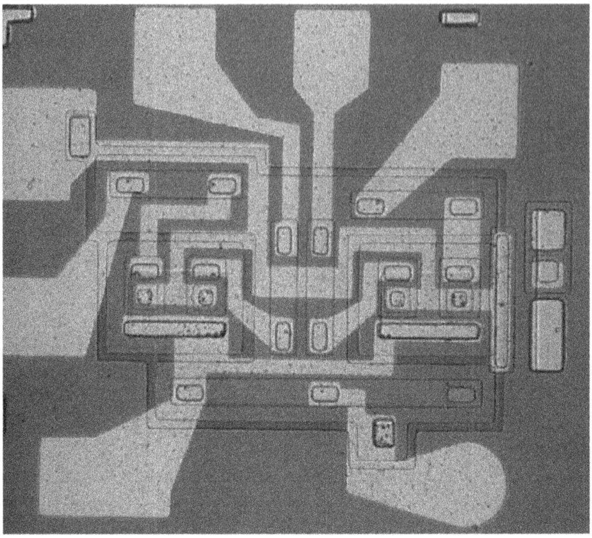

Fig. 4.43. Fairchild four input NAND/NOR gate FμL 907

Although the Fairchild management was not a typical example of such a situation, there were significant mishaps. Everybody who ever worked with Bob Noyce felt immediately Bob's charisma. Bob was approachable and laughed easily. He had grace and strength. He melted women easily. Bob, however, also did, not want to make a decision and he very seldom wanted say no. He wanted to please everybody and he frequently said different things to different people.

Gordon Moore said in [3] that: *"I had become increasingly frustrated with difficulty in transferring new products and technology into the manufacturing organization of the company."*

Who else should fix this problem if the Director of R&D and VP and General Manager of the Semiconductor Division were not able to do it?

Robert Noyce was involved with the general company management and, contrary to common myth, played only a very small part in actual design or fabrication of the first integrated circuits. This prompts a question as to who was behind the Fairchild patents # 2,981,877 "Semiconductor device and lead structure" and # 3,117,260 "Semiconductor circuit complexes." Both patents belong to Robert N. Noyce and were filed on July 30, 1959 and September 11, 1959. They were considered as proof that the planar integrated circuit was invented by Bob Noyce. Gordon Moore stated in [2] that *"Noyce assembled a meeting to brainstorm how Hoerni's invention might be extended into the integrated circuit realm."* Neither Moore nor Hoerni were at the meeting and Jay Last doubted that any such meeting was held.

Noyce later said that as he began thinking about Kilby's circuit *"all the bits and pieces come together one day."* The entry in Noyce's notebook dated

January 23, 1959 is entitled "Methods of isolating multiple devices." This is the only entry which is related to integrated circuits. All previous notes dealt with various transistors and diode matters. Noyce wrote: "*in many applications now it would be desirable to make multiple devices on a single piece of silicon in order to be able to make interconnections between devices as part of the manufacturing process, and thus reduce size, weight, etc., as well as cost per active element.*"

The key development that made the integrated circuit feasible was the invention by Jean Hoerni of the planar transistor in December 1957, which Noyce signed as "read and understood." Noyce also signed Hoerni's disclosure "Selective control of electron and hole lifetime in semiconductor device" on January 20, 1959 when planar process certainly come into the discussion. In November 1958, Noyce learned details about Aluminum sputtered interconnect approach developed at DOFL. Very likely Noyce learned more about this technique when he interviewed Jim Nall in January 1959. When Noyce wrote "*all the bits and pieces*" in his entry on January 23, 1959 he combined Hoerni's planar process idea with Nall's and Lathrop's sputtered Aluminum interconnect ideas.

Noyce's later comments explaining how he conceived the idea by "thinking about Kilby's circuit" are irrelevant to his entry from January 23, 1959. Kilby's application was submitted on February 6, 1959 and TI did not reveal a single word about their approach until the IRE Convention in March 1959, two months after Noyce recorded his notes.

Noyce's notebook entry discusses in detail the isolation between multiple devices (Fig. 4.46), however, Noyce's patent application filled on July 30, 1959, which become U.S. Patent # 2,981,877, did not contain any mention about junction isolation as described in Noyce's notebook.

An additional puzzling entry in Noyce's notebook dated January 30, 1962 where he discussed bonding Pyrex glass to the surface of the wafer as a support for subsequent silicon etch from the backside of the wafer. This is essentially the technique used in the very first Fairchild integrated circuit designed by Last's group almost three years earlier.

Noyce's reasoning for this type of isolation was based on the thought that this type of isolation would eliminate the extra capacitance introduced into a circuit by the isolating PN junction, which would deteriorate the circuit performance. The fact that Noyce considered this type of device isolation of some importance is supported by a highly unusual note that Gordon Moore, on December 12, 1962, added to Noyce's notebook where he wrote "*I do remember discussing this idea with RNN early this year.*"

If Noyce's dates are correct, he knew a better solution to the problems that Last's group tried to solve. Why was this idea not passed to Jay Last? Why did Noyce, at the time when Micrologic with PN junction isolation was already in production, consider an idea that was filed to the Patent Office on August 15, 1960 by Jay Last?

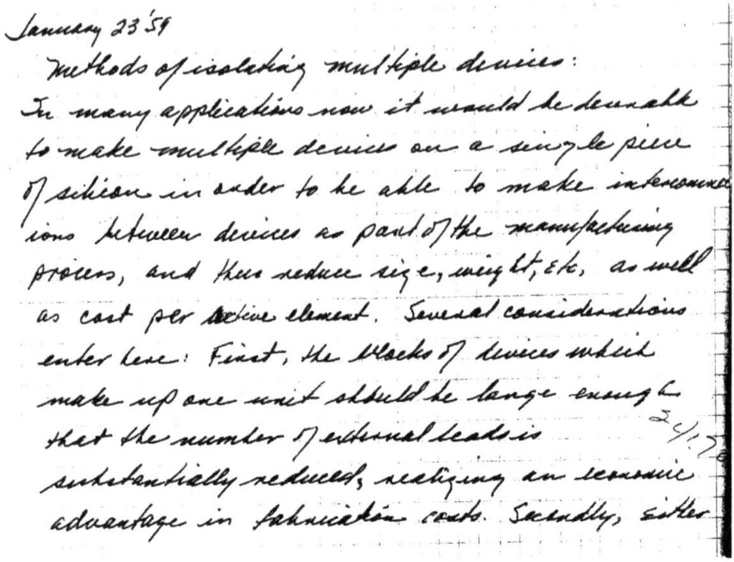

January 23 '59
Methods of isolating multiple devices:
In many applications now it would be desirable
to make multiple devices on a single piece
of silicon in order to be able to make interconnect-
ions between devices as part of the manufacturing
process, and thus reduce size, weight, etc, as well
as cost per active element. Several considerations
enter here: First, the blocks of devices which
make up one unit should be large enough
that the number of external leads is
substantially reduced, realizing an economic
advantage in fabrication costs. Secondly, either

Fig. 4.44. Page 70 of Noyce's Lab Notebook

Fig. 4.45. Robert N. Noyce in 1959

The entry mentioned in Noyce's notebook as "connections are made by evaporating metal through the holes in the oxide to interconnect the diodes as desired for a particular circuit" also cannot withstand scrutiny. This note correlates with other previously known facts. In May 1957, the Microelectronics work group of Diamond Ordnance Fuze Laboratory (DOFL) began investigation of the application of extremely miniaturized components that

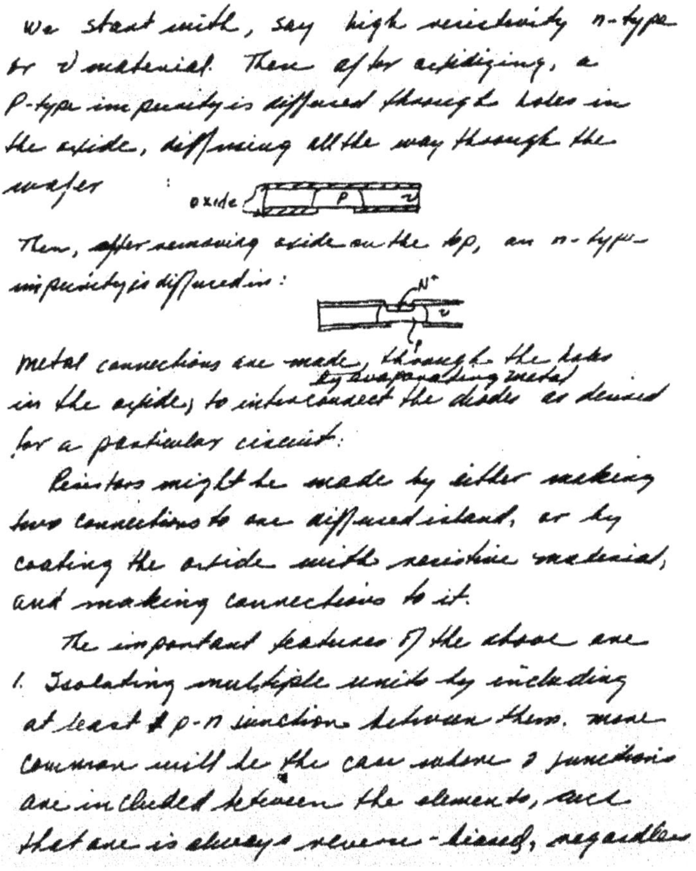

We start with, say high resistivity n-type or ν material. Then after oxidizing, a P-type impurity is diffused through holes in the oxide, diffusing all the way through the wafer:

oxide P ν

Then, after removing oxide on the top, an n-type impurity is diffused in:

N+ ν P

Metal connections are made, through the holes (by evaporating metal) in the oxide, to interconnect the diodes as desired for a particular circuit.

Resistors might be made by either making two connections to one diffused island, or by coating the oxide with resistive material, and making connections to it.

The important features of the above are

1. Isolating multiple units by including at least 1 p-n junction between them. More common will be the case where 2 junctions are included between the elements, such that one is always reverse-biased, regardless

Fig. 4.46. Page 72 of Noyce's Lab Notebook

had recently become available on the market. Members of the group were James R. Nall, Jay W. Lathrop, Thomas A. Prough and Robert S. Marty. This group developed two very important processing steps. They used transistors and diodes only as pieces of germanium wafer with vacuum-deposited aluminum interconnect (Fig. 4.48.) The second process was photographic masking of transistors (Fig. 4.49.)

James Nall joined Fairchild in February1959. Robert Noyce who was at that time manager of Fairchild Semiconductor Research and Development Department interviewed Nall in January the same year. Nall received B.S. degree from George Washington University in 1952. While conducting undergraduate study he worked in the Department of Pharmacology at the University. He joined the staff of the Electron Tube Laboratory at the National Bureau of Standards in 1952. In 1956, he transferred to the Diamond Ordnance Fuze Laboratories where he was project leader in device development.

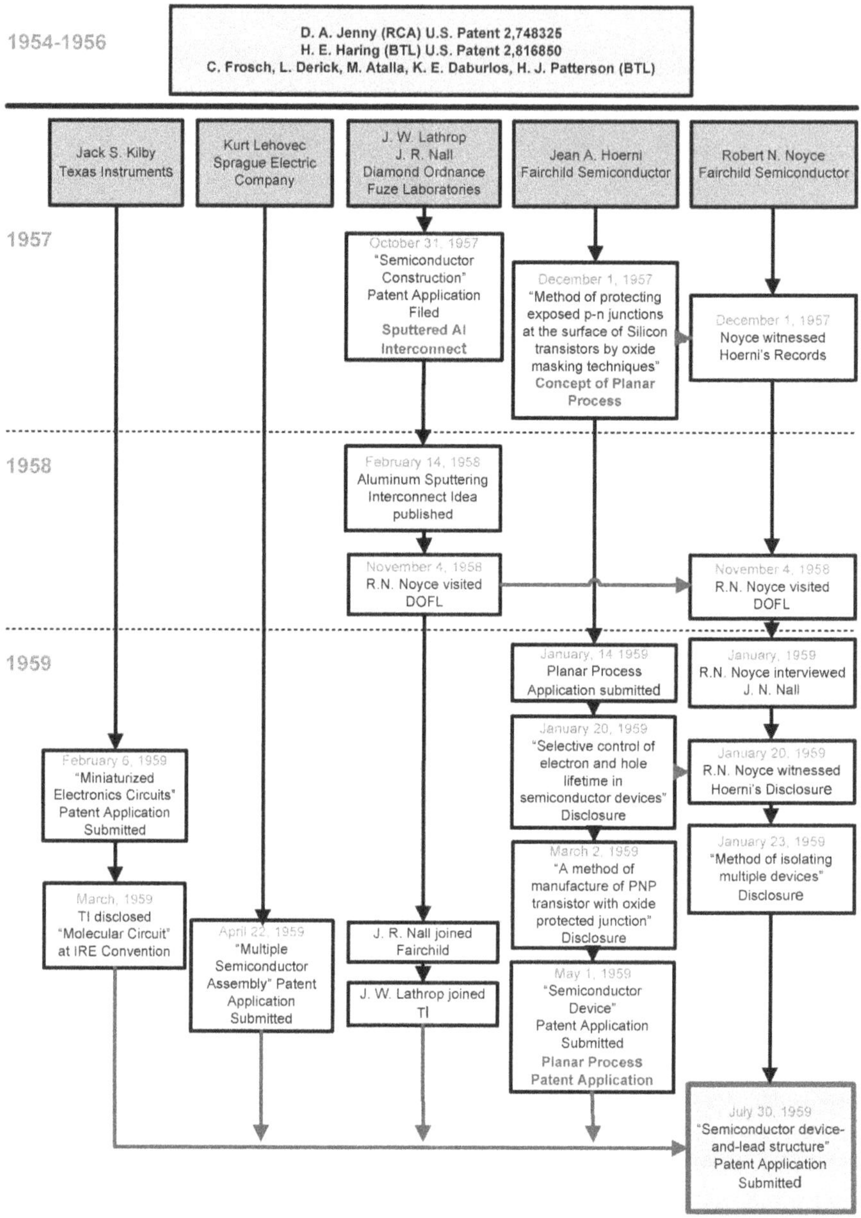

Fig. 4.47. "*All the bits and pieces*" used in Noyce's Patent Application submitted on July 30, 1959 which became U.S. Patent 2,981,877

Fig. 4.48. The first transistor made with photolithography and with evaporated aluminum (Nall, Lathrop, Prough, DOFL, January 1958)

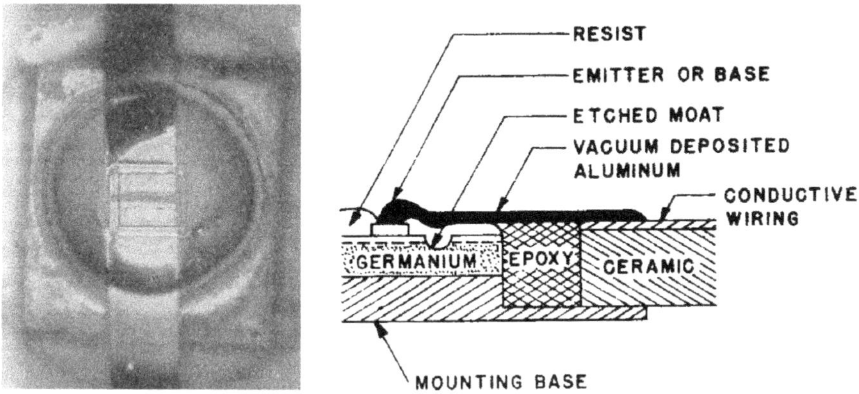

Fig. 4.49. Photographs of deposited aluminum contacting emitter and base of Ge mesa transistor (vertical stripe) and cross-section of germanium die assembled in ceramic wafer (DOFL 1958)

I asked Bob Noyce if he was aware of Nall's previous work on deposited aluminum interconnects. Bob first asked if I worked with Nall, and after that, he answered that he knew about Nall's work and said, *"this was definitely one of the hints."*

I found in DOFL papers list of visitors and licenses of DOFL technologies prepared by T. M. Liimatainen, Deputy Chief of Branch 340 "Electron Tube." Bob Noyce visited DOFL at Van Ness Street in Washington, D.C., on November 4, 1958.

COMMERCIAL COMPANIES
VISITING THE DIAMOND ORDINANCE FUZE LABORATORIES
ON MICROMINIATURIZATION AND PHOTOLITHOGRAPHIC TECHNIQUES
9/10/57 THROUGH 7/15/59

NAME OF COMPANY	VISITS	VISITORS
Federal Telecommunication Labs	2	3
General Electric Company	5	18
* Radio Corporation of America	10	19
Bell Telephone Laboratories	2	3
* Motorola	4	6
* Sylvania	19	40
Zenith Radio Corporation	2	4
Electro-Optical Systems	3	3
Telemeter Magnetics	1	2
Texas Instruments	10	18
Philco Corporation	12	30
* Melpar, Inc.	6	12
ACF Electronics	3	5
Universal Transistor Products	1	2
* Varo Corporation	2	4
Gahogan, Inc.	1	1
* Sprague Electric Company	9	16
* Bosch Arma Corporation	6	16
Ford Instruments	2	3
AVCO	1	1
General Chemical	1	1
* Westinghouse	6	18
Western Electric Company	1	1
* Fairchild Semiconductor	1	2
Ohio Semiconductor	1	2
Reuter-Stokes	1	1

* Commercially using DOFL "2D" Concept and Solid Circuit Concept.

Fig. 4.50. List of visitors and licenses of DOFL "2D Solid Circuit Concept"

Bob's name is also on the distribution list of DOFL internal memo "The use of Photolithographic Techniques in Transistor Fabrication" by J. R. Nall and J. W. Lathrop. This was only report available at that time. The main difference in later (spring 1959) practice over what was described in this report is use of moat etching in place of pedestal etching.

PRODUCTION TECHNIQUES

Transistors Are Made By Photolithography

By J. R. Nall and J. W. Lathrop
Diamond Ordnance Fuze Laboratories
Washington, D. C.

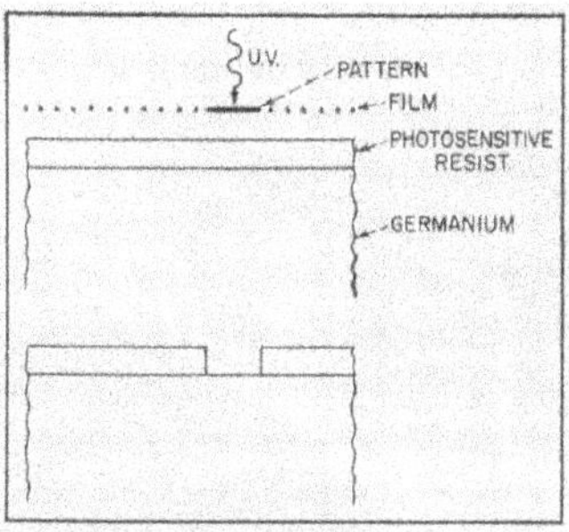

Schematic diagram of exposure process for germanium blanks

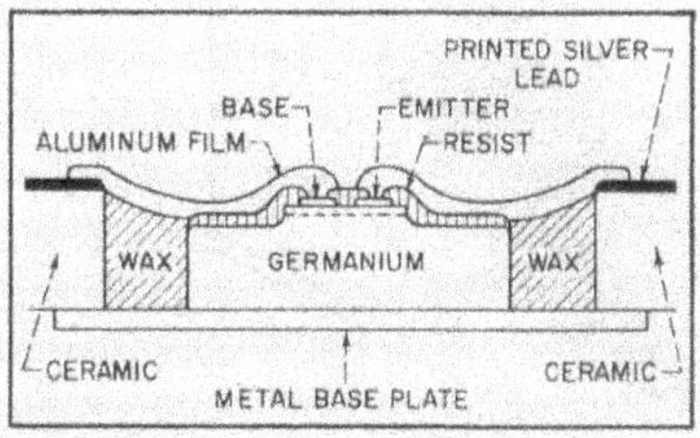

Cross section shows finished transistor fixed in hole in ceramic printed circuit board. Final aluminum film makes leads

Fig. 4.51. ELECTRONICS Magazine, Engineering Edition, February 14, 1958

I am not familiar with any single publication which would mention Jim Nall's or Jay W. Lathrop's contribution to the development of planar integrated circuit. Although Diamond Ordnance Fuze Laboratory published several papers describing their process and technology several years before Bob Noyce's patent application was filed, yet, the Fairchild boss of Nall is the only one to whom belongs all credits.

Politicians, many business leaders a CEO's often *say* that innovation is critical to the future of civilization, our country, their company, etc. But in practice, these same people often *act* as if innovation is an evil that must be suppressed, or at least tightly controlled. In Fairchild Semiconductor, only the Director of Research and Development and later vice-president, Robert N. Noyce, decided which ideas would be patented and which would be suppressed. When Last asked Fairchild's management where his project fit into the company plan, nobody said, "Integrated Circuit is a brilliant idea" and we will do it and support you. Of course, when Jay's group accomplished the goal, they forgot their earlier predictions and the lack of the vision and they started claiming how their contribution was important.

It is sad that the invention of the planar integrated circuit at Fairchild is credited to Robert N. Noyce only, and the monumental accomplishment of Jean Hoerni and others are forgotten. It was Jean Hoerni's creativity and Jay Last's persistence that made the integrated circuit a reality. Robert Noyce was in charge of the company where Jean Hoerni, Jay Last, Isy Haas, Lionel Kattner, Robert Norman, James Nall, Gary Tripp, Robert Marlin, Chester Gunter, James Wilkerson, Jerry Lessard, and Melvin Hoar created the integrated circuit, which they called at that time a *"pea on a dinner plate"* [6].

References

[1] Eight Leave Shockley to form Coast Semiconductor Firm, Electronic News, October 21, 1957

[2] G. E. Moore, "The Role of Fairchild in Silicon Technology in the Early Days of "Silicon Valley," Proc. IEEE, Vol. 86, (1998), pp. 53–62

[3] G. E. Moore, "Intel – Memories and the Microprocessor"', Daedelus 125, No. 2. 1996

[4] J. R. Nall, J. W. Lathrop, "Photolithographic Fabrication Techniques for Transistors which are an Integral Part of a Printed Circuit," 1957 Electron Devices meeting of the IRE PGED, November 1957

[5] Transistor made by Photography, Electronic Week, November 11, 1957, p. 5

[6] R. Norman, J. Last, I. Haas, Presentation at Solid State Circuit Conference, February, 1965.

5

Driving the Company Out of Business

"Almost all of our qualifications on MIL devices have lapsed, and we are thus excluded from shipping to the high-priced and profitable military market unless we continue to lie. There are many reasons for not doing so. First, it is fraud and, if exposed, could be almost a death blow to our business. Second, when we are lax enough so that everyone knows we shall ship product regardless of the quality level, no one get excited about fixing the problem; soon the quality level degrades so that even commercial customers refuse our product."

C. Lester Hogan

Fairchild's epitaxial Micrologic product line was introduced in August 1963 at the Western Electronic Show and Convention (WESCON); Fairchild added new elements (4-input gate, half shift register without inverter) to their line which replaced completely the original Micrologic line.

Element	1–99		100–999	
	Old	New	Old	New
Buffer	$26.00	$24.60	$20.80	$16.50
Counter	70.00	48.10	56.00	32.20
Flip Flop	40.00	26.60	32.00	17.15
Gate	20.00	20.00	16.00	13.40
Half Adder	60.00	37.00	48.00	24.80
Half Shift	70.00	53.40	56.00	35.75

The product manager for Micrologic, Mel H. Phelps announced at WESCON *"an order of magnitude increase in deliveries of the new Micrologic."* The reduced time of isolation diffusion, from 22 hours to 6 hours, resulted in higher yield due to reduced wafer warpage and breakage.

Bob Noyce announced that by January 1964 the Mountain View facility would be converted to production of integrated circuits mainly, moving transistor work to the Portland, ME., plant. The problem was that the Fairchild parts became obsolete. In 1972 Bob Noyce said *"we pushed RTL much longer than we should have."* The introduction of DTL by Signetics had an enormous impact on the industry, because designers were quickly accustomed to working with the concept behind DTL.

The original four founders of Signetics had no circuit expertise. They hired Orville Baker as the manager of circuit development, and he unilaterally decided to go into DTL. Signetics came to the 1962 IRE show in New York with exactly one circuit, a NAND gate which sold for $125.00. Signetics never started in the discrete component business but quickly began a serious challenge to Fairchild in digital products.

When there is no competition, almost anyone can run a company. From the start the Fairchild management had an advantage to be ahead of the crowd in a market at least by a year and a half (Table 5.1 and Table 5.2.) When Signetics began winning the digital customers, and Fairchild digital revenue started to decline, Bob Widlar's analog products maintained Fairchild revenue. Fairchild reported total sales $109.7 in 1965; just over $40 million was generated by higher priced linear parts. The problem of mismanagement of Fairchild was for another year or two hidden by Widlar's success.

The Improved Minuteman ICBM (WS-133B) called for the largest integrated circuits procurement. The prime contractor Autonetics defined the integrated circuit specifications and made it available to the industry in mid 1963.

The required reliability testing involved putting each unit through 1000 hours of electric life test at high temperature, vibration test, etc. An explosive demand for integrated circuits which was triggered by a large quantity of Minuteman contracts caught the industry with demand in excess of capacity. Exact numbers were classified, but very reasonable estimates called for tens of thousand of parts every month. Everybody tried to climb on the bandwagon at once. The cost of manufacturing, cycle time, and overall manufacturing efficiency suddenly became a big issue.

The circuits were used in the inertial guidance, computer, flight control and ground support for Minuteman II. In addition to the current four suppliers (Texas Instruments – 18 circuits, Westinghouse – 18 circuits, RCA one circuit, General Electric one circuit) four other companies were considering to apply for Autonetics qualification. Texas Instruments was the chief supplier at $9,875,649, Westinghouse Electronics Corporation $3,787,900; RCA $1,300,920; and the General Electric Co., $153,042. The four other companies were Pacific Semiconductors, Fairchild Semiconductor, Signetics and Melpar.

A new serious competitor arrived on the scene just after Autonetics' deadline expired. In August 1963, Sylvania announced the world's fastest silicon

Table 5.1. Lag Time before adoption of planar process

Company	**Lag Time [months]**	**Transistor Sales in Time of adoption**
Fairchild		$30 [millions]
Texas Instruments	17	$70
RCA	17	$25
General Instruments	17	$10
General Electric	20	$20
Rheem	22	$7
Motorola	27	$60
Sperry	29	$5
Raytheon	29	$7
Honeywell	29	$5
Sylvania	29	$10
Philco	29	$15
Bendix	35	$5
Transitron	35	$23
Amperex	37	$5
Tung-Sol	37	$5
Sprague	44	$10

Table 5.2. Lag Time before adoption of Planar Diffused Integrated Circuit (not Texas Instruments Functional Electronics Blocks) [*Source: R. E. Freund, Union Carbide Corporation 1968*]

Company	**Lag Time [months]**
Fairchild	
Signetics	7
Westinghouse	11
Motorola	17
Texas Instruments	17
Radiation	18
General Instrument	18
Sylvania	19
Transitron	19
Amelco	24
Honeywell	26
Raytheon	28
Sprague	35

switching transistor (2N2784) with high beta in the microamperes range, with gradual falloff beyond 10 mA, low saturation voltage, typically 0.2 V and switching speed of 12 nsec.

The abilities of management are tested when a company must compete. To manage a cost-effective manufacturing company with timely delivering of parts, requires the utmost attention of management. When there is no competition or the customer is the government, almost any one could be the CEO of the company. Fairchild's Semiconductor management was now facing such a test.

The first indications of managerial problems at Fairchild organization surfaced in the beginning of 1965. However, the phenomenon and success of Widlar's parts delayed bigger visibility of company difficulties.

A major reorganization of Fairchild Semiconductor took place in April 1965. Fairchild Camera and Instruments Corporation re-grouped into two divisions: Fairchild Instrumentation and Fairchild Semiconductor. Fairchild Instrumentation produced oscilloscopes, test equipment and invented dual-slope integrating digital voltmeters. Robert Noyce, vice-president and general manager of the Semiconductor division was named to a new post of group vice-president, with responsibility for both the Semiconductor and Instrumentation divisions. The new Instrumentation division was headquartered in Clifton with John S. Auld as general manager. Charles E. Sporck, former operations manager, succeeded R. N. Noyce as a general manager of the Semiconductor division.

What Fairchild needed was firm and steady leadership. In 1967, Bob Noyce was fighting lawsuits and wrestling with the Fairchild Camera and Instruments stock option committee over stock options. Noyce began to spend most of his time at Fairchild's headquarters on the East Coast.

Charlie Sporck was a well-organized manager with a clear vision how semiconductor manufacturing should be organized. He benefited highly from his tenure with Ford and General Electric where he learned mass production techniques. Although he was a mechanical engineer by training, he surrounded himself with engineers of great technical expertise and very soon Sporck's group was more competent than G. Moore's R&D organization. This situation created tension between the R&D and manufacturing groups. As Moore put it: *"They were less willing to listen to the authorities in the R&D laboratory of how it had to be done and wanted to kill the process and start all over again so it was theirs instead of something that was transferred to them."* The real situation was not that simple. As Don Kobrin, one of Sporck's aides recalls *"when people in manufacturing would ask for help from the R&D organization, no one would show up."*

Under Noyce's leadership, Fairchild Semiconductor was basically controlled by the Marketing Group. The Marketing Group had power over key decisions, such as what products to make, and when and where. Charlie Sporck attempted to overcome these problems between R&D and production by establishing several Product Managers with responsibilities to coordinate production of specific devices. Sporck, a long admirer of Alfred Sloan's management techniques at General Motors gave each manager of Fairchild's six semiconductor plants wide authority to decide which products he would make. Sporck was arguing that a product oriented organization – such as that at Motorola – with product managers; marketing, manufacturing, and their own engineering, is the correct way to run a semiconductor business. G. Moore vehemently rejected such a plan, and when Noyce and John Carter refused to go along with the reorganization, a frustrated Sporck realized that Bob Noyce would be unable to coordinate operations as the company needed. At the beginning of February 1967, Sporck submitted his resignation to Noyce. Robert N. Noyce said that *"the loss of five top men won't hurt Fairchild's operations. Fairchild is not just a few people; we are not scraping the bottom of the barrel for talent."* Talent is certainly very important, but more important is to make money. In less than six months Noyce's talents were driving the company performance into red ink.

Fairchild failed to bring to market new and competitive products because the R&D group produced nothing. Successful competitors forced Fairchild to lower prices and very quickly it became apparent that the company was in trouble. In the fourth quarter of 1966, Fairchild's profits dropped significantly due to *"problems encountered in the introduction of new processing facilities and new semiconductor devices."*[1].

Texas Instruments, the second largest producer of integrated circuits, was gaining on the leader, Fairchild Semiconductor. TI's sales had climbed more than 60% from their 966 level, while Fairchild had increased only slightly. With Widlar gone, the very profitable linear business of the 700-series began to erode as a result of competitive products. Fairchild's fortunes plummeted quickly. In July 1967, the semiconductor operation lost money for the first time since 1958. The situation needed action and as always in similar situations, some of the personal scores may be adjusted. Roswell L. Gilpatric, Fairchild board member, the New York attorney and former Deputy Secretary of Defense, who did not like John Carter, the chairman of Fairchild Camera, suggested that a committee should study the company's performance. Carter was a man who was primarily a financial man. He was hired by Fairchild from Corning Glass. He was a very affable guy but liked to live the high life and didn't pay much attention to the semiconductor business. He came out to Mountain View maybe once or twice a year to see the group, but his main interest was to build the company up and other activities, in printing, graphic arts and instrumentation.

The committee consisted of Gilpatric, Walter Burke (Sherman Fairchild's personal financial advisor) and Joseph B. Wharton, Jr. (who was a financial and tax advisor.) The committee attacked Carter's acquisitions policies and after a bitter conflict broke out, Carter, sensing a power play, resigned.

The Carter resignation caught Sherman Fairchild by surprise. For a while Richard Hodgson, the President of Fairchild Camera, acted as the company CEO. Hodgson was a good manager and he favored the semiconductor division; the problem was that the committee, which was only financially oriented, did not approve his plan for company growth. In spring of 1968, the committee, persuaded Sherman Fairchild that he had made a mistake in mak-

Fig. 5.1. John Carter

ing Hodgson chief executive and S. Fairchild removed Hodgson as CEO of the company. Instead the committee suggested running the company as a team with S. Fairchild and R. Noyce, group vice-president in charge of the semiconductor and instrumentation division, acting as operating officer. Sherman Fairchild needed time to find a new chief executive.

In the meantime, the chaos in Mountain View turned from bad to worse. Thomas Bay, previous manager of Fairchild Instrumentation Division, became general manager replacing Charles Sporck, in February 1967. Under Bay's restructuring, Fairchild Semiconductor formed three semi-autonomous units – discrete components, integrated circuits, and special products. Donald E. Yost was named the head of the discrete devices department. The integrated circuits group was headed by John Sentous, formerly manager of IC manufacturing. The special product department was headed by John Ready. Two former Motorolans with a strong ego – Donald T. Valentine and W. J. Sanders – challenged each other. When sufficiently bruised, Don Valentine cleaned his office on Sunday July 7, and left Fairchild. Monday July 8, former aerospace and defense marketing manager W. J. Sanders became Fairchild Semiconductor's director of marketing.

On August 7, 1967 W. Jerry Sanders, 30, the flamboyant and aggressive director of marketing, decided alone, what new products the division would develop and sell. As a part of his gigantic marketing makeover, Sanders decided to introduce a new product every week (Fig. 5.2.) With a slow manufacturing department, such a promise was certainly a big risk, but a confident Sanders in his thirties was sure that he could manage any risk.

In January 1968 Thomas Bay laid off about 250 employees, about half of them technical personnel. Tom Bay said *"we were doing too many things at Mountain View that we could not afford to do."*

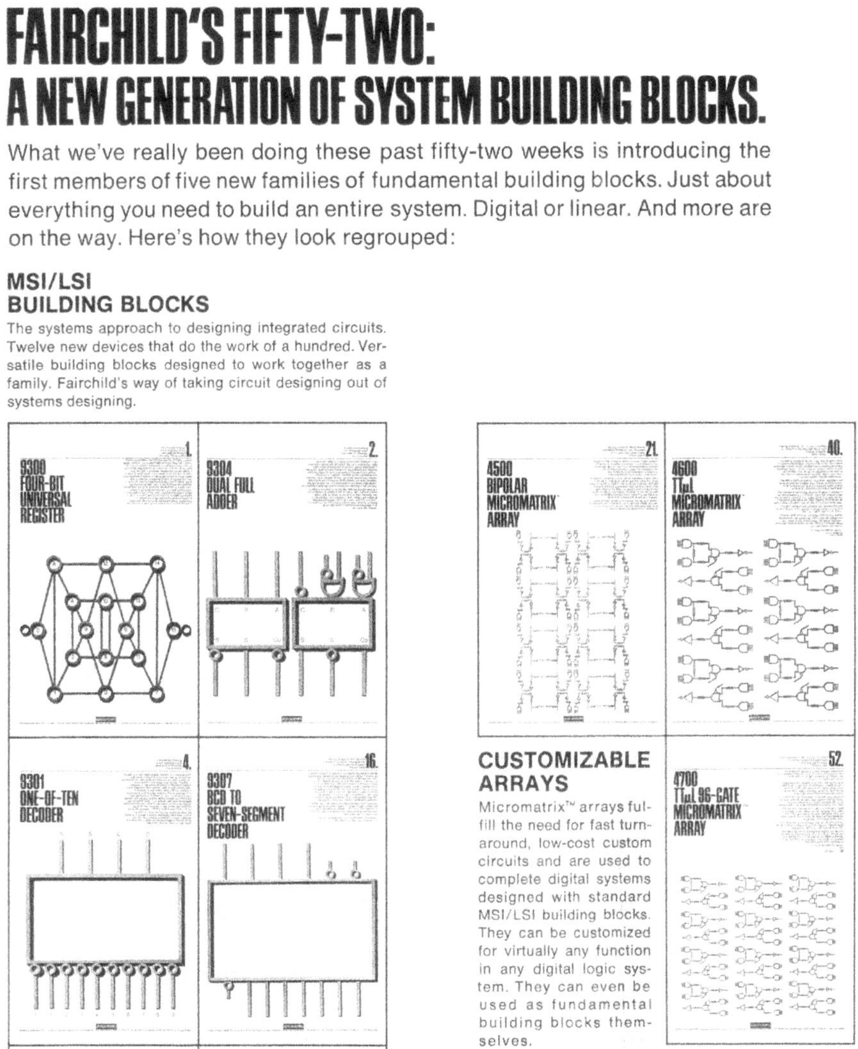

Fig. 5.2. Jerry Sanders "Fairchild Fifty-Two" plan

In May 1968 after a week of hemming and hawing (as become typical for Fairchild Semiconductor) Fairchild announced that Victor Grinich was leaving the company for a teaching "sabbatical" of a year or so. With half of the "Eight Traitors" gone and two working on their new company, the semiconductor division deteriorated even more.

Bob Noyce was never considered to be CEO of Fairchild Camera and Instruments. Sherman Fairchild knew that Noyce did not like to made decisions. Noyce was telling different people, different stories, and people who worked closely with Noyce agreed that Fairchild Semiconductor's problems were a result of Noyce's indecisiveness. I think the best characterization of Bob Noyce come from Harry Sello, one of Noyce's subordinates; Sello said *"Bob was too nice to too many people."* Robert N. Noyce assessed his chances after he realized that he was not going to be considered for Fairchild's presidency and considered a new start up. Benefiting from his experience in Shockley Semiconductor Laboratory, Noyce talked to Moore and in March 1968 they called Arthur Rock again. They said they wanted to form a company, and they needed to raise $2.5 million. Noyce said that they wanted to manufacture semiconductor memories. *"I thought that was possible and a good idea and agreed on the terms. There was no negotiation; everything was done very quickly. I made a few phone calls and raised the $2.5 million"*, Rock said.

Sherman Fairchild changed strategy, and considered spinning off the semiconductor division as a separate company. As hinted by Sporck's original suggestion, Sherman Fairchild turned the search to C. Lester Hogan. Hogan, 47, was the division executive vice-president and general manager of Motorola Semiconductor in Phoenix.

Hogan joined the start-up Motorola Semiconductor as a general manager in 1958. Hogan built up the operation from sales of less than $5 million in 1958 to over $230 million in 1967. Hogan was good in his engineering judgment and was always optimistic; people enjoyed working with or for him. Hogan, contrary to Noyce, could make decisions. Most of his decisions were good. And Hogan's bad decisions were better than no decisions. Under his leadership Motorola in 1968 passed Texas Instruments as the biggest U.S. producer. Worldwide; Texas Instruments' sales were still bigger than Motorola's.

Fig. 5.3. Dr. C. Lester Hogan in 1968

On June 25, 1968 Noyce, who knew Hogan, called him and made an appointment for Walter Burke to visit Hogan in Phoenix. At that time Motorola was the only major semiconductor manufacturer who did not license Fairchild's Planar Process. Hogan feared that Fairchild might sue for patent infringement and agreed to meet Burke. In a several-hour talk, Burke mentioned a job offer to Hogan. Hogan, though receptive, turned it down.

Noyce's lawyer gave an ultimatum to Noyce to resign – if he still wants to go ahead with his plan to start a new semiconductor company. A conflict-of-interest situation could prevent the forming of a new company, which Noyce and Moore called N-M Electronics.[1] On June 28, 1968 Noyce submitted his formal resignation to Sherman Fairchild.

A week later, Sherman Fairchild himself flew to Phoenix and visited Hogan in his home. The job offer extended to Hogan was for President of Fairchild Camera and Instrument. Hogan said he was dedicated to his career at Motorola and just as a side issue mentioned that *"the stock options were pie in sky, if one didn't have the money for them."*

A few days later, when Hogan called to decline the job, Sherman Fairchild offered a personal loan to pay for the stock option. The Hogan deal was a) a salary of $120,000 a year, $30,000 more than he got at Motorola, b) an interest-free loan for $5.4 million to enable him to exercise an option for 90,000 shares of Fairchild stock at $60 a share, c) an additional 10,000 shares of restricted stock at $10 a share.

On August 3, Hogan flew to New York. He met Sherman Fairchild and Walter Burke in a private room at the Sky Club atop the Pan Am building. Hogan accepted the job. Sherman Fairchild was surprised, as he had been convinced that Hogan would not leave Motorola. James F. Riley, president of Signetics, was considered as another candidate for the vacant presidency of Fairchild Semiconductor.

Lester Hogan turned in his resignation to Daniel E. Noble, who had hired Hogan 10 years ago. Noble was disappointed and urged Hogan to reconsider and to talk to Galvin. Galvin arrogantly started the conversation by discounting Hogan's contribution to Motorola's success. When Galvin ordered Hogan not to pick up his personal belongings he had left in his office, the meeting was over. Motorola immediately filed a suit charging the Fairchild with unfair competition, antitrust violation, and unjust enrichment. Galvin quickly forgot how he lured and hired 18 engineers from the semiconductor division of General Electric.

The 40-page suit was filed by attorney Mark Wilmer of the Phoenix law firm Mueller, Aichele & Rauner of Chicago, and Townsend & Townsend, San Francisco. The Defendants along with The Fairchild Camera and Instruments are Lester Hogan, Leo E. Dwork, Wilfred J. Corrigan, Eugene A. Blanchette, Thomas D. Hinkelman, George M. Scalise, Andrew A. Procassini, William L.

---

[1] Because people joked about the name of the company as "M*ore Noisy Electronics*", they changed name to Intel.

Lehner, and the defendants' wives. The wives were named because the financial benefits allegedly "wrongfully gained" fall under Arizona community property law. Fairchild, on hearing of the action, stated that "we feel we are not guilty of any wrongdoing and we intend to defend ourselves." After spending a lot of money on legal fees, Motorola won nothing.

Les Hogan was authorized by Sherman Fairchild to offer Fairchild's stock options and invited seven Motorola top managers to go with him to California. The team was called "Hogan's Heroes" because at the time, a very popular television series was called Hogan's Heroes. Of course, because his name was Hogan, this became Hogan's Heroes at Fairchild.

When Hogan arrived at Mountain View on August 10, 1968 he found the company in chaos. A short confrontation with Thomas Bay resulted in Bay's resignation. Then Hogan insisted on moving the corporate headquarters from Syosset to Mountain View. The next monumental job was to align the former Moore R&D organization with manufacturing divisions. Hogan defined the areas in which he thought Fairchild would lead the industry:

- Gallium Arsenide for microwave and light emitting devices
- MOS Technology
- LSI Components (complex parts with 800–1000 devices)
- Computer Aided Design to speed up the mask-making operation
- Schottky TTL Circuits
- High Frequency Transistors

Hogan's assumption was that with this technology and with the re-organized manufacturing division, he would be able to repeat his success story once again. The problem was that he did not have time to see the production lines yet. Fairchild Semiconductor's manufacturing was under-invested. Hogan's heroes were puzzled that there was no capital investment, that all of the assembly and test equipment, all of the diffusion equipment, was ancient. Fairchild had not fully addressed the epitaxy approach. *"It was very backward to haven't done any of the things that we'd been doing at Motorola,"* and so Hogan's initial conclusion was to put a massive infusion of capital into the company to revive it.

Hoggan assigned the responsibilities as following: Wilf Corrigan – Discrete Devices; Gene Blanchette – Integrated Circuits; Andy Procassini Q.A & Reliability; George Scalise – Manufacturing; Bill Lehner – Equipment Engineering and Facilities; Tom Hinkelman – Planning.

After Hogan's people became familiar with Fairchild's situation, Hogan called The Board of Directors to a meeting (Fig. 5.4.).

Hogan started the meeting and said:

> *"The purpose of our meeting is to discuss the status and future of Semiconductor Division, introduce the management team, and decide how the company can turn for the better. First off, the semiconductor*

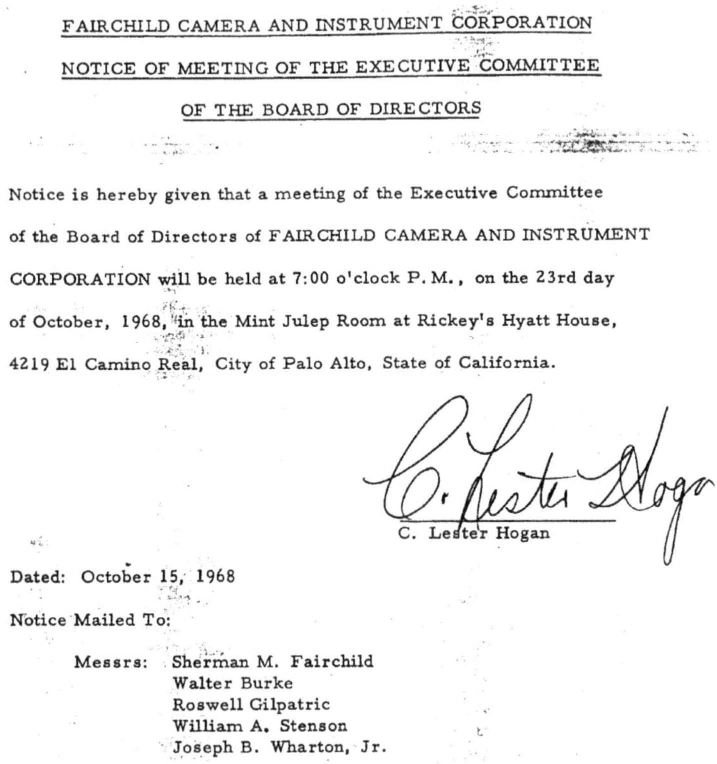

FAIRCHILD CAMERA AND INSTRUMENT CORPORATION

NOTICE OF MEETING OF THE EXECUTIVE COMMITTEE

OF THE BOARD OF DIRECTORS

Notice is hereby given that a meeting of the Executive Committee of the Board of Directors of FAIRCHILD CAMERA AND INSTRUMENT CORPORATION will be held at 7:00 o'clock P. M., on the 23rd day of October, 1968, in the Mint Julep Room at Rickey's Hyatt House, 4219 El Camino Real, City of Palo Alto, State of California.

C. Lester Hogan

Dated: October 15, 1968

Notice Mailed To:

Messrs: Sherman M. Fairchild
Walter Burke
Roswell Gilpatric
William A. Stenson
Joseph B. Wharton, Jr.

Fig. 5.4. Notice of the Fairchild's The Board of Directors meeting from October 15, 1968

division was very sick. The gradual decay began about 1965 and it is apparent now in looking at some of the detailed financial statistics.

It is very difficult in business to perform an autopsy and describe all the illnesses that killed the patient when you were never really familiar with the patient in bright years. Luckily we found the patient before he died and I can assure you that the patient won't die.

I think, however, that in order to give our presentation some meaning it is worthwhile for us to list the symptoms and guess at the ailments.

First, the cost of producing parts is extremely high; so high, in fact, that there is no possibility of standing toe-to-toe with our biggest competitors, Motorola and Texas Instruments, and slugging it out with them for business today. The cost of producing parts is high because yield in diffusing devices is extremely low, the assembly costs are too high because nothing was invested in even simple automation.

> *There were and still are other ills in the Division. The greatest problems leading to the deterioration of the Division in the past years was lack of a management willing to make a decision. Young, bright, but totally inexperienced engineers were running almost all production areas. They either lacked the courage necessary to make the changes that would have improved the situation, or they looked upward for approval of their proposed change and received no answer back.*
>
> *In addition, the inventory is so much too low to maintain the sales level but, even so, has not been kept clean. Somewhere between $1 to $2 million of the present inventory is completely un-salable.*
>
> *It was common to cheat to ship products that had never been tested as required or, if tested and failed, the product was shipped anyway and we just lied about the results. Almost all of our qualifications on MIL devices have lapsed, and we are thus excluded from shipping to the high-priced and profitable military market unless we continue to lie.*
>
> *There are many reasons for not doing so. First it is fraud and, if exposed, could be almost a death blow to our business."*

The picture of the company was even grimmer when W. Corrigan and Gene Blanchette finished their presentation. Perhaps the one strength the company had, was the image of being leaders in Semiconductor Technology and because of this, the company could attract and retain many customers. Fairchild was still the glamour leader able to attract bright, young engineers due to the Fairchild Semiconductor name.

G. Blanchette listed two pages of weaknesses in order of priority. On the top of the list is: *"For the last three years, R&D has done no product development as such. This lack of product development in 1966 and 1967 has led to a very serious lag in our ability to service the market in Current Mode Logic and TTL."*

The list ends with the paragraph: *"Our production capacity is inadequate, our technology is about two years behind the leaders, our yields are about the lowest in the industry, and even on those products which we can produce, we are not cost competitive due to our lack of mechanization and the absence of a plastic package."*

The corrective actions (Fig.5.7) were widely discussed and approved. As always, it is important to define problems and set up the roadmap leading to the defined goal; however, more important is to implement and execute corrective action.

Hogan's revitalization program caused a personal turnover that ran as high as 40% in the marketing organization. Fairchild Marketing Director Jerry Sanders, following the recent departure of Donald Rogers, former Fairchild international manager, to Intersil, Inc., reduced in July 1968 the number of people reporting directly to him. The aero-space-defense marketing group was headed by Chaz Haba; Marshall Cox, former Computer Market-

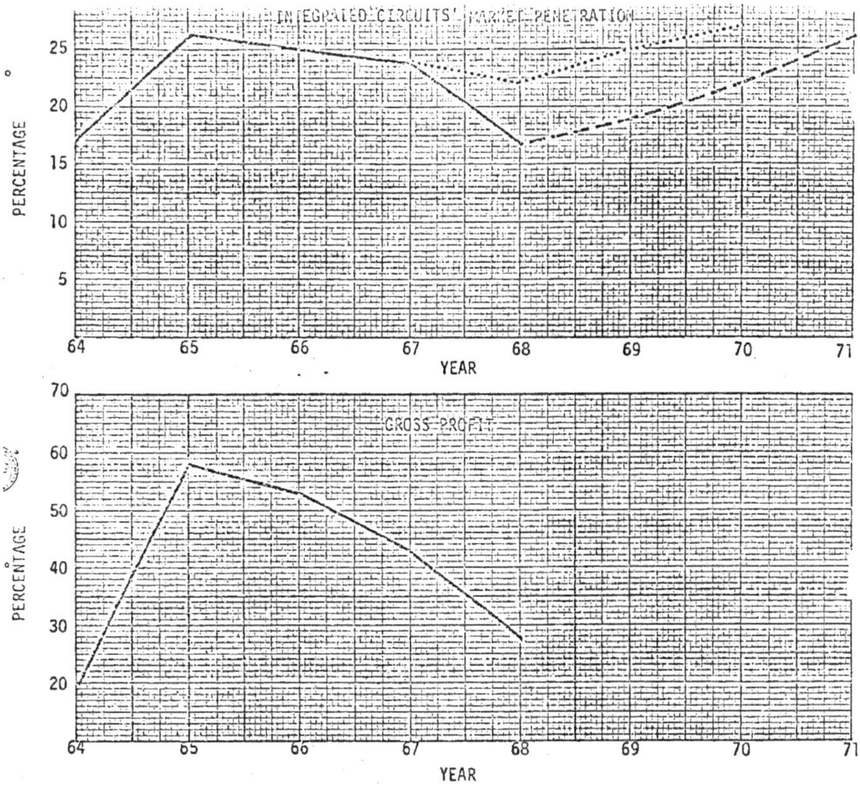

Fig. 5.5. Fairchild Semiconductor IC Market penetration and Gross Profit

ing manager became manager of the commercial marketing group; Bernard Marran succeeded Don Rogers as manager of the international Marketing; and Ben Anixter became manager for all semiconductor products.

Hogan had a very different revelation on how to deal with customers. Hogan realized that Sanders "manipulated" several customers with questionable, if not unethical, practices. Hogan and his team called the Fairchild sales force the "*Suede Shoe Boys*" because they were very flamboyant. A few of them wanted to impress customers with several gold chains and open-necked shirts. Jerry Sanders was wearing white suits and was at that time very "Hollywood like". Except for George Scalise and Wilf Corrigan, Sanders' boys were almost half a generation younger then the other conservative ex-Motorola managers. Les Hogan had decided to bring in F. Joseph Van Poppelen, Jr., from Signetics, for the job of marketing director, and this, of course, ended the party style of "The Suede Shoe Boys."

The marketing organization under Sanders' leadership responded only to orders worth \$5-million to \$30-million. It ignored any customer with an order

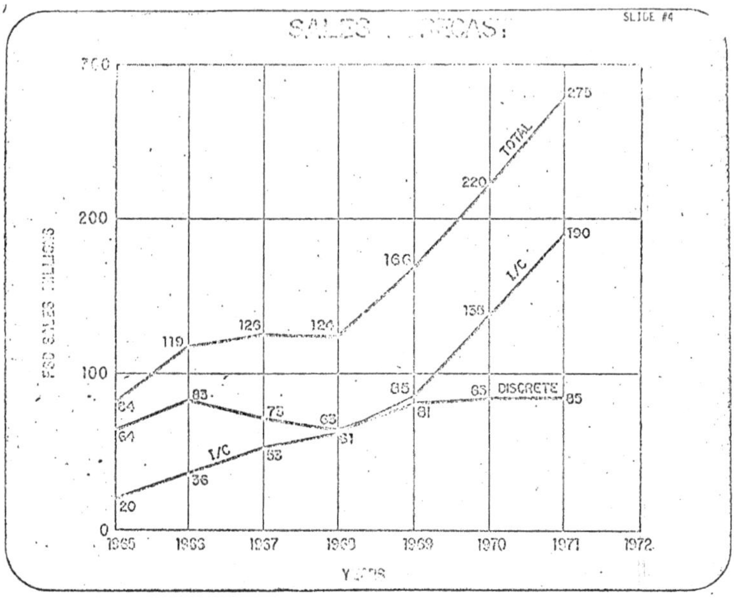

Fig. 5.6. Actual Fairchild Semiconductor Sales 1965–1968 and Projected sales 1969–1972

CORRECTIVE ACTION

1. Build up highly competent I/C force in Portland, Maine. Plan to move nearly all high volume products to Portland. Portland would be capable of development work as well.

2. Set up R & D to develop product. Move Senior Product Design Team from Mountain View to Palo Alto.

3. Begin to mechanize manufacturing lines. In the interim, use foreign plants for low cost labor.

4. Organize such that Manufacturing Manager identify with and be available to customers with problems of delivery and reliability.

5. Organize manufacturing such that individual Product Managers are measured by:

   A) Profit and Loss

   B) New Product Introductions

   C) Customer Satisfaction

6. Begin intensive program to attain Management depth in three (3) years. Will train present employees and hire outsiders.

7. Assist in setting up strong Q. A. operation to monitor product quality.

Fig. 5.7. List of the corrective actions approved by the Fairchild's Board of Directors meeting on October 15, 1968 that was less than $1-million. Fairchild got all its sales from less than 300 customers, while for example Texas Instruments had nearly 1000.

Hogan appointed Jerry Sanders to the new position of assistant vice president of marketing development of Fairchild corporate staff. Jerry Sanders should report to Joseph Van Poppelen, Fairchild's corporate vice president of marketing. Douglas J. O'Connor resigned as marketing manager for Texas Instruments' semiconductor components group and became group director of marketing for Fairchild Semiconductor Division. The changes worked out perfectly as planned. Jerry Sanders resigned from Fairchild Semiconductors and temporarily moved south to Malibu beach.

Douglas J. O'Connor replaced Jerry Sanders as marketing director of Fairchild in January 1969. O'Connor had split marketing activity into only two units: European and non-European. The European unit was headed by Chaz Haba. Responsibility for all non-European was assigned to Dedy Saban, one of the second batch of "Hogan's Heroes" from Motorola. Bernard T. Marren took charge of all U.S. sales. Reporting to Bernard were Gordon Russel, Ben Anixter, John Richardson, and Jack Gifford, now head of computer marketing.

The alumni had grown so numerous and so bitter that their bad-mouthing hurt Fairchild in the marketplace. Shortly before the Christmas of 1970, some former employees of Fairchild Camera & Instrument Corporation held a "Fairchild Alumni" party at a Los Altos restaurant. The invitation suggested *"If you wish, bring a small gift for L. C. Hogan's Care Package."*

Because of the ultimate failure of Fairchild Semiconductor, Hogan's efforts to fix problems are not always seen in the right perspective. Hogan joined a company that was a technology leader in several fields but whose production system was chaotic. Fairchilds manufacturing facilities were not adequate; they were dirty and antiquated, with yields 20% at the best. The new line of products announced in Sanders' "Fifty-Two" plan could not be produced with a yield greater than 1%. 90% of shipments were late and had specification defects. Hogan rushed to correct the situation with a massive infusion of new equipment. In December 1969, the company ramped production to 3000 two-inch-diameter wafers per week. In August 1968 the hand-operated device testers were operated by 250 women; in December 1970 only 20 women handled the bigger volume of testing with a computerized systems. The newly mechanized Fairchild discrete transistor line produced 3.5-million devices a week with production costs about 30% cheaper than Fairchild's transistor line in Hong Kong.

Fairchild Semiconductor Division improved significantly; however, Fairchild System Division become the number one loser. Hogan had recruited Eugene R. White from Honeywell Information System as general manager of the System Division. They planned to produce the calculators with custom chips from the Semiconductor Division. Hogan split the semiconductor operation among Wilfred J. Corrigan as vice-president and general manager for domestic operation. George M. Scalise headed Far East operation, Douglas J. O'Connor run the European sector, and Leo Dwork was responsible for MOS and memory products.

Hogan expected that by investing in the production operation and by increasing yield, the company would have the ability to be more aggressive than competitors with respect of prices. The things did not work out like planned. The focus on production problems resulted in slow development of new technology. During 1969 and 1970 the company did not introduced any significant new product. Hogan himself was going through health problems and the industry was hit by recession in the beginning of the seventies.

Hogan made a deal with his buddy – Wilf Corrigan. In 1974, Hogan became the vice-chairman of the board and Corrigan was named president and chief executive officer of Fairchild Camera and Instrument Corporation. Once Corrigan got in, the operation did not work as it was planned, and Corrigan tried to get rid of Hogan.

Corrigan was at that time an extremely unpopular person at Fairchild; members of the R&D group called him "Liverpool Butcher" because of the insensitive decisions he made. Corrigan cut off several important R&D projects and pushed the VP of R&D, Jim Early, aside. Corrigan decided *"We've got to cut the number of products we make in half,* " and he gave to Bob Ulrickson the job of arbitrating which products would go and which ones would stay. Ulrickson was dealing with division vice presidents who did not want to lose their products and he got to tell them what to do. Corrigan shut down the ASIC Micromosaic effort and everybody had to find a position or be laid off. In the short term Corrigan balanced the financial sheet; however, his destructive policy of only cuts without establishing a new productive roadmap with delivery, brought Fairchild to the end. The company plunge was gaining speed.

Today, a majority of people involved in the Fairchild saga agree that one of the major failures of Fairchild was misunderstanding MOS technology, and a failed transition from bipolar to MOS. Gordon Moore did not recognize importance of MOS in the early sixties. Jerry Sanders in August 1967, when Philco-Ford cut prices of its MOS product, ridiculed Philco's product marketing manager Don Richard. Sanders said *"Prices are still outrageous. Who will pay $13 for a J-K flip-flop when I can sell one for $3?"* The problem was that Fairchild did not have any visionary who would ask, for example, a question, "Can I build a 100-bit register in bipolar?" Fairchild had a Roy Pollock in charge of MOS development.

My father and life lectured me about judgment: the older I am, the slower in judgment I become. In the case of Pollock, I have only one characterization no matter how old I am: Pollock was very simply The Idiot who not only knew nothing about engineering; he also had as many possible flaws as any human could have. His achievement of destroying MOS at Fairchild was not enough; Pollock later went on to destroy the same stuff at RCA [2]. How Corrigan could fire so many excellent engineers and keep Pollock in an important position is today difficult to understand.

Fairchild declined to the point where it was almost impossible to survive against the competition of newcomers, mainly Mostek and Intel. Fairchild was losing money even as the rest of the semiconductor industry was booming. Fairchild Camera and Instruments was sold in 1979 to a French oil-field company Schlumberger Limited. Schlumberger, a French global oil-field service company incorporated in the Dutch Antilles, purchased Fairchild for $425 million as part a diversification effort.

Schlumberger fired Corrigan instantly and installed Tom Roberts as a head of the new company. Roberts was a West Point graduate, starting out in sales at Schlumberger and then at IBM. He began to move up in the financial area and he finally became CFO. Roberts did not know the difference between AC and DC current, and had never heard about semiconductors. It did not seem to bother him. His unlimited arrogance lead him to the conclusion that he could manage any industry. Tom Roberts was wrong and he could not manage any business.

In 1985 Donald W. Brooks, a former Texas Instruments engineer who started his career with Jack Kilby became president of Fairchild. Don Brooks finally in 1986 decided that the best solution was to try to sell Fairchild to Fujitsu.

Fujitsu Limited was founded in Japan in 1923 when German Siemens & Halske and Furukawa Electric created Fuji Electric to produce electrical equipment. Fuji created a communications division, Fujitsu, in 1935. Originally a manufacturer of telephone equipment, it made antiaircraft weapons during World War II. Shortly after the war, Fujitsu returned to telecommunication, becoming one of four major suppliers to Japan's state owned telecommunications monopoly, Nippon Telephone and Telegraph. Fujitsu grew very quickly during Japan's rapid economic development in the end of fifties and in the sixties. The company developed the Japan's first commercial computer in 1954, the FACOM 100.

During the 1980's Japanese companies flooded market with cheap chips, and prices were dropping precipitously. In DRAM, Japanese companies increased their market share from five percent to a ninety percent. The U.S. semiconductor industry lost more than forty thousand jobs. The competitive hostilities were building up, Charles Sporck said: "*We are at war with Japan – not with guns and ammunition, but in the economic war with technology, productivity and quality.*"

AMD's Jerry Sanders wanted Washington to declare that the U.S. semiconductor industry was *"scared ... and we will not permit it to be destroyed."* Wilfred Corrigan said: *"The Japanese business strategy is to obliterate competition. The bid for Fairchild was just a probe being observed by other major Japanese corporations. If Fujitsu succeeded, by the end of 1987 all the ma-*

jor Japanese companies would be coming in to acquire U.S. semiconductor manufacturers."

To Don Brooks, Fujitsu was not the enemy. Schlumberger was losing money due to the global oil recession and needed curtail future financial loses. Fairchild had lost $175 million in its 1985 fiscal year and was expected to lose another $75 million in fiscal year 1986. Schlumberger had tried to sell Fairchild to a U.S. firm for several months, but there was no interested buyer. Don Brooks was concerned that Schlumberger might completely shut down Fairchild.

After the Semiconductor Trade Agreement was signed, Fujitsu realized that their 3% share of the $8.7 billion U.S. semiconductor market was in peril. The least expensive and quickest way to fix this problem was to acquire an existing U.S. Firm with manufacturing facilities located in the U.S.

After several weeks of negotiations, Fujitsu offered in October 1986 to buy eighty percent of Fairchild and agreed to allow Schlumberger to keep twenty percent. The deal entitled Fujitsu to expand Fairchild's semiconductor operations by integrating them with Fujitsu's U.S. semiconductor business and certain other Fujitsu European semiconductor operations. Brooks worked on the merger where the products would be marketed in the western world under the name Fairchild, in the Asian world under the name Fujitsu. Fairchild would manufacture at lower cost and would design and market geographically. Fairchild would have an organization that would embody what was Fairchild in California, and in Europe. And Fujitsu would do the same thing in Asia. Fujitsu planned to begin integrated production of VLSI circuit in the U.S. by making the most of Fairchild's plants in California, Washington, Maine, and South Carolina. The takeover was expected to increase sales of the Fujitsu Semiconductor Division to $2.7 billion, a level equivalent to those of Toshiba and Hitachi.

Don Brooks left for Washington confident that, with the Fujitsu deal, he had saved ailing Fairchild company – once the flagship of the American semiconductor industry. When the merger negotiation was completed and ready to be executed, several Fairchild Alumni headed by Bob Noyce flew to Washington, D.C. to argue against the merger. They claimed it was a military business like Cray computers, and the deal would give Japanese control over these primary areas. Brooks discovered that he was subject of political controversy.

The Department of Commerce, Department of Defense, and National Security Agency raised objection to the Fairchild – Fujitsu deal. Secretary of the Commerce Malcolm Baldrige offered the following three objections:

1. Japan is a market closed to U.S. supercomputer manufacturers. Therefore, the U.S. should not allow Japan's largest producer of supercomputers to acquire a distribution mechanism in the U.S. for its supercomputers. Baldrige said *"It is a way for Fujitsu to launder its nationality, a way to look American."*

2. Allowing one U.S. semiconductor firm to be acquired by a Japanese company would create domino effects. This would deplete the U.S. ability to manufacture semiconductors domestically, increase the erosion of the U.S. technological base, and render the U.S. dependence on the Japanese for critical high-technology products.
3. The combination of a Japanese takeover of the semiconductor and supercomputer markets would further widen the U.S. – Japan trade deficit.

As is typical in politics, the Secretary of State, George Schultz, and Secretary of the Treasury, James Baker, both maintained the position that an open investment practice in the United States was sound policy. The Treasury Department also feared that the Japanese would retaliate against the U.S. by not attending Treasury bill auctions, and the State Department feared that intervention would further sour the already tense diplomatic relations with Japan. However, the Deputy Under Secretary of Defense, Stephen Bryen opinion was that: *"If one of the semiconductor companies of our semiconductor industry falls into the hands of the Japanese, we could end up with no semiconductor industry. We can lose technology race by default."*

The Fairchild-Fujitsu debate gained speed when senator James Exon wrote letters to several administration officials warning that the U.S. technological lead over the Communist bloc was being eroded and that the semiconductor industry was being crucial to nation defense. Exon called to *"take action to preserve at least a minimum and prudent level of America's vital defense industry within our borders and jurisdiction."*

Had it not been for political intervention, Fujitsu would have completed the Fairchild acquisition by the January 1987 target date. But it became clear to Fujitsu that the deal could not be saved. On March 17, 1987 Fujitsu's President Takauma Yamamoto announced in Osaka that *"The U.S. Government objection led me to decide not to go through forcibly with the acquisition."* The deal to acquire Fairchild Semiconductor was off. The Justice Department responded to Brooks with 18-wheelers worth of documents that had been sent there and they still had to make a decision.

The irony of the end is that in September 1987, Fujitsu announced that it planned to invest $70 million to build a semiconductor plant in Oregon. The company planned to manufacture application specific integrated circuits (ASIC), CMOS and ECL devices, and market them in the United States [3].

The premier semiconductor company in the world was sold to Sporck's National Semiconductors in 1987. National Semiconductors only paid a hundred million cash. Schlumberger had a leveraged buyout offer in at a hundred and eighty to two hundred million dollars. But Schlumberger never accepted the leveraged buyout because the board never wanted to be embarrassed; they wanted to get it off the books. After all, to the field oil company, fifty or hundred million does not make any difference. When National Semiconductor faced financial constraints in 1988, Fairchild's defense-related semiconductor operation was shut down and Fairchild's former Washington facility was sold to Matsushita. This time, neither the Pentagon, nor the Commerce Department intervened.

All what happened to Fairchild after Noyce's flight to Washington was irrelevant; there was no way to bring Fairchild glory back. Many suggested that Wilf Corrigan would bring ill Fairchild down anyway [4]. In my opinion Corrigan deserves a lot of credit for the destruction of Fairchild; and then the Shlumberger people, helped by politicians, completed it.

References

[1] Fairchild Camera and Instrument Annual Report 1966
[2] R. Ulrickson, J. Nichols, Interview with R. Walker, April, 19, 1995
[3] Japan Economic Journal, September 12, 1987, p.14
[4] L. C. Hogan, Interview with R. Walker, January 24, 1995

6

Integrated Circuits outside Fairchild Semiconductor

"We do not exist to make a profit. We exist to create useful products and services and it is ***Opportunity*** *to operate at a profit that is the* ***Incentive*** *to create, make, and market these useful products and services for our society."*

Patrick E. Haggerty, February 1965

The incredible financial success of Fairchild Semiconductor could not go unnoticed. From 1959 to 1962 the net profit of Fairchild Semiconductor rose by two million dollars per year. In the early sixties there was no barrier sufficient to deter entry into this industry where the expectation of possible rewards out-weighted the risks.

The large boom of the semiconductor industry in the beginning of the sixties was a consequence of the increase of R&D funding by the Government. Between 1945 and 1960 the R&D spending in the semiconductor industry increased by about 7% per year. Government support of R&D activities increased nine times. The advantages of research and the innovative changes led to enthusiasm in society at large. The idea of innovation and newness became a part of our national character. After WWII the value of R&D was highly regarded and political leaders were served by visionary and responsible intellectuals. Vannevar Bush, aide to president Truman, wrote in his book published in 1946 [1]:

> *"Advances in science when put to practical use mean more jobs, higher wages, shorter hours, more abundant crops, more leisure for recreation, for study, for learning how to live with the deadening drudgery which has been the burden of the common man for ages past. Advances in science will also bring higher standards of living, will lead to the prevention or cure of diseases, will promote conservation of our limited national resources, and will assure means of defense against*

aggression. But to achieve these objectives and to secure a high level of employment, to maintain a position of world leadership, the flow of new scientific knowledge must be both continuous and substantial."

New firms have three fundamentally different paths to entry into an industry:

1) Introduce new and technically advanced products from scratch and finance costly R&D expenditures
2) Become a second-source supplier to an already established supplier and pay a license fee and royalty
3) Develop new technology quietly and secretly under "sponsorship" of an established company, and then "steal" technology and start up a new company

The introduction and development of new technologies requires a large R&D effort, which is not only costly but also time-consuming. A second source supplier enters a market with delay. This delay does not allow taking an advantage of high prices which may be charged in the case of a sole supplier. For example, Fairchild sold its mesa silicon transistor 2N697 in March 1958 for $150. In August 1958 the price dropped to $75. In January 1959 the price further dropped to $45. In July 1959 to $28.50; and in February 1960 to $22.70.

Theoretically, by definition, a patent should provide protection to technically innovative companies, but it did not. Semiconductors companies apply for thousands of patents each year; however, from the historical point of view, only three patents are noteworthy: (1) the original Bardeen, Brattain, and Shockley transistor patents owned by Bell Telephone Laboratories, (2) Jean Hoerni's planar technology patent owned by Fairchild, and (3) the integrated circuit patent.

Bell Laboratories' patents have never been challenged. Bell Telephone sells patents rights to all manufacturers for a flat fee of $25,000 and approximately 1.5% royalty on all non-military sales. Hoerni's patent of the planar process, filed in 1959, was a major technical breakthrough in the industry. By year 1964, 75% of the industry had adopted this approach. Nevertheless, by January 1965 fewer than ten manufacturers had signed license agreements with Fairchild.

The integrated circuit patent was first awarded to Texas Instruments after bitter litigation with Fairchild. Upon appeal, the patent was granted to Fairchild and the Nobel Prize was awarded to Jack Kilby of Texas Instruments.

At that time the industry had minimal patent rights enforcement. Expenditure to build a new manufacturing facility had been "reasonable" and the barriers for new industry entrants were relatively low.

Table 6.1. The transient nature of leadership of the Semiconductor Industry 1955–1996

VACUUM TUBES	TRANSISTORS	PLANAR TRANSISTORS	INTEGRATED CIRCUITS	INTEGRATED CIRCUITS (VLSI)	SUBMICRON INTEGRATED CIRCUITS
1945	**1955**	**1965**	**1975**	**1985**	**1995**
RCA	Hughes	Texas Instruments	Texas Instruments	Motorola	Intel
Sylvania	Transitron	Fairchild	Fairchild	Texas Instruments	NEC
GE	Philco	Motorola	National	NEC	Toshiba
Raytheon	Sylvania	General Instrument	Intel	Hitachi	Hitachi
Westinghouse	Texas Instruments	GE	Motorola	National	Motorola
Amperex	GE	RCA	Rockwell	Toshiba	Samsung
National Video	RCA	Sprague	General Instrument	Intel	Texas Instruments
Rawland	Westinghouse	Philco	RCA	Philips	Fujitsu
Eimac	Motorola	Transitron	Philips	Fujitsu	Mitsubishi
Lansdale	Clevite	Raytheon	AMD	Fairchild	Philips

In the late fifties, Hughes Aircraft Co. and Transitron Corporation dominated the market with germanium transistors. Hughes had entered the semiconductor business early, and by late 1959 had a dominant position in Ge and Si diodes

Texas Instruments consistently held a leadership position in the industry and always maintained a complete line of advanced products. Motorola, Inc., before 1960 concentrated on a narrow product line for sale mostly to other Motorola divisions. A aggressive investment in 1960, when production facilities grew from 68,000 sq. ft. to 1,118,000 sq. ft. in 1964, and employment increase in the same period of time from less then 500 to 3200 brought Motorola up between the industry leaders. By 1965 Motorola, Fairchild, and Texas Instruments were the first three sales leader of the industry and maintained this position through 1970.

General Electric had a spotty record of semiconductor product development. The company who claims today "*We bring good things to life*", was not considering transistors as "good things" for a very long time. Fig. 6.1 is perhaps the best illustration of GE's record and their marketing philosophy in the end of the fifties. General Electric's major contribution to semiconductor technology was the automated manufacturing of low-cost epoxy-encapsulated transistors. RCA, while reporting many new products, the first MOS transistor and CMOS circuits among them, has never had a major marketing and production operation to capitalize on its R&D achievements.

IBM, the prominent semiconductor company today, was very soft on semiconductor research and manufacturing in sixties. The management of IBM at that time did not consider a planar integrated circuit as a reliable device, and IBM concentrated their effort on the development of thin-film based products. IBM was awarded in 1961 a large contract from the Bureau of Ships to establish automatic fabrication of advanced thin-film circuits.

Beside Shockley Semiconductor Laboratories and Fairchild Semiconductors, companies such Amelco, Signetics, Texas Instruments, Sprague Electric Company, Sylvania Electric Products, Advanced Micro Devices, and Intersil established new industry policy or improved in a significant way the technology of integrated circuits.

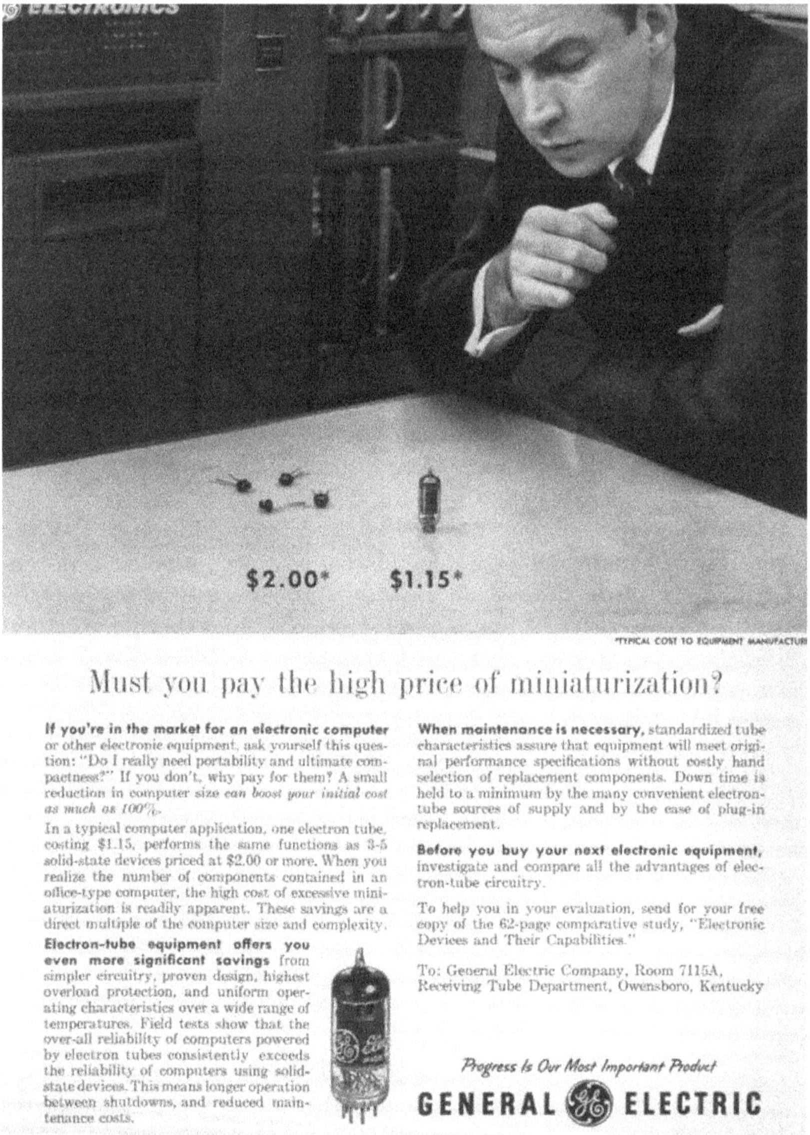

Fig. 6.1. General Electric Advertisement from the early 1950's with GE's mission statement "*Progress is Our Most Important Product.*"

Amelco Semiconductor, Division of Teledyne, Inc.

By the end of 1960, Fairchild Semiconductor had become a major manufacturer of silicon planar transistors and diodes with device performance surpassing by far any other competitor. Fairchild's main effort focused on the building of new production facilities and expansion of existing production lines. With enormous market demand for planar devices, the new integrated circuit developed by Last's group played little importance in Fairchild's plans.

When Jay Last realized that many managers at Fairchild Semiconductor were pre-occupied with the temporary success of planar devices and had none or very little interest in the new integrated circuits technology, he considered all his alternatives. Jean Hoerni was in a similar "blasé" mood because Noyce and Moore never acknowledged his contribution to the planar process.

Last talked to Arthur Rock at Hayden Stone and, as often in similar situations, a solution emerged by chance. Henry Singleton and George Kozmetsky, two former executives of Litton Industries, with Rock's help, formed Teledyne in the summer of 1960. They wanted to form a company specializing in military electronics with in-house manufacturing of integrated circuits. Rock contacted Jay and Jean, and urged them to meet Singleton.

There was a significant risk in leaving Fairchild Semiconductor at the end of 1960 to try to design and produce integrated circuits elsewhere. Last, who earned his Bachelor's degree in Optics at the University of Rochester in 1951 and his Ph.D. from MIT in 1956, was head of the Micrologic section. Hoerni, with two Ph.D. degrees, was head of the Physics section. They could have had a comfortable career as just a parasite on the large company. This, however, is not a characteristic of engineers like Last or Hoerni.

Arthur Rock set up a meeting for Jay and Jean with Singleton. Singleton made a very attractive offer to set up the new operation in Mountain View, and to promote both Jay and Jean into vice-presidents of Teledyne.

On January 31, 1961, a chimpanzee, Ham, was rocketed into space in the U.S. Mercury capsule. The same day Jay Last and Jean Hoerni left Fairchild. Ham proved that men and animals would be able to survive in weightlessness and paved the way for future travels in space. Jay and Jean set up Amelco Semiconductor at Terra Bella Avenue in Mountain View, the company that would design and manufacture very specialized circuits for NASA and military space operations.

Fig. 6.2. Jay Last, co-founder of Amelco Semiconductor and former manager of Fairchild's group which designed and realized the world's first planar integrated circuit (1960)

The Amelco facility was just a few blocks from Fairchild. Amelco's founders were aware that they must avoid any conflict with Fairchild if they did not want to follow the path of Rheem Semiconductors. Contrary to Ed Baldwin, Jay and Jean were the key persons in the development of Fairchild's integrated circuits with detailed knowledge of every aspect of the new technology. When C. Sheldon Roberts with Isy Haas joined them, Amelco Semiconductor had the nucleus of the expertise needed to build up the new company. A few weeks later, E. Kleiner become Amelco's consultant, and formed the service department. Half of the original Fairchild eight founders were now associated with Amelco Semiconductor.

After Last's and Hoerni's resignations, Gordon Moore moved Fairchild R&D group into new fields: superconducting materials and microwave devices. This was clearly a wrong decision and both efforts ended with a fiasco. Noyce viewed Jay's and Jean's departure as betrayals, and because another group of capable engineers: Allison, Kattner, James, and Weissenstern, defected from Fairchild and formed Signetics just a few months after Last and Hoerni departed, Noyce become obsessed with the effort to destroy any new start-up. Signetics was his focus for a while. Last and Hoerni at Amelco were developing devices that were not directly competing with Fairchild, however, the same was not true in the case of Signetics. Noyce did not hesitate to use any tactics to drive Signetics out of business. He tried to lure Allison to re-join Fairchild, he announced Fairchild's devices that were not in production yet, he dumped prices of DTL parts, he sued; but Signetics miraculously avoided going under.

Cecil A. Lasch, Jr., a technician who worked at Fairchild with Gordon Moore and Jean Hoerni on the development of diffusion furnaces, left Fairchild also, and set up in 1960 the first semiconductor equipment manufacturer, Electroglas Incorporated, in Menlo Park. Amelco benefited from Lasch's work. After twelve months of fevered work, Amelco's 45,000 square-foot facility was equipped with the state-of-the-art Electroglas DF-3 diffusion furnaces (Fig. 6.3) and ultra-clean water wet benches (not DI water which was industry standard at that time.)

Fig. 6.3. Amelco Semiconductors clean room (1963)

Sheldon Roberts developed the epitaxial process. J. Last with optical engineer R. E. Lewis, developed at that time the most advanced and monstrous, 20 feet long camera for accurate mask production. The design sophistication was going beyond conventional standards. A large granite pedestal was equipped with temperature compensation that regulated the temperature based on the ambient and number of persons present in the room.

In addition, Amelco abandoned Fairchild's method of "mask indexing" and for the first time introduced mark-in-mark overlay targets (Fig. 6.5).

Fig. 6.4. Doping dilution system of Electroglas DG series diffusion furnaces

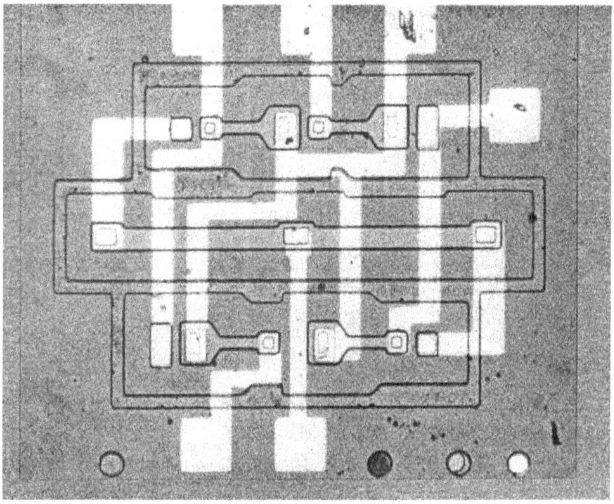

Fig. 6.5. Mark-in-mark overlay targets on Amelco's integrated circuit

Such photolithography technology enabled production of smaller devices with reduced tolerances.

Jean Hoerni focused on the process and the development of the device which Shockley's team did not complete – a junction field effect transistor. Hoerni's planar technology was ideal for the junction FET device, and he developed the junction FET concept that is used without major changes up to this date. Hoerni's process eliminated problems with previously manufactured JFETs developed by Crystalonics in Cambridge, MA. Crystalonics, which was founded in 1958, succeeded in JFET construction which was very similar to the concept that Bob Noyce's group was pursuing at Shockley Laboratories. Crystalonics announced six devices of "amplifier types" designations (C610-C615) and four devices "switching types" (C650-C653). All parts were supplied in TO-5 packages. Maximum anode (drain) currents were 50 mA. The not well-controlled transconductances of Crystalonics devices varied from 100 to 1200 at room temperatures. Inter-electrode capacitances were very high, 35 to 50 pf, and limited the design to frequencies below 1 kHz.

Fig. 6.6. Hoerni planar junction FET device (February 1963)

JFETs were probably Amelco's most successful parts commercially, and they eventually resulted in the industry first JFET matched pair and JFET input operational amplifier. There is a nice story going with Amelco's matched JFET's: Amelco's salesperson, Ed Barrett, came by to see Robert Pease at Philbrick with a "*new, improved*" matched JFET pair. Ed said, "*Everybody wants to evaluate this FET by putting it on the curve-tracer. Some customers think it's the best matched pair they ever saw*". Bob asked Ed if he would let him evaluate this part. Ed replied, *"It wouldn't work well for you because you would plug it into the front end of an op amp – and it would not work."* Ed admitted, "*That is because it is really just one FET inside, bonded out to both sides of the six-pin package. It looks pretty good on a curve-tracer. But if you tried to use it, it wouldn't amplify.*"

Jay Last recognized the importance of linear integrated circuits in the pre-Widlar era and Amelco became the first major supplier of linear integrated circuits. The parts were contracted by Westinghouse, Honeywell and by others. In 1964 Amelco with about 700 employees offered about 30 standard integrated circuits, the D13001 monolithic low drift differential amplifier

AMELCO
FIELD EFFECT TRANSISTORS

TENTATIVE SPECIFICATIONS
OCTOBER, 1964

2N3066 2N3067 2N3068 2N3365 2N3366 2N3367

MAXIMUM RATINGS

	Premium	Industrial
Storage Temperature	−65° C to 300° C	−65° C to 175° C
Operating Junction Temperature	200° C Max.	150° C Max.
Power Dissipation	300 MW	300 MW
Derate	1.7 mw/° C	2.4 mw/° C

DESCRIPTION
Amelco N-Channel Field Effect Transistors are Silicon Planar Diffused devices that combine high transconductance with high input impedance and low capacitance. Because of this they are useful in many applications including low level amplifiers and choppers.

ELECTRICAL CHARACTERISTICS AT 25° C

SYMBOL	CHARACTERISTICS	TEST CONDITIONS	PREMIUM 2N3066	PREMIUM 2N3067	PREMIUM 2N3068	INDUSTRIAL 2N3365	INDUSTRIAL 2N3366	INDUSTRIAL 2N3367	
BV_{DGO}	Drain to Gate Breakdown Voltage	1	50	50	50	40	40	40	Volts Min.
I_{GSS}	Total Gate Leakage Current	2	1.0	1.0	1.0	5.0	5.0	5.0	nA Max.
$I_{GSS\ Hi\text{-}Temp}$	Total Gate Leakage Current	3	1.0	1.0	1.0	1.0	1.0	1.0	μA Max.
I_{DSS}	Saturation Current	4	0.8-4.0	0.2-1.0	.05-.25	0.8-4.0	0.2-1.0	.05-.25	Ma
g_m (f = 1.0kc)	Transconductance	4	400-1000	300-1000	200-1000	400-2000	250-1000	100-1000	μ mhos
V_p	Pinch Off Voltage	5	10	5	2.5	12	7	2.5	Volts Max.
C_{DG} (f = 140kc)	Drain to Gate Capacitance	6	1.5	1.5	1.5	2.0	2.0	2.0	pf Max.
C_{SG} (f = 140kc)	Source to Gate Capacitance	7	3.0	3.0	3.0	3.0	3.0	3.0	pf Max.
C_{os}	Small-Signal, common-source output capacitance (input shorted) Freq = 1 MC	4	1.5	1.5	1.5	-	-	-	pf Max.
g_{os}	Small-signal, common-source output conductance (input shorted) Freq = 1 MC	4	50	20	5	-	-	-	μ mhos Max.
C_{is}	Small-signal, common-source input capacitance (output shorted) Freq. = 1 MC	8	10	10	10	-	-	-	pf Max.
NF	Noise Figure	9	3.0	3.0	3.0	-	-	-	db Max.

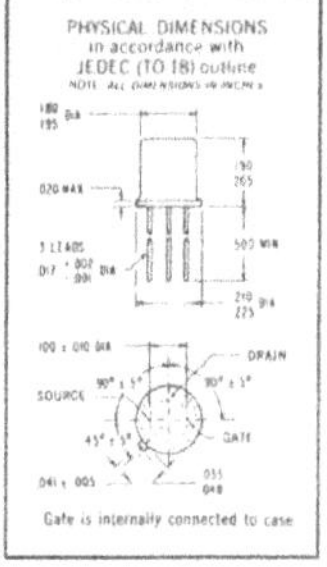

AMELCO SEMICONDUCTOR
DIVISION OF TELEDYNE, INC. / 1300 TERRA BELLA AVE., MOUNTAIN VIEW, CALIFORNIA
MAIL ADDRESS: P. O. BOX 1030, MOUNTAIN VIEW, CALIF. / PHONE: (415) 968-9241 / TWX: (415) 969-9112

Fig. 6.7. Preliminary data sheet for Amelco junction FETs

with tight Vbe thermal coupling, a series of JFET devices and competitive DCTL logic.

Amelco's start up was remarkable for other reasons. Contrary to Fairchild, Amelco was under-funded and the lack of the capital forced Amelco founders to operate at a maximal effectiveness. The original Last and Hoerni plan to support the system company with internally produced parts showed as not sustainable because military and government frequently asked for low quantities of parts that must pass severe procurement procedures. With a small volume of production, it is very difficult to achieve high yield. Singleton ran the company on a shoestring for a long time, balancing frequently a small volume of specialized government parts without the benefit of the production of high volume commodity products.

Amelco gradually expanded more into commodity products, and also manufactured discrete transistors. However, the financial pressure took a toll.

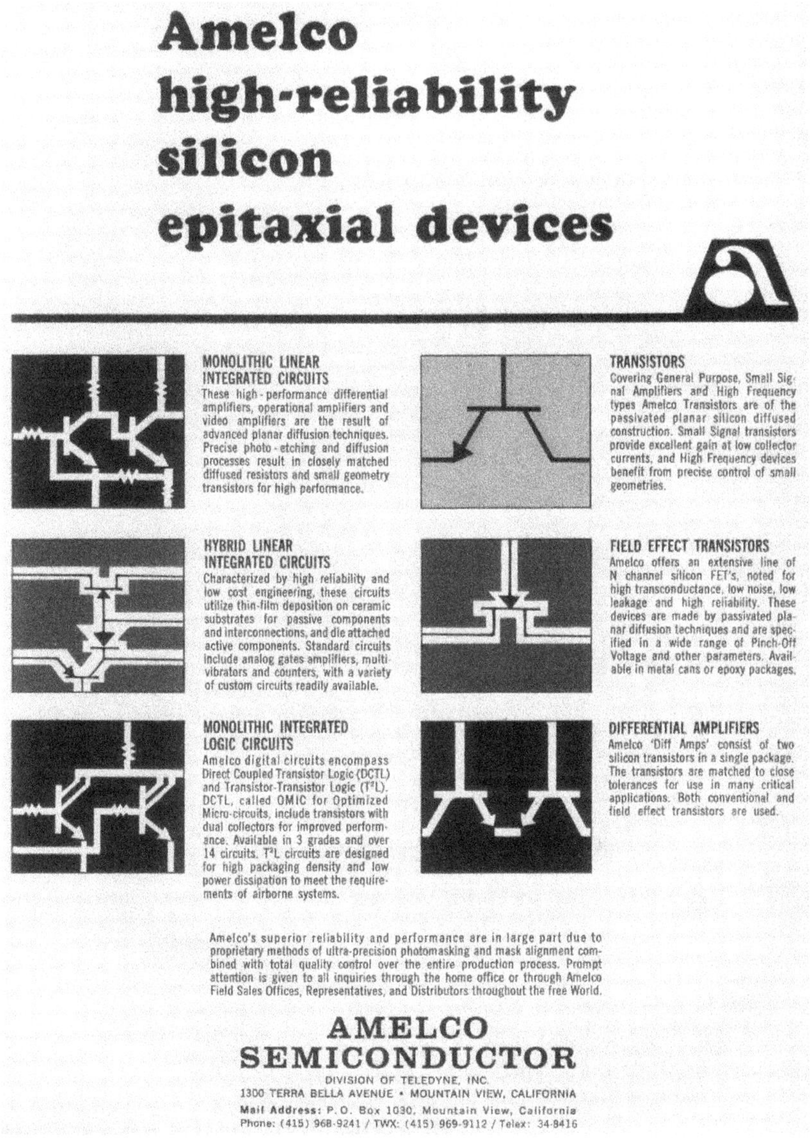

Fig. 6.8. Amelco advertisement released in May 1966

Hoerni, who was in the position of General Manager got into an argument with Kozmetsky. Kozmetsky knew very little about manufacturing of integrated circuits and sadly, he was already on a mission. Kozmetsky made a deal with James F. Battey, who directed Shockley Laboratories for Clevite Transistor. To avoid possible conflict between Clevite and Teledyne, Battey resigned from Clevite in March 1963. For Hoerni it was difficult to accept

Kozmetsky's argumentation and he abruptly, as was his style, left Amelco. Roberts followed him the same week. Battey became the General Manager of Amelco in September 1963.

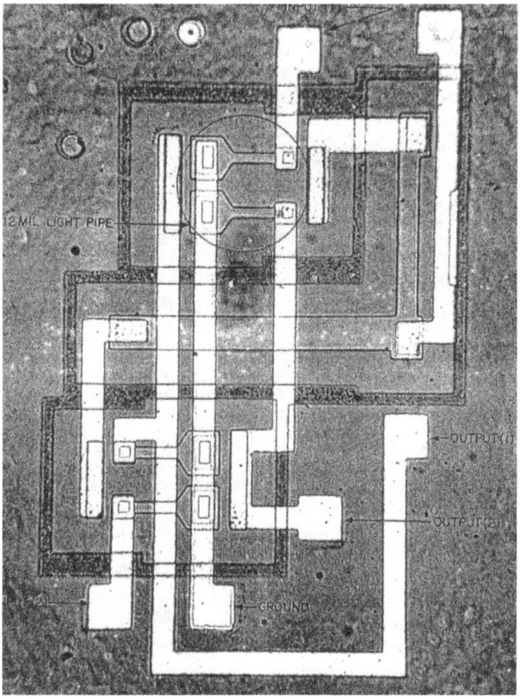

Fig. 6.9. Last's and Gilleo's Flip-Flop driven through optical interconnect (1963)

The hybrid and integrated op amps were a huge success. It generated sales; it generated new designs; and it generated competition. Amelco acquired Philbrick and Nexus in 1966, creating Teledyne Philbrick Nexus. Philbrick had designed for Amelco already before 1966, A13-251/351 op amp and D13/43 diff amp, and were quite successful on the market before the Widlar revolution. Bob Pease designed Philbrick's hybrid op amps Q25AH and Q85AH, which were manufactured by Amelco. As a response to Widlar's 709, Amelco introduced the 805BE. The 805 was comparable in complexity to the μA 709, but was packaged in a 10-pin package. Later on the Amelco 809BE (which was the same as Philbrick T52) competed with LM101. Amelco introduced also the first integrated ADC using Pease's V-F circuitry.

Jay Last and M. A. Gilleo developed the concept of interconnect which will be again re-discovered in the near feature. In 1963 they developed an optical interconnect (Fig. 6.9) where coupling between individual components is based on the radiation generated by carrier recombination and re-converted back to current by photovoltaic effect.

Although many do not remember Amelco's accomplishments today, their achievements are definitely a significant part of the semiconductor saga. Amelco's parts were used for high-end weapons systems, submarines, space launchers, satellites, space stations, and many other applications. Amelco, like not many other companies, survived all mergers. In 1966 it merged into Teledyne Philbrick Nexus, which became Teledyne Philbrick, which became Teledyne Components, which became TelCom Semiconductor, which became part of today's Microchip.

Texas Instruments, Inc.

In 1941 Cecil Green, J. Erik Jonsson, Eugene McDermott, and H. B. Peacock bought Geophysical Service Inc. which was founded in 1920. In 1951, Dallas-based Geophysical Service became Texas Instruments Incorporated. Patrick E. Haggerty, president of TI, decided to create a major semiconductor company. In 1951 TI fully committed to the development, manufacture, and marketing of active semiconductor devices, and licensed transistor technology from Western Electric in the spring of 1952. In 1952 the company had sales of just $ 20.5 million and net profit after taxes of $0.9 million.

Texas Instruments was a pioneer in the marketing of germanium transistors. By persuading Regency Radio to produce a radio receiver using these components, a project which cost Texas Instruments almost $2 million as its part of the investment, TI gained a strong hold in the market.

During 1952–55 Haggerty took a significant risk and put $3 million into the new plant, equipment and development of the silicon transistor. Haggerty recruited Gordon K. Teal from Bell Laboratories as research director [2]. Teal joined TI's 1770 employees in the end of 1952. The solid state research and development laboratory was established on January 1, 1953 and TI's engineering team began work on a small-signal transistor which would meet military environmental conditions. Teal hired Willis A. Adcock, along with Boyd Cornelison, Morton E. Jones, J. W. Thornhill, and E. D. Jackson, who designed an improved version of Bell Labs' crystal grower.

A little over 12 months later, in early 1954, they had succeeded and they demonstrated a working silicon device similar to the germanium junction transistor designed earlier by Morgan Sparks at Bell Labs. Starting with pure silicon, the initial melt was doped by antimony, and this part of the ingot formed the collector region. An additional part of the ingot, approximately 10 μm, was doped by gallium and then quickly by a much larger amount of arsenic or antimony to make the low-resistivity emitter. Upper and lower N-type doped parts were cut off and an approximately 5 mm thick wafer with N-P-N layers was sliced from the ingot. The wafer was divided into rectangular bars which were chemically etched to remove surface damage and then mounted on a header by the collector and emitter ends. A thin gold wire was pulse bonded to the edge of the base layer.

As in the case of Bell Laboratories transistors, moisture was the main problem, causing high reverse leakage current. The manufacturing of grown junction transistors was a very labor-intensive effort and difficult to automate.

Because of their large investment into grown-junction technology, Texas Instruments was very slow in introducing diffusion to transistor manufacturing. According to Kilby, in 1958 there was still no working silicon mesa transistor [3]. It was only after Fairchild's success with the diffused mesa transistor that TI introduced TI's version of the 2N696 and 2N697 "diffused base" devices (Fig. 6.10).

electrical characteristics at 25 °C ambient (unless otherwise noted)

	PARAMETERS	TEST CONDITIONS		min.	max.	unit
I_{CBO}	Collector Reverse Current	V_{CB} = 30v	I_E = 0	—	1.0	μa
	at 150°C	V_{CB} = 30v	I_E = 0	—	100	μa
BV_{CBO}	Collector-Base Breakdown Voltage	I_{CBO} = 100μa	I_E = 0	60	—	v
BV_{CER}	Collector-Emitter Breakdown Voltage	I_{CER} = 100ma	R_{BE} = 10 ohms	40	—	v
BV_{EBO}	Emitter-Base Breakdown Voltage	I_{EBO} = 100μa	I_C = 0	5	—	v
h_{FE}*	D-C Forward Current Transfer Ratio	I_C = 150ma	V_{CE} = 10 v (2N696)	20	60	—
			(2N697)	40	120	—
V_{BE} (sat)*	Base-Emitter Saturation Voltage	I_C = 150ma	I_B = 15ma	—	1.3	v
V_{CE} (sat)*	Collector-Emitter Saturation Voltage	I_C = 150ma	I_B = 15ma	—	1.5	v
h_{fe}	A-C Common Emitter Forward Current Transfer Ratio	I_C = 50ma	V_C = 10v f = 20mc	2.5	—	—
C_{ob}	Collector Capacitance	I_E = 0ma	V_C = 10v	—	35	μμf

*Pulse conditions: length = 300μs ; duty cycle < 2%.

maximum ratings at 25 °C ambient

Collector-Base Voltage	60v
Collector-Emitter Voltage (R_{BE} = 10 Ω)	40v
Emitter-Base Voltage	5v
Total Device Dissipation	0.6w
Total Device Dissipation at case temperature 25°C	2w
Storage Temperature Range	−65°C to +175°C

Fig. 6.10. The electrical characteristics of TI "difused base" 2N696 and 2N697 mesa transistors

There are plenty of stories describing how Jack Kilby developed his integrated circuit in 1958. Jack Kilby graduated in Electrical Engineering from the University of Illinois in 1947 and joined The Centralab Division of Globe Union, Inc. in Milwaukee, as a product engineer working on hearing aid amplifiers. In 1952 Centralab acquired a transistor license from Western Electric and started work on germanium transistors. Kilby attended the first Western Electric transistor technology symposium in 1952 (Fig. 6.12) and headed Centralab's project to build transistors. Centralab was heavily involved in the Micromodule Components program directed and funded by the Signal Corps. The main contractor, RCA, worked with other sub-contractors on modification of electronic components for possible "integration" into micromodule blocks. The program objective was a 10-fold reduction in military electronic equipment size. Prototype modules had component density up to 350,000 parts a cubic foot. Centralab had resistor technology (screened components – thick film circuits) and used transistors inserted into a substrate and connected them to the substrate using a photolithographically defined pattern using technology developed in 1957 by Lathrop and Nall at the Army Diamond Ordnance Laboratories (Fig. 6.16.)

Centralab, however, was too small, and their work on transistors was progressing very slowly. Kilby got an offer from Willis Adcock to work for Texas

Fig. 6.11. The world's largest semiconductor manufacturing facility in 1960 – Texas Instruments components division plant

Fig. 6.12. Kilby (back row, center) at the first transistor technology symposium in 1952. Jack Morton stands at front, left

Instruments in the spring of 1958. Obviously, at the time when TI hired Jack Kilby, the company had no program or plan to research an integrated circuit. In a recent interview Kilby said: *"After several interviews, I was hired by Willis Adcock of TI. My duties were not precisely defined, but it was understood that I would work in the general area of microminaturization. Soon after starting at TI in May 1958, I realized that since the company made transistors, resistors, and capacitors, a repackaging effort might provide an effective*

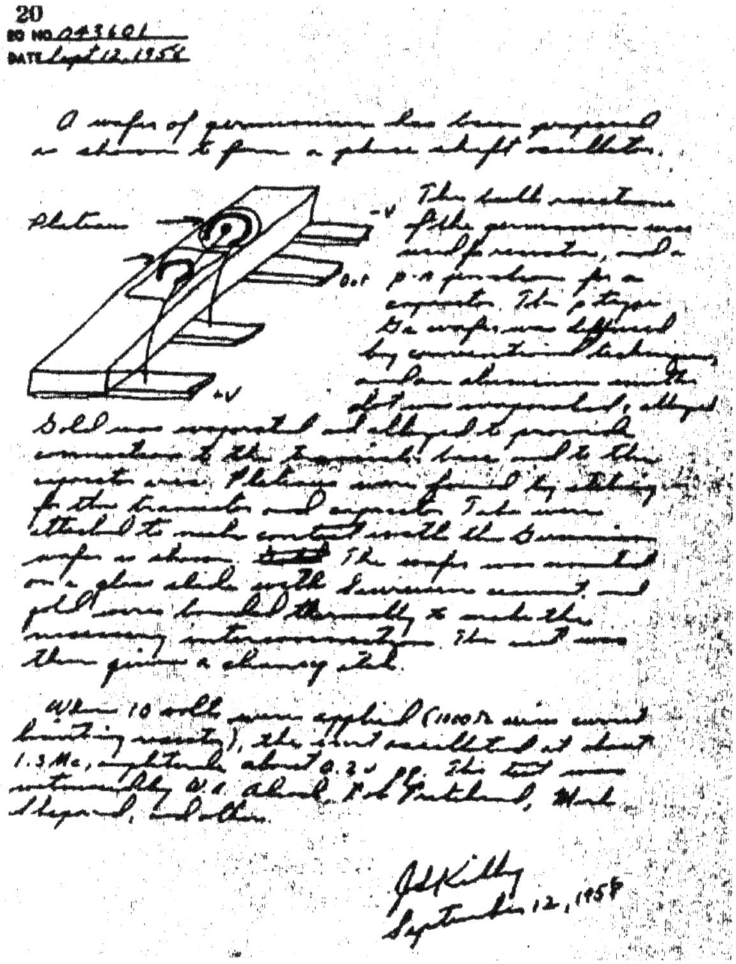

Fig. 6.13. Phase shift oscillator which Kilby built in August and September 1958

alternative to the Micro-Module. I therefore designed an IF amplifier using components in a tubular format and built a prototype. We also performed a detailed cost analysis, which was completed just a few days before the plant shut down for a mass vacation. As a new employee, I had no vacation time coming and was left alone to ponder the result of the IF amplifier exercise. The cost analysis gave me my first insight into the cost structure of a semiconductor house. The numbers were high – very high – and I felt it likely that I would be assigned to work on a proposal for the Micro Module program when vacation was over unless I came up with a good idea very quickly. In my discouraged mood, I began to feel that the only thing a semiconductor house could make in a cost-effective way was a semiconductor. Further thought led me to the conclusion that semiconductors were all that really required – that resistors and capacitors, in particular, could be made from the same material as the active devices.

Although the test showed that circuits could be built with all semiconductor elements, it was not integrated."

Because of the unavailability of silicon diffusion transistors at TI, Kilby used the germanium transistor. Texas Instruments used square, 10 by 10 mm, germanium wafers. In five rows and five columns there were 25 mesa transistors with emitter and base contacts evaporated through metal masks. Kilby used a P-type wafer with resistivity about 3 Ω-cm and with Antimony diffusion created an N-type base layer. From this wafer he cut a bar 1.6 mm wide and alloyed contacts to the back of this semiconductor bar to provide contacts to the bulk resistor. Aluminum evaporated through an appropriate mask was used to form an emitter alloyed region. The gold-alloyed contact was formed to the N-type base. Wax was applied by hand to mask the mesa of the transistor and the region forming a distributed RC network. Loose gold wire was then thermally bonded to the appropriate areas to complete the interconnection. The transistor used by Kilby was identical to Lee and Shockley's transistor developed in 1955 in BTL.

In fact, the note about the inductor in Kilby's patent: "Small inductances, suitable for high frequency use, may also be made by shaping the semiconductor as evident by Figure 5a" (Fig. 6.15) indicates that Kilby was still considering microminiaturization as used in the Micromodule program under the Signal Corps. Kilby's patent did not solve or suggest any isolation between multiple devices excepting "appropriate semiconductor shape" and "air gap." Kilby's network did not resemble the integrated circuit as used today at all.

The importance of Kilby's invention, was that it proved that all electronic components can be produced from semiconductors. It was Texas Instruments and its "Texas style" that sold "Solid Circuits" to the Air Force. TI's aggressive management, under the presidency of P. E. Haggerty, did not hesitate as Fairchild's Moore and Noyce did. Patrick E. Haggerty, president of Texas Instruments, recognized the possible potential of Kilby's work for military applications. Haggerty, an electrical engineer with a degree from Marquette University, started his career with the Badger Carton Company in Milwaukee. During World War II he served in the Navy, where he was responsible for the procurement and production of airborne electronic equipment. He joined Geophysical Services in 1945 and served on the Government Council of Science and Technology. The "Solid Circuit" program launched by Texas Instruments at the IRE Convention in March 1959, put in question the established status of miniaturization of electronic assemblies.

Jay W. Lathrop joined TI's Transistor Product Group at almost the same time as Kilby. Although he did not work directly with Kilby, he set up a small photomask shop and passed on to Kilby a major piece of information which

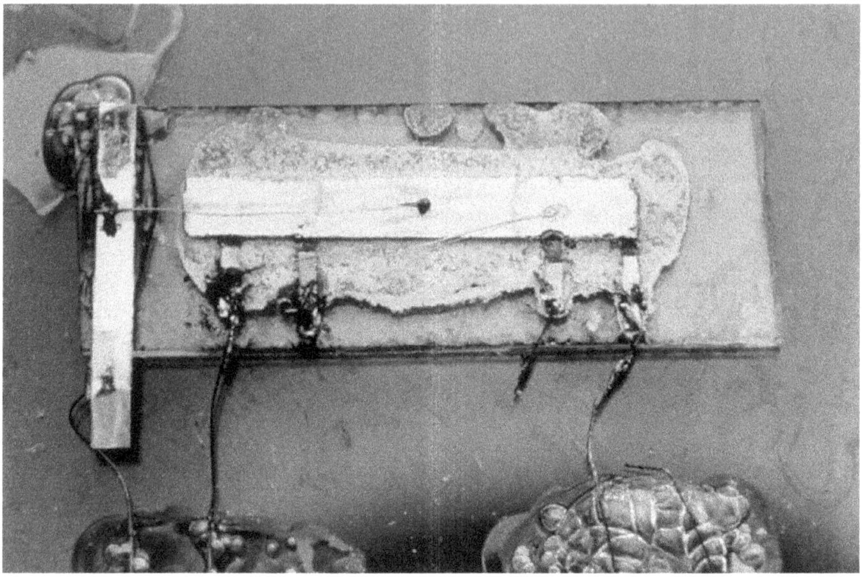

Fig. 6.14. Jack S. Kilby's FIB-phase shift oscillator with a germanium mesa transistor, bulk resistor, and distributed RC network (Texas Instruments, September 12, 1958)

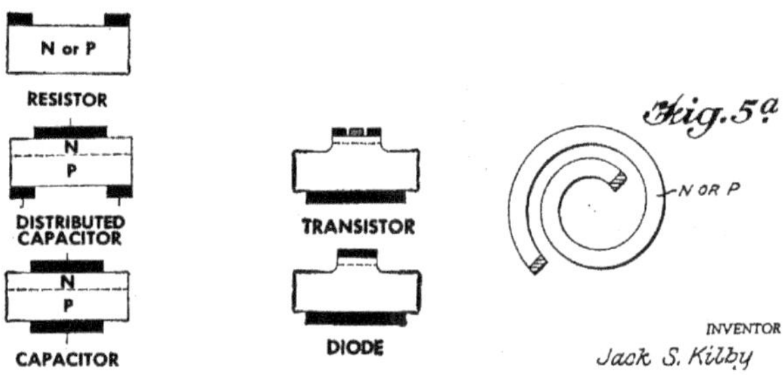

Fig. 6.15. Semiconductor components including "Spiral semiconductor inductance" as suggested by Kilby in 1959 [4]

was missing in Kilby's ideas described in his notebook from September 1958. Jay Lathrop and Lee Barnes designed and prepared masks for the circuit using "air isolation" which Kilby, as sole author, submitted to the Patent Office in February 29, 1959.

The Air Force granted TI a contract to build a digital computer in January 1961. TI had eighteen months to build a pilot production line that could turn out 500 integrated circuits daily for at least ten consecutive days. The job to build a chip fabrication line fell to Kilby. Harvey Cragon from the Equipment Division developed the computer architecture. In October 1961 Texas Instruments showed the "Molecular Computer" consisting of 587 sili-

2,890,395

SEMICONDUCTOR CONSTRUCTION

Jay W. Lathrop and James R. Nall, Silver Spring, Md., assignors to the United States of America as represented by the Secretary of the Army

Application October 31, 1957, Serial No. 693,788

3 Claims. (Cl. 317—234)

(Granted under Title 35, U.S. Code (1952), sec. 266)

The invention described herein may be manufactured and used by or for the Government for governmental purposes without the payment to us of any royalty thereon.

This invention relates to a novel semiconductor construction, especially as applied to transistors.

It is the object of this invention to provide a semiconductor construction which permits of very great miniaturization and provides a high resistance to very large shocks and vibrations.

It is a further object of this invention to provide a semiconductor construction which is integrally incorporated with a printed circuit plate.

A typical form of the invention essentially comprises a transistor unit mounted in a printed circuit plate with its junction contacts electrically connected to appropriate printed circuit leads by means of metal deposited therebetween. This construction eliminates the need for the spring loaded or pressure bonded leads utilized in the prior art thus providing the important advantages of permitting a very considerable reduction in size, and greatly increasing the resistance of the unit to shock and vibration. It should be noted that although some of the techniques utilized in producing the novel semiconductor construction of this invention are similar to the techniques employed in printed circuitry, there has been no previous attempt in the prior art to incorporate these techniques with semiconductor construction so as to provide the very important advantages achieved by this invention

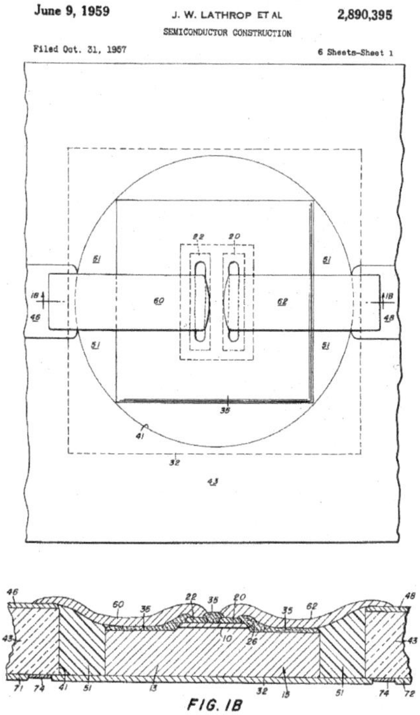

Fig. 6.16. "Planar" circuits with evaporated metal patterned by photolithography developed by Lathrop and Nall (DOFL, October 1957)

con semiconductor networks and occupying 6.3 cubic inches. Air Force and TI spokesmen emphasized that the small computer was developed only as a vehicle to prove the capability of molecular electronics. P. E. Haggerty, TI's president, said that each semiconductor network consists of a silicon functional block capable of performing the functions of a Flip-Flop, NOR gate, and driver. Hermetically sealed networks measuring $6.35 \times 3.175 \times 0.75$ mm were assembled into rigid modules (Fig. 6.17.) The computer was made from 47 such modules, and the memory capacity was 300 bits. Mr. Haggerty said that TI's effort in the development of the circuits was difficult to separate from other company work, but he estimated \$2 million had gone into the program.

The publicity campaign worked well for Texas Instruments. Fashion magazine's for women printed articles about "Solid Circuits" and described how this would change our lives. Texas Instruments, however, was not leading the microelectronics revolution. At the end of the fifties and at the beginning of the sixties they were at least 12–24 months behind Fairchild and Tom Longo's Sylvania know-how.

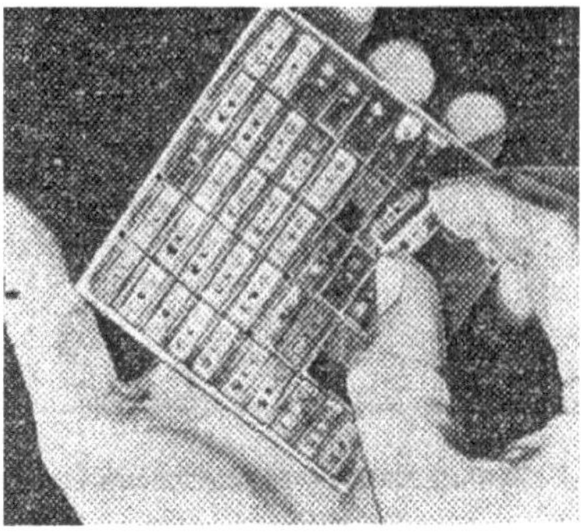

Fig. 6.17. Molecular Computer demonstrated by Texas Instruments in October 1961

Jay Last of Fairchild still has in his possession one of the first Fairchild products with a label "Open Here". Fairchild marketing people knew that TI bought some quantity of each new Fairchild part. As a joke, Jay Last labeled some of these parts "Open Here" instead of with a product label.

It took Texas Instruments another five to eight years before they consolidated their R&D and production activity to surpass Fairchild.

A very good illustration of the status of integrated circuit development at TI, can be derived from the company re-organization announced in May 1961 [5]. The company had an Equipment Division, which would handle mostly the government contracts. Texas Instruments Semiconductor and Components division combined its germanium and silicon transistors groups into one new transistor product group headed by J. R. McDade. Jay R. Reese, former manager of the silicon products group, had been named head of the components group which included diodes, rectifiers, capacitors and resistors. Robert Bronson was manager of the mesa-planar transistor department. Howard Moss was heading the transistor department. The special computer products department had been renamed the alloy transistor department and continued to be managed by Jim Reese. There was no "integrated circuits department"

In the Transistor Anniversary issue of Transactions of Electron Devices [3], Kilby described twenty years later, that he omitted to include into the patents a more sophisticated version of the circuit with diffused resistors and oxide capacitors. There is absolutely no evidence that Texas Instruments had at that time more sophisticated devices.

The fact that almost the same engineering ideas and solutions originated at different development organizations, is not unusual. However, the unethical and unprofessional legal tactics were the reasons why Texas Instruments lost in litigation with Sprague and Fairchild. Although it is clear that Jack Kilby was pushed by Texas Instrument's management to participate in such séances, he, unfortunately, did nothing to set the record straight.

The legal practices of Texas Instruments, which later T. J. Rodgers, the CEO of Cypress Semiconductor, called "The Dallas Legal Firm" established a precedent in the semiconductor industry and resulted in many frivolous suits used as tactics to slow down or discredit the competitors.

Sprague Electric Company

In the nineteen fifties all transistor manufacturers resided on the east coast. The manufacturing of pointcontact and alloyed germanium transistors was very similar to the assembling of vacuum tubes. Vacuumtube companies such as RCA, Westinghouse, General Electric, Sylvania, Raytheon and others saw the transistor as a similar product. The management of vacuum-tube companies had no knowledge of semiconductors and they were not able to extend the design of the transistor beyond the knowledge they bought with license rights from Western Electric. Gradually all of them quit the semiconductor business.

Sprague Electric Company is an example of a slightly different kind of business decay. The two men in charge of the company were two brothers, R. C. Sprague, the founder, and his brother Julian Sprague, who was in charge of sales. They each struggled for control of the company. The brothers and their sons, Julian and his son Peter Sprague, and R. C. Sprague and his son John were each trying to put his son into the company's leadership. As could have been predicted, neither of them achieved his goal. However, this internal fighting drove the company slowly out of business. Kurt Lehovec, not aware of Sprague's internal struggles, joined the company as the Research Director and launched the Sprague semiconductor business in 1952. Lehovec was lured by the promise of complete freedom to pursue the semiconductor device project of his choice.

Czech-born Kurt Lehovec started his career during World War II at Charles University in Prague as a Ph.D. student of Prof. B. Gudden (B. Gudden graduated and worked under Prof. Pohl). Kurt's Ph.D. study focused on lead-selenide as an infrared detector. During the course of this research, he noticed that Thallium diffused into Selenium very rapidly and the "blocking effect" of the rectifier was significantly improved. Süddeutsche Apparate Fabrik in Nüremberg supported Lehovec's study. This association let Lehovec attend a secret scientific conference about "Material X" in 1942 in Munich. The code name, Material X, was used for Germanium, and Lehovec did not know what Material X was.

German military research maintained two research groups working on semiconductors in occupied Czechoslovakian "Böhmen und Mähren" territory: The Prague group headed by Prof. B. Gudden worked on the rectifying diode; a second and larger group worked in the small town of Tanvald in the North of Bohemia on microwave point contact diode and radar research. Many members of the German research groups, including Prof. Gudden, were killed in May 1945 during the liberation of Czechoslovakia by the Red Army. A major part of the Tanvald research facility was confiscated by the Soviet army and transferred to the Soviet Union; the rest of the inventory was set on fire and destroyed.

Several kilograms of Germanium survived the fire and were used by Jan Tauc and Helmar Frank to build the first European transistors in 1949. Jan Tauc later immigrated to the United States. Helmar Frank, who was also a student of Prof. Gudden at Charles University, directed post-war semiconductor research sponsored by the Czechoslovakian Army at the former Philips Laboratories in Prague.

During the war the British Intelligence Assault Unit 30AU identified prominent scientists involved in German research. At the top of the list was Helmut Walter, German genius and aircraft designer of the world's first jet plane, the Messerschmit 262. At the end of the war, the CIA and KGB quickly joined the British effort and transferred several researchers from the Prague and Tanvald research groups to their respective countries. The CIA with U.S. Joint Intelligence Objectives Agency (JIOA) executed the Project Paperclip, and picked up Kurt Lehovec. Under very dramatic circumstances, when the Red Army began the liberation of Prague, and when hours not days made a difference, Lehovec peddled on a bicycle with his tennis racquet from Prague to the American occupied zone. Later, he was deported as a Paperclip item in a U.S. Navy ship to the Squier Laboratory at U.S. Signal Corps in Fort Monmouth in New Jersey.

The thinking behind project Paperclip was exemplified in a letter Major General Hugh Knerr, Deputy Commanding General for Administration of U.S. Strategic Forces in Europe, wrote to Lieutenant General Carl Spatz in March 1945: "*Occupation of German scientific and industrial establishments has revealed the fact that we have been alarmingly backward in many fields of research, if we do not take this opportunity to seize apparatus and the brains that developed it and put this combination back to work promptly, we will remain several years behind while we attempt to cover a field already exploited.*"

Everything in Kurt Lehovec's life was extraordinary. His start with the Signal Corps was not an exception. Lehovec did not speak English, and other than the clothes he carried he owned nothing. The contracting officer, who worked for British Intelligence, gave Kurt a few boxes of cigarettes and suggested he go to the black market and buy some clothes and luggage. Lehovec arrived at the U.S. Signal Corps with physical chemists, Dr. Rudolf Brill and Ernst Baers, physicists; Horst Kedesti, Georg Hass, Georg Goubau, Günter Guttwein, and electrical engineers; Hans Ziegler, Edvard Gerber, and Richard Guenther. From 210 Paperclip scientists The Signal Corps utilized 24 scientists, all involved in radar and communication research. The rocket scientists ended up in White Sands.

The project Paperclip is a textbook example of the irony of life and politics. Dr. Kedesti, born with the German sense to be "pünktlich[1]," expressed disappointment with the organization of the Signal Corps just after he passed the entrance gate – the electric poles at the Fort Monmouth base were not aligned straight and they varied in height. Dr. Hans K. Ziegler received his degrees from the Technical University in Munich and carried out research in German industry as an expert in ceramic insulators for power transmission lines, and was one of the few Paperclip scientists who received the golden badge of the German Nazi Party (NSDAP). Ziegler was a skilled politician. After his arrival to Fort Monmouth he immediately established contact with the "right" friends, and he become the "spokesman" for Signal Corps' Paperclip group. His countrymen called him an "apple polisher" and distanced themselves from him. In 1959, Ziegler became the Chief Scientist of U.S. Army Signal Corps' Laboratories at Fort Monmouth, N.J. After the Army's re-organization, he was appointed in 1963 to Deputy for Science and Chief Scientist of the US Army Electronics Command and in 1971 to Director of the U.S. Army Electronics Technology & Devices Laboratory. Dr. Ziegler was also a Fellow of the IEEE and the American Academy of Science. The Army recognized his achievements with two Meritorious – and the coveted Exceptional – Civil Service Awards. As such, Dr. Ziegler is one of a few, if not the only one, who can pin up on his jacket medals from the Nazis and the Americans.

It was only natural that a large community of Jewish workers in Signal Corps was openly hostile to the Paperclip scientists. Lehovec's co-worker in micro-optical laboratory, Ben Levin, organized slogans for his forces: "*We would not offer them a seat, they need to sit on the floor*," or "*Germany should be converted to the dessert and attached to the Stalin's territory.*" It did not take a long time for Joe McCarthy to fire Levin with his leftist views. Then, the Paperclip scientists, including highly decorated NSDAP heroes, sat firmly in American chairs.

Lehovec asked the chief scientist of Squier Laboratory, Dr. Golay, what his research project should be, and he received the answer: "*you can do whatever you want, we need everything.*" Lehovec was one of the youngest members of the Paperclip team, and he worked very hard from day one. His group was assigned to the "Institute of Advanced Study" and had about 10 members. In 1948, Lehovec was sent to Purdue University to visit Prof. Lark-Horovitz's group to discuss ongoing research with the Signal Corps. With Dr. Harold A. Zahl, the Chief Scientist of the Signal Corps, Lehovec visited Bell Laboratories and started the Army's research on silicon and germanium devices, primarily on the problems of light emission from solids.

From his Prague studies he was familiar with the work of Russian scientist O. Lossew on "cold" light emission, first observed in silicon carbide in 1907. This effect, which was manifested by passing current through selected crystals of silicon carbide, showed a type of light emission which remained

---

[1] punctilious

Fig. 6.18. Kurt Lehovec (left) and Helmar Frank in Santa Monica, CA (2003)

unexplained. Dr. Carl A. Accardo remembered when he was approached by Lehovec with an inquiry "*if I would be interested in studying the effects with him and his co-worker Edward Jamgochian.*" Their research indicated that the produced light was due to the recombination of electrons and holes across a so-called p-n junction. The two articles authored by Lehovec, Accardo and Jamgochian appeared in Physical Review [6] and were presented at the American Physical Society in New York. The articles provided the theoretical background and an explanation of what later become known as LED's, or light emitting diodes.

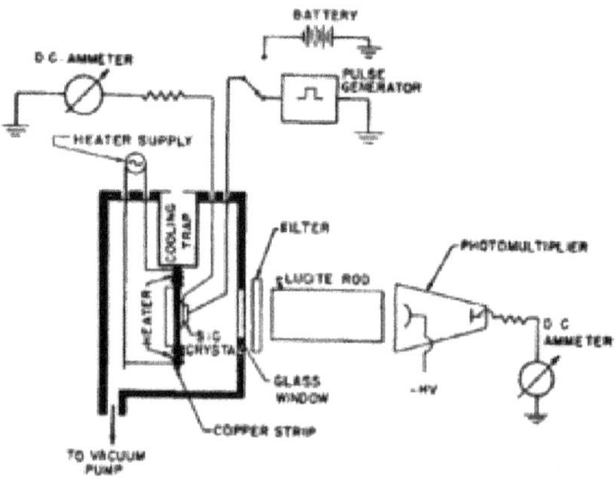

Fig. 6.19. Experimental setup for measurement of the light emission from SiC used by Lehovec group in 1948 [6]

Lehovec was a principal and reviewer of Semiconductor Research Contract DA36-0390SC 71131 between the U.S. Signal Corps and Prof. Karl Lark-Horovitz's group at Purdue University. During 1949–1950 Lark-Horovitz and R. Bray investigated the effect of strong electric fields on the observed spreading resistance of metal-germanium point-contact rectifiers. Some of these experiments closely resembled the work carried out simultaneously at Bell Telephone Laboratories by Bardeen, Brattain, and Shockley.

Prof. Lark-Horovitz offered Lehovec a position at Purdue University which Kurt turned down. Today, Kurt sees this as one of his worst decisions.

There is one more, rather tragicomic story, about Paperclip scientists. It was illegal, and U.S. law explicitly prohibited Nazi officials from immigrating to America. Although not all Paperclip scientists were Nazis, as many as three-quarters of the scientists in question were allegedly committed Nazis. As Orwell, and later Communists and modern American Doctrines, taught us, the law applies only to some people. After some wrestling between the military, the State Department, and the Executive Branch, the legal status of the Paperclip scientists was finally resolved in 1952. The Paperclip scientists who were brought to the U.S. illegally could apply for U.S. Citizenship or they could return to their home countries (not all Paperclip scientists were Germans). Kurt Lehovec was driven to Niagara Falls and directed to walk over the bridge toward Canada. Then, he was instructed to turn around at the Canadian side of the bridge and return to the USA. He obtained a stamp of legal entry. Five years later, when he obtained U.S. citizenship he needed to explain to the judge in Massachusetts how he got into the USA. *"From where did you enter the USA?* the judge asked, Lehovec answered "*From Canada.*" The judge asked again *"From where did you come to Canada?"* Lehovec truthfully answered *"From the USA."* Judge was satisfied with this explanation and Lehovec became a U.S. citizen. In early 1952, R. C. Sprague sent Dr. Preston Robinson, Sprague's Director of Research, to visit Kurt Lehovec in Squier Laboratory with the intention of hiring Lehovec and bringing the Sprague Company into the semiconductor business. Young Lehovec who was just planning to "tie the knot," was impressed with New England scenery and turned down offers from Dr. Shive to join Bell Labs, and from Dr. North to join Pacific Semiconductors. Instead, he left for the church and got married on May 23, 1952. He joined Sprague Electric Company.

During his interaction with Philco, RCA, GE, and other companies at the Signal Corps, Lehovec learned very quickly that in the U.S., top executives grow into their positions by chance, or by birth, or by friends they know. Lehovec appropriately characterized this phenomenon as the "3M effect". "*The workers at the lowest level handle the Material, the managers at the middle level handle the Men, and the top management handles the Money.*" Lehovec liked Robinson and R. C. Sprague's interest in new technology and innovation. Robinson was a key man in the development of the tantalum capacitor and he had respect for research work. Once they were together, they discussed the possibility of a Government contract. R. C. Sprague commented, *"If it is worthwhile doing it, I will pay for it. And if it is not worthwhile doing it, I do not want to do it, even if the Government pays for it."* Kurt was impressed.

Lehovec set up laboratory production in record time and he was on a mission. His goal was not to duplicate Bell Telephone Laboratories devices. He wanted to be an innovator, not a follower. He noticed already at the Signal Corps that the only drawback of the silicon devices was lower mobility of free carriers, which would make the devices somewhat slower compared to the germanium devices. Kurt Lehovec realized that the only way to overcome this disadvantage was to make the device smaller. However, to manufacture small devices in the beginning of the nineteen fifties was not an easy task. In addition, Sprague's management philosophy was, *"if this is any good, Bell would have done it already."*

Lehovec found two basic problems with the Bell point contact transistor which the Sprague Company licensed:

1) Tedious labor to manipulate under a microscope two tiny wires to the correct position.
2) The possibility of sliding of the wires during the handling the device.

Lehovec come up with an ingenious solution to these problems which Armen Fermenian put into practice, a narrow sheet of beryllium copper was guided over a ceramic frame and glued to it. Two metal chisels cut the sheet within the frame providing the triangular shapes with sharp points facing to each other. A plunger then bent the tips of the triangles into vertical position at the desired spacing of the point contacts.

Lehovec visited Bell Labs and showed a transistor to Jack Morton. Morton was very impressed with Sprague technology and considered an order of a large number of devices for Western Electric. The deal was, however, aborted

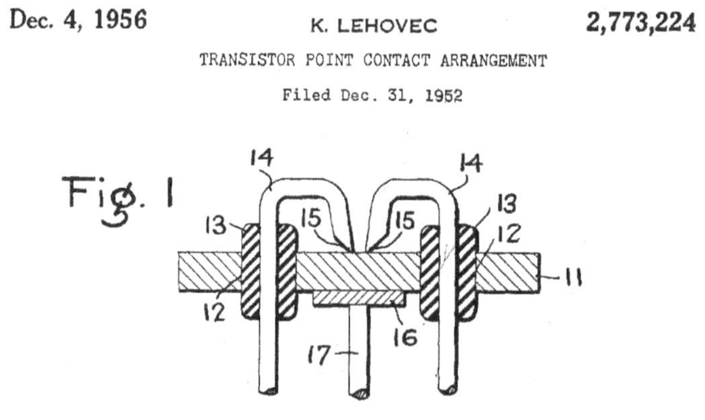

Fig. 6.20. The 1952 Sprague point contact transistor with a production cost of one tenth of the production cost of the BTL transistor produced by Western Electric

by Sprague management. The next major improvement of transistor technology was the introduction of new "capillary alloying" for junction transistor. This method allowed to calculate the amount of germanium dissolved in the indium from the germanium-indium phase diagram at the given temperature, and thus to control the junction depth. In addition, the surface-melt technique enabled the production of multiple junctions in a single slice of semiconductor. The method was superior to the BTL and TI grown-junction transistor process, which could produce only PNP or NPN structures.

The promising success of Lehovec' group was hampered by intrigues organized by R. C. Sprague's brother, Julian. Julian was vice-president of the company, in charge of sales. His son, R. C. Junior was vice-president and in charge of the personnel department. Julian opposed the Sprague Electric Company entrance into the semiconductor from the beginning. The reason behind the intrigues was the uncertainty related to anything new. One brother could handle such risk and see the future of the company, the other saw it as a path to destruction. Julian Sprague, filled with concern over an unwanted situation, once commented: "*My father was a genius, but always broke. I decided to make money.*"

To make money, Julian hired consultant from the Arthur D. Little company with the "goal" to streamline the R&D group. The consultant's advice was quite simple "*... transistors are so complicated and small that they are a job for watchmakers, not for Sprague Electric Company.*"

Kurt Lehovec's boss, Preston Robinson, was fired in 1954 for malicious reasons by Julian K. Sprague and was replaced by drug chemist Wilbur Lazier who was previously fired by DuPont and Pfizer. Lazier knew nothing about semiconductors and approached Lehovec with a question "*is the transistor a capacitor?*" His idea was to boost the capacitor business of Sprague Electric Company and cut off everything else. Lazier's stupidity embarrassed company officials many times. When IBM engineers visited Sprague to negotiate a contract and crystal growth was discussed, Lazier exploded into laughter and commented "*flowers are grown, not crystals ....*"

Another of Lehovec's major inventions, junction isolation, had a very interesting background. Lehovec attended a workshop at Princeton University in the end of 1958. Torkel Wallmark of RCA presented in this workshop a visionary presentation about the next generation of electronics. Wallmark listed problems that needed to be solved before integrated circuits could be designed. One of these limitations was device isolation. Lehovec, educated from his work on the surface-melt multiple-junction devices found a solution while driving his red Corvette – PN junction isolation.

The engineering problem was solved; the bigger problem remained to be overcome. At that time nobody had heard anything about integrated circuits, so only visionaries could have appreciated the importance of Lehovec's invention. The Sprague Company did not have visionaries and they did not want to file a patent application. Sprague's patent attorney, Hillary Sweeney, explained at great length to Lehovec, that a patent needed to have a docket number and must be in sequence with previous disclosures, and that he was too busy to do it. Lehovec wrote the patent application himself and mailed it to the Patent Office. After months, a persistent Lehovec finally prevailed. His patent was approved and issued by the Patent Office without a major objection. Immediately after the patent 3,029,366[2] was issued, Al Bower, of the patent firm Connolly & Hutz that handled the Sprague patents, called Lehovec that Texas Instruments is suing Sprague for patent interference.

In a panic, R. C. Sprague sent an internal memo to Hillary Sweeney, asking him to settle the case and avoid the interference hearing. Lehovec called Connolly & Hutz, and was told that Mr. Hutz would travel to Dallas to represent Sprague against Texas Instruments. Kurt asked if he could travel at his own expense with him. Hutz agreed.

Texas Instruments set up the interference proceeding hearing in a large conference room and equipped Kilby with double digits of TI's attorneys. Sprague Company was represented by Attorney at Law, Mr. Hutz, and Kurt Lehovec. TI called as an expert witness, Professor Robert J. Maurer, director of the Atomic Energy Commission and ARPA, (which jointly established the Materials Research Laboratory at the University of Illinois.)

Texas Instruments presented three arguments in the case:

1. The NPN transistor is an N-P emitter diode isolated from the collector N collector region by the PN collector junction.
2. Professor Maurer argued that when a large number of components should be placed on piece of semiconductor, the P-N junction isolation is automatically an obvious solution.
3. TI claimed that they actually produced a practical circuit assembly which was using P-N junction isolation prior to Lehovec patent filling date, and which was documented by page 19 in 1958 Kilby's Notebook (Fig. 6.21).

Kurt Lehovec was not allowed to present a counter argument personally, but Mr. Hutz was allowed to talk. After a brief discussion with Kurt, Hutz argued that "transistor action requires closeness of the emitter and collector junctions, so the current flow occurs across the collector junction."

To examine the second argument, Lehovec quickly drew on sheet of paper a circuit consisting of three resistors connected in a series with three capacitors terminating at the diode. The sheet of paper was handed to Prof. Maurer and Hutz repeated what Kurt was whispering into his ears.

> *"How many types of components are in the circuits?"* asked Hutz.
> "*Three*" Maurer reply.
> *"What kinds of devices are in the circuit?"*
> *"Resistors, capacitors and a diode"* was Maurer answer.

[2] Multiple Semiconductor Assembly, U.S. Patent Application filed on April 22, 1959.

"How many components in total are in the circuit?" asked Hutz.
"Seven."

Then Lehovec draw a schematic with "fork like structure" where three branches of the fork being semiconductor resistors consisting of the N-regions, each branch separated from the P-region of the fork by P-N junction capacitances and PN diode at the bottom of fork. *"Where is the PN junction isolation in this circuit?"* Mr. Hutz asked Professor Maurer. Maurer threw up his hands and said *"There is none."* Maurer, also, could not explain why, in his definition, such "*automatic and obvious solution*" was not obvious when TI designed the circuit for the Minuteman.

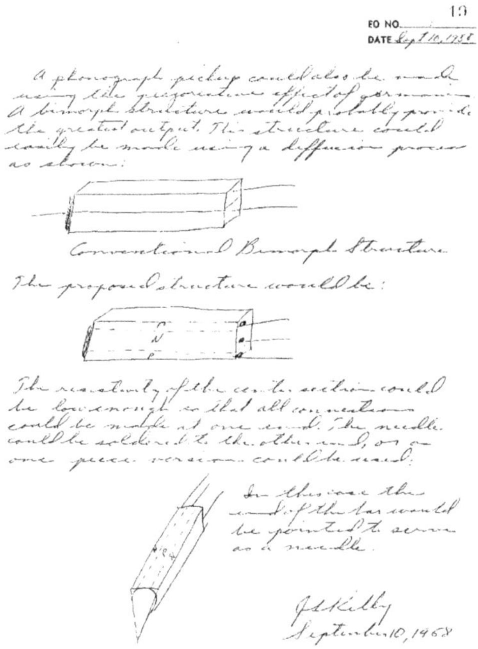

Fig. 6.21. Page 19 of Kilby's 1958 notebook used in the interference hearing Sprague vs. Texas Instruments in March 16, 1966

Finally, a TI attorney described page 19 of Jack Kilby's notebook (Fig. 6.21). The structure shown on the page was P type of material with N-layers on the top and bottom of the sample. On one end of the sample all three regions were connected by the contact. The N-layers acted as two resistors in series. A stylus attached to one end of the sample, while clamping the opposing end with contacts to the N-layers, provided a gramophone pick-up. By bending the sample, one N-layer expanded and the opposite layer compressed. Using these two layers in a balanced bridge circuit the signal proportional to the bending of the sample may be sensed.

Texas Instruments screenwriters set up a process line with all necessary equipment at the conference room, demonstrated each individual manufacturing step, and compared it to the Kilby's notes. Smart Lehovec, who was not only an excellent theoretician, but also a good practitioner, noticed that N-layers produced by gaseous diffusion would create N-layer at both ends of the sample. According to TI's description the N-layer across the narrow face of the strip was ground off. Why grind off the N-layer which already connected both top and bottom N-layer and replace such a natural connection by the metal contact? Kurt Lehovec was allowed to question the technician who was running the demo line. The confused technician acknowledged that the face N-layer was indeed not ground off. Thus, there were not two resistors isolated by P-N junction, but just one N-layer.

The whole Texas Instruments séance ended as a complete debacle, and attorney Hutz was asked to deliver a written summary of arguments presented at the hearing. Three weeks later the U.S. Patent Office decided that Lehovec was entitled to all patent claims and was awarded patent priority (Fig. 6.22).

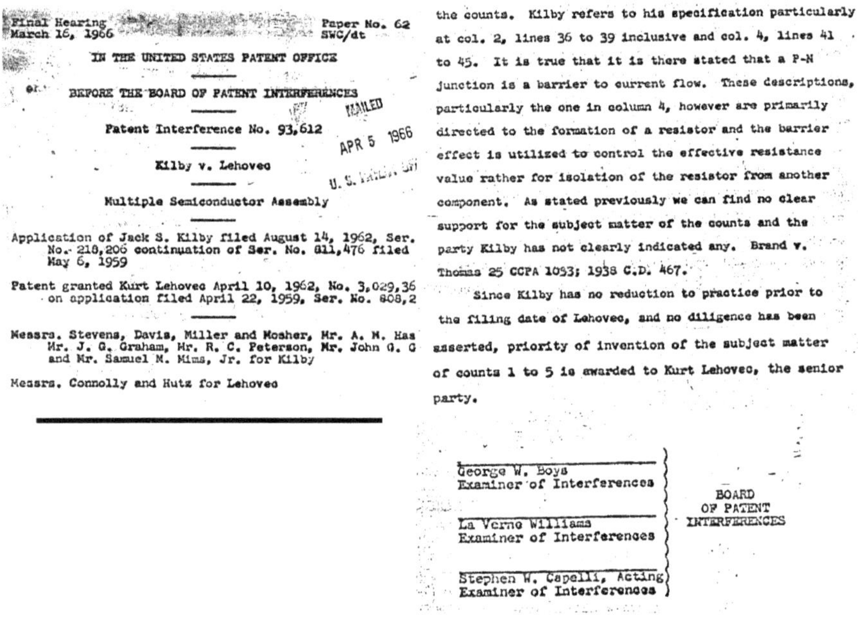
Final Hearing
March 16, 1966

Paper No. 62
SWC/dt

IN THE UNITED STATES PATENT OFFICE

BEFORE THE BOARD OF PATENT INTERFERENCES

MAILED
APR 5 1966
U. S. PATENT OFFICE

Patent Interference No. 93,612

Kilby v. Lehovec

Multiple Semiconductor Assembly

Application of Jack S. Kilby filed August 14, 1962, Ser. No. 218,206 continuation of Ser. No. 811,476 filed May 6, 1959

Patent granted Kurt Lehovec April 10, 1962, No. 3,029,36 on application filed April 22, 1959, Ser. No. 808,2

Messrs. Stevens, Davis, Miller and Mosher, Mr. A. M. Has Mr. J. G. Graham, Mr. R. C. Peterson, Mr. John G. G and Mr. Samuel M. Mims, Jr. for Kilby

Messrs. Connolly and Hutz for Lehovec

the counts. Kilby refers to his specification particularly at col. 2, lines 36 to 39 inclusive and col. 4, lines 41 to 45. It is true that it is there stated that a P-N junction is a barrier to current flow. These descriptions, particularly the one in column 4, however are primarily directed to the formation of a resistor and the barrier effect is utilized to control the effective resistance value rather for isolation of the resistor from another component. As stated previously we can find no clear support for the subject matter of the counts and the party Kilby has not clearly indicated any. Brand v. Thomas 25 CCPA 1093; 1938 C.D. 467.

Since Kilby has no reduction to practice prior to the filing date of Lehovec, and no diligence has been asserted, priority of invention of the subject matter of counts 1 to 5 is awarded to Kurt Lehovec, the senior party.

George W. Boys
Examiner of Interferences

La Verne Williams
Examiner of Interferences

Stephen W. Capelli, Acting
Examiner of Interferences

BOARD OF PATENT INTERFERENCES

Fig. 6.22. Parts of the Transcript from Kilby vs. Lehovec, final U.S. Patent Office hearing, March 16, 1966

R. C. Sprague asked Hillary Sweeney to remove and destroy his previous internal memo attached to the Lehovec files, and Kurt was paid $1.00 for his invention. Several months later he received a letter from R. C. Sprague stating *"your patent has turned out to be a valuable piece of property"* (Fig. 6.23).

Lehovec was never reimbursed for expenses related to writing the patent application, the postage fee to mail it to the Patent Office, nor for his trip to Dallas.

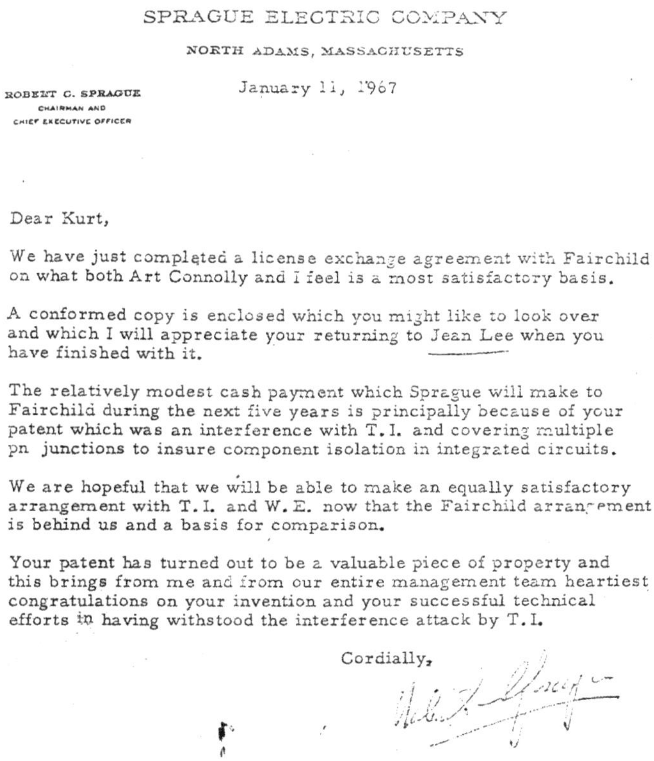

SPRAGUE ELECTRIC COMPANY

NORTH ADAMS, MASSACHUSETTS

ROBERT C. SPRAGUE
CHAIRMAN AND
CHIEF EXECUTIVE OFFICER

January 11, 1967

Dear Kurt,

We have just completed a license exchange agreement with Fairchild on what both Art Connolly and I feel is a most satisfactory basis.

A conformed copy is enclosed which you might like to look over and which I will appreciate your returning to Jean Lee when you have finished with it.

The relatively modest cash payment which Sprague will make to Fairchild during the next five years is principally because of your patent which was an interference with T.I. and covering multiple pn junctions to insure component isolation in integrated circuits.

We are hopeful that we will be able to make an equally satisfactory arrangement with T.I. and W.E. now that the Fairchild arrangement is behind us and a basis for comparison.

Your patent has turned out to be a valuable piece of property and this brings from me and from our entire management team heartiest congratulations on your invention and your successful technical efforts in having withstood the interference attack by T.I.

Cordially,

Fig. 6.23. R. C. Sprague letter to K. Lehovec issued shortly after the interference hearing "Kilby vs. Lehovec"

The Sprague Electric R&D group successfully completed the design of their first integrated circuit during 1961 and 1962. The technology, after further modification, was transferred into the Semiconductor Division of Sprague Electric Company in Concord, NH where the UNICIRCUIT series was manufactured. Sprague Company was, in the early sixties, in the lead of ion-implantation technology, thanks to engineers J. D. Macdougall, K. E. Manchester, Hans Schier, P. W. Anderson, P. E. Roughan, and Leo Fedotovsky. Their implanter was constructed semi-illegally in a hallway under constant smoke contamination. Hans was a chain smoker, who could not smoke only one cigarette at the time; he needed to smoke two cigarettes at the same time. Because it was difficult to smoke two cigarettes and enter notes simultaneously into a research notebook or assemble some part of equipment, Dr.

Schier constructed a special quartz cigaretteholder. The holder was basically quartz tubing which split a mouth-piece into two parts, each of them could hold one cigarette. The implanter was later "discovered" by MOSTEK (who was using the Sprague Fab to manufacture MOSTEK integrated circuits) to adjust the threshold voltage of MOS transistors.

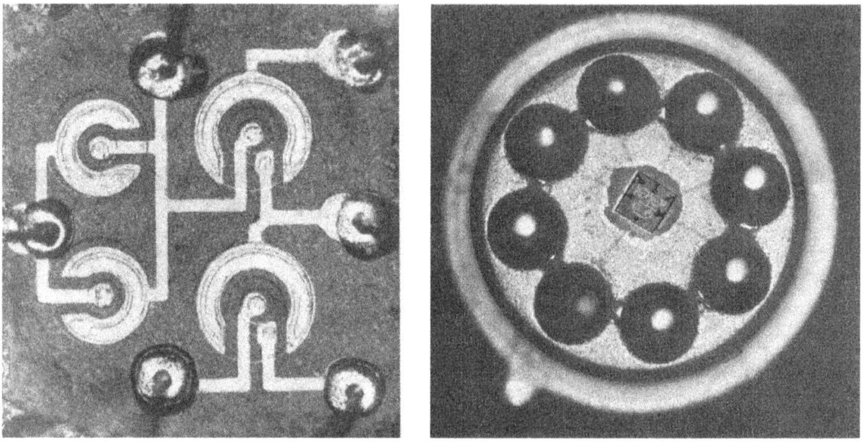

Fig. 6.24. The first Sprague Integrated circuit (1962)

Robert Sprague's son, John Sprague, after graduating from Stanford, did not want to return to the Sprague Company. R. C. Sprague asked Kurt to fly to California and bring John back. Kurt assumed that John would easily recognize Lazier's deficiencies and that R. C. Sprague would put John in charge of the R&D group. R. C. Sprague instead ordered John to report to Kurt Lehovec. John Sprague wanted to hire Prof. von Hippel of MIT as a consultant and asked Kurt for his opinion. Kurt was a friend of von Hippel, but he hesitated. This was another wrong decision, which Kurt regrets today. In the meantime, political W. Lazier hired Dr. Burns of Bell Labs as a consultant, and the doom of Sprague Semiconductor business was only a matter of time.

Sprague Electric licensed Fairchild Semiconductor to use the Lehovec patent in exchange for the use of Hoerni and Noyce's patents, and they were well positioned to be a major player in the semiconductor business. A similar agreement between Sprague and TI was signed by Robert Sprague and Mark Shepard in August 1967.

The creative engineer, himself, however, was not able to overcome poor company management and the incompetence of W. Lazier. Sprague Electric Company failed, and today, nobody even remembers, that Sprague Electric was in the semiconductor business.

In 1966, Kurt saw Francois Truffat movie "Four Hundred Blows." In the movie, an outstanding but rebellious young athlete, representing the reform school, beat easily and repeatedly athletes from other prestigious and private schools. In his last race, he led by a large margin and stopped just before crossing the finish line to let others win. The movie pre-occupied Kurt's mind for a very long time and resulted in his abrupt separation from Sprague Electric Company. A top-notch skier, Kurt Lehovec moved temporarily to Austria and when the Alps and skiing saturated his mind, he moved to Los Angeles and retired.

Lehovec's inventions and works contributed to advances in CV techniques, solar cells, solid-state batteries, and LED diodes. His theoretical work "Space Charge Layer and Distribution of Lattice Defects at the Surface of Ionic Crystals" [7], is still referenced, fifty years after its publication. In the last part of his life, Lehovec became a poet and a play writer. He published a half dozen books of poetry adoring nature and the beauty of women. He remains a loner and enjoys lone walks during the night, putting no emphasis on his previous career as a physicist.

Sylvania Electric Products, Inc.

Thomas A. Edison invented the modern incandescent light bulb in 1879. Use of the light bulb spread, spurred on by achievements in electrical engineering led by Germany's Werner von Siemens, whose company perfected the light bulb in 1880.

Fig. 6.25. Frank Poor facility renewing burned-out light bulbs (1902)

Sylvania traces its roots back to 1901, when a young entrepreneur Frank Poor became a partner in a small company in Middleton, Massachusetts, that renewed burned-out light bulbs. The company would buy an old bulb for a few cents, cut off the glass tip, replace the filament, and reseal the bulb. Poor soon moved the business to Danvers, bought out his partner, and called his new company the Bay State Lamp Company. His brothers soon joined him in the enterprise. The Poor brothers started the Hygrade Incandescent Lamp Company in 1909 to sell new carbon-filament light bulbs, and by 1911, tungsten-filament light bulbs. The Bay State refilling operation was soon discontinued. In 1916, Hygrade opened a new plant headquartered in Salem, Massachusetts, which could turn out 16,000 lamps a day. After several transitions, the company was finally purchased in 1922 by entrepreneur Bernard Erskine and two associates, who founded the Nilco Lamp Works. In 1924, Nilco formed the Sylvania Products Company to make receiving tubes for another new invention that would change the world: the radio.

William Pietenpol, former director of Semiconductor Device Development at Bell Labs, joined Sylvania in 1958 to establish a transistor facility at Sylvania Electric Products, where diode manufacturing capability already existed. The R&D Division was directed by Dr. J. Earl Thomas, another BTL alumnus. Device Technology was headed by Dr. John O. Percival who was pivotal in developing Sylvania's epitaxially-grown silicon mesa transistor. In 1959 Sylvania merged with General Telephone & Electronics (GT&E) and become a subsidiary of GT&E.

Thomas Anthony Longo started his career at General Telephone Labs in 1958. Tom Longo received the best education in semiconductor physics available at the time. He received the B.S. in Chemistry and Physics, the M.S. degree in Physics, and as a member of Professor Karl Lark-Horovitz's group he graduated with a Ph.D. degree in Physics from Purdue University in 1957. His group contained such heavyweights as Professor H. Y. Fan and students W. G. Spitzer and A. K. Ramdas. Longo was going along with "difficult" Lark-Horowitz very easily and fully used the academic freedom he had. Longo had a choice to work on Ge or Si; he choose silicon and used radiation as a tool to study the physical properties of Si. The title of his Ph.D. thesis was "Nucleon Irradiation of Silicon Semiconductors." When Professor Lark-Horowitz died, Longo left Purdue and joined General Telephone Laboratories in North Lake, IL., as Head of the Research Semiconductor Laboratory.
Shortly after GT&E merged with Sylvania Electric Products in Woburn, MA., Longo become the divisional director of R&D and managed the Advanced Devices Development. Sylvania Electric Products was one of the original companies licensing transistor technology from BTL and producing alloy transistors. By 1959 the transistor technology was well evolved and Tom Longo's group developed the first Sylvania mesa double-diffused transistor.

In the early days of integrated circuits, process development was not driven by device requirements. The circuit designer had to use devices with parameters given by available processes at the time. Robert H. Norman joined Fairchild in 1959; he later became a section head for device evaluation. He developed the industry's first monolithic resistor-transistor logic (RTL Mi-

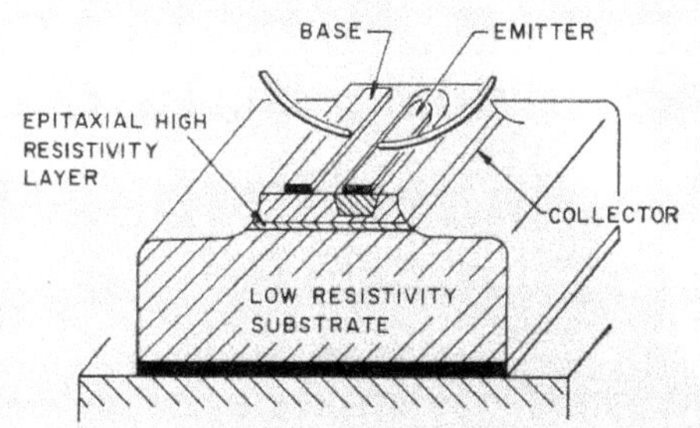

EPITAXIAL DIFFUSED TRANSISTOR CONSTRUCTION

Sylvania epitaxial transistor is fabricated by diffusing a base region into a thin, high resistivity layer. This thin layer is vapor grown in epitaxial fashion onto a low resistivity collector substrate. The single crystal structure of the substrate is continued into the thin film.

PNP EPITAXIAL DIFFUSED-BASE GERMANIUM MESA TRANSISTORS

ABSOLUTE MAXIMUM RATINGS: (at 25°C)	SYL-2300	SYL-2301	UNIT
Collector to Base Voltage	−15	−12	V
Collector to Emitter Voltage	−15	−12	V
Emitter to Base Voltage	−2.5	−1.0	V
Collector Current	−50	−50	mA
Power Dissipation (free air)	150	150	mW
Power Dissipation (case at 25°C)	300	300	mW
Storage Temperature	−65 to +100	−65 to +100	°C
Junction Temperature	+100	+100	°C

ELECTRICAL CHARACTERISTICS AT 25°C:

		SYL-2300		SYL-2301		UNIT
Symbol	Conditions	Min.	Max.	Min.	Max.	
BV_{CBO}	$I_C = -100\ \mu A,\ I_E = 0$	−15		−12		V
BV_{EBO}	$I_E = -100\ \mu A,\ I_C = 0$	−2.5		−1.0		V
BV_{CES}	$I_C = -100\ \mu A,\ V_{BE} = 0$	−15		−12		V
h_{FE}	$I_C = -10$ mA, $V_{CE} = -0.22$ V	25				—
	$I_C = -10$ mA, $V_{CE} = -0.25$ V			20		—
V_{BE}	$I_C = -10$ mA, $I_B = -0.4$ mA	−0.34	−0.44	−0.34	−0.50	V
I_{CBO}	$V_{CB} = -5$ V, $I_E = 0$		−3.0		−3.0	μA
V_{CE} (Sat.)	$I_C = -10$ mA, $I_B = -0.4$ mA		−0.22			V
	$I_C = -10$ mA, $I_B = -0.5$ mA				−0.25	V
	$I_C = -50$ mA, $I_B = -2.5$ mA		−0.25		−0.35	V
$t_d + t_r$	$V_{BE(p)} = 0.5$ V, $I_{B(1)} = -1$ mA		70		90	mμsec
t_s	$V_{CC} = -3.5$ V, $R_C = 300$ ohms		45		60	mμsec
t_f	$I_{B(2)} = 0.25$ mA		70		95	mμsec

Fig. 6.26. Sylvania's epitaxially-grown silicon mesa transistor (1959)

crologic) [4]. The reasoning for this choice was simple: Fairchild's transistors at that time were slow, large and with big variations of electrical parameters. The resistor was considered a simple device.

When Orville Baker, a former Fairchild employee, moved to Signetics as employee number 5, he developed the DTL series SE100. David F. Allison (former head of the Device Development section at Fairchild), Mark Weisenstein, Lionel Kattner, and David James wore badges numbered one through four. The Signetics DTL series featured, for the first time, a metal-over-oxide capacitor, with a propagation delay of approximately 25 nsec with a power consumption of 6 mW per gate. The process was an improved version of Fairchild's planar process with only "stripe-type" transistors. Although the SE100 series was a significant improvement compared to TI's RTL and Fairchild's RTL Micrologic, the circuitry still needed large resistors.

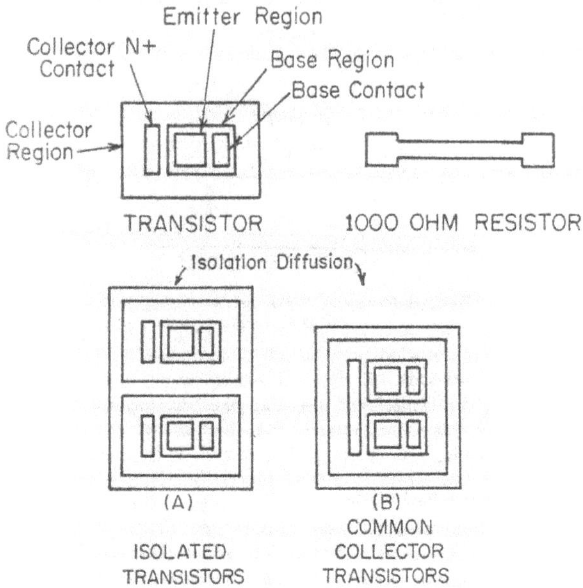

Fig. 6.27. Comparative sizes (in scale) of a transistor and a 1000 Ω resistor, and two transistors with individual and with shared isolation

Tom Longo's motivation for high speed device performance originated from his previous experience. Tom Longo served in the U.S. Navy from 1944 to 1950 and joined Navy Reserve in 1950. During his duty he became connected with the National Security Agency which sponsored the development of high speed computers in a project called "Lightning." There he met Seymour Cray[3] and learned about requirements for components used in computer architecture.

At that time James Buie at Pacific Semiconductor, Inc. (later TRW Electronic Components Group) developed logic constructed from discrete devices that became later known as Transistor-Transistor Logic, or TTL. When Buie's idea became known, Fairchild's Ruegg and Beeson published an application paper describing the TTL gate constructed from discrete Fairchild transistors. Tom Longo advanced Buie's and Fairchild's ideas and developed a logic concept which became the most popular logic family product for more than three decades. Tom Longo was the first to recognize that the old myth, that the resistor is the cheapest electronic component, was no longer true in monolithic circuits. But more importantly, Tom Longo departed from the design concept used at that time. Fairchild, TI, and others designed their integrated circuits around the same transistors which were similar to discrete parts. Longo for the first time designed a circuit where transistor requirements

[3] Seymour Cray co-founded Control Data Corporation in 1957, where he built the first computer using transistors instead of vacuum tubes.

were determined by their function in the circuitry. Multiemitter input transistors, load transistors, inverters, level shifters, Darlington pair and cascade – totem pole, were all sized differently. In 1961, Sylvania, with help from GCA, installed the first advanced stepper into production and introduced masking technology with an unprecedented registration. Sylvania's advanced epitaxial process and photolithography techniques resulted in a total gate area no bigger than the area of a typical Fairchild transistor. For example, the dual NAND TTL gate had a die size of 355 × 355 μm, the emitter diameter of the popular Fairchild's 696 transistor was over 700 μm.

Tom Longo showed his first low-level TTL circuit at the 1962 WESCON in Los Angeles. The biggest advantage was device speed; the noise margin was poor, and the circuit could not drive any capacitance load. There was no significant excitement about the speed, but people did want the TTL flexibility. Tom and his colleagues R. Bohn, R. Sirrine, and I. Feinberg came up with an improved design which emerged as Sylvania's SUHL in late 1962.

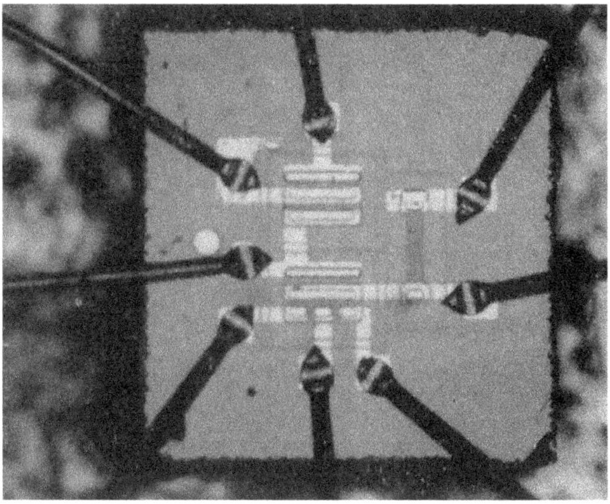

Fig. 6.28. Sylvania's NAND gate with less than 10-nanosecond propagation delay

The first four logic blocks (single/dual NAND, OR, and RS Flip-Flop) were announced by Sylvania's general marketing manager Ernest H. Ulm in early 1963. The circuits were based on an advanced form of a silicon epitaxial planar process with unprecedented density – 500 circuits on a 1-inch diameter wafer. Typical propagation delays were 8 nsec with less than 10 mW of dissipation per stage. The units were priced at $200.00 in quantities of one to nine.

There is a very interesting story which clearly demonstrate the differences how business was conducted forty years ago. Thermocompression bonds with fine gold wires had been the industry standard for attaching connecting leads between the metallized aluminum at the transistor die and the gold plated Kovar post of package leads. The problem was that at elevated temperatures aluminum-gold forms an intermetallic compound which may significantly deprede the device reliability. This problem was known as the so-called Purple Plague. Tom Longo and B. Selikson eliminated this problem by the use of aluminum wire bonding. They published the results of their findings and the whole industry freely applied this method. There was no law suit, no patent infringement litigation – something difficult to imagine today.

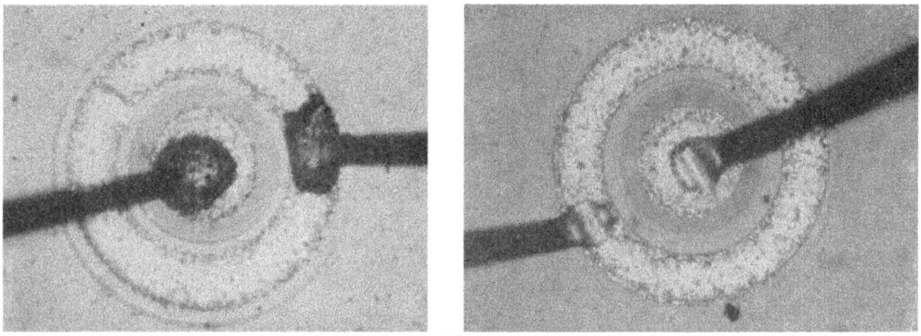

a) b)

Fig. 6.29. The transistor after stress at 300°C for 24 hour bonded **a**) with gold wire (Purple Plague) and similar transistor bonded with aluminum wire **b**)

The Sylvania logic products were first introduced to Litton Guidance and Control in Woodland Hills, CA as Sylvania Ultra High-Level logic circuit (SUHL) in January 1963. Litton was skeptical, but thought there might be a place for Longo's baby in the Phoenix Missile which was built by Hughes Aircraft. Litton wanted to see the circuit diagram before they would make a decision. The Capricorn-born Tom Longo refused, and gave an ultimatum to Litton: Purchase Order at $300.00 a part first; circuit diagram second. Litton turned around and requested bids from the whole industry. They received 7 bids and made awards to Fairchild, Texas Instruments, Transitron and Tom's Sylvania. Once the Purchase Order was received, Tom Longo released the circuit diagram.

In November of the same year, Tom Longo presented a paper at the IRE group meeting (PGED – Professional Group on Electron Devices) in Washington. It was from that paper that Texas Instruments copied this device and introduced two years later the Texas Instruments TTL series 5400 with only a single difference – different pin out.

Actually, Tom Longo visited TI after he presented the paper. TI's Howard Moss who was in charge of the 5400 series development said to Tom: *"we are not too far behind you."* However, TI added to Tom's circuit a very important feature: TI developed the new plastic DIP package. This was the reason why the very-lowpriced logic series became the most popular family of logic circuits.

Fig. 6.30. Thomas A. Longo, the semiconductor memory pioneer. (1962)

The case of Sylvania TTL is a typical example of advances in integrated circuits. Except for Jean Hoerni's idea of the planar transistor, which stood out, most of the achievements of monolithic integrated circuits are the results of contributions of many participants: James Buie sparked the idea; Thomas Longo made it functional and manufacturable; and Texas Instruments lowered the price to make the product favorable over the existing devices on the market.

By early 1965, Sylvania had developed the largest selection of compatible digital circuits. The Sylvania SUHL, unlike any other logic series at that time, offered the best speed-power product for saturated logic and provided the highest level of noise immunity.

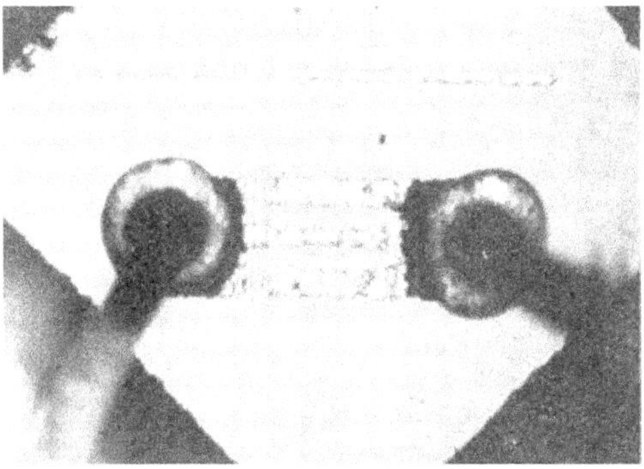

Fig. 6.31. Sylvania planar epitaxial transistor 2N2784 – the fastest device in 1963 with $fT = 1$ GHz@5 mA

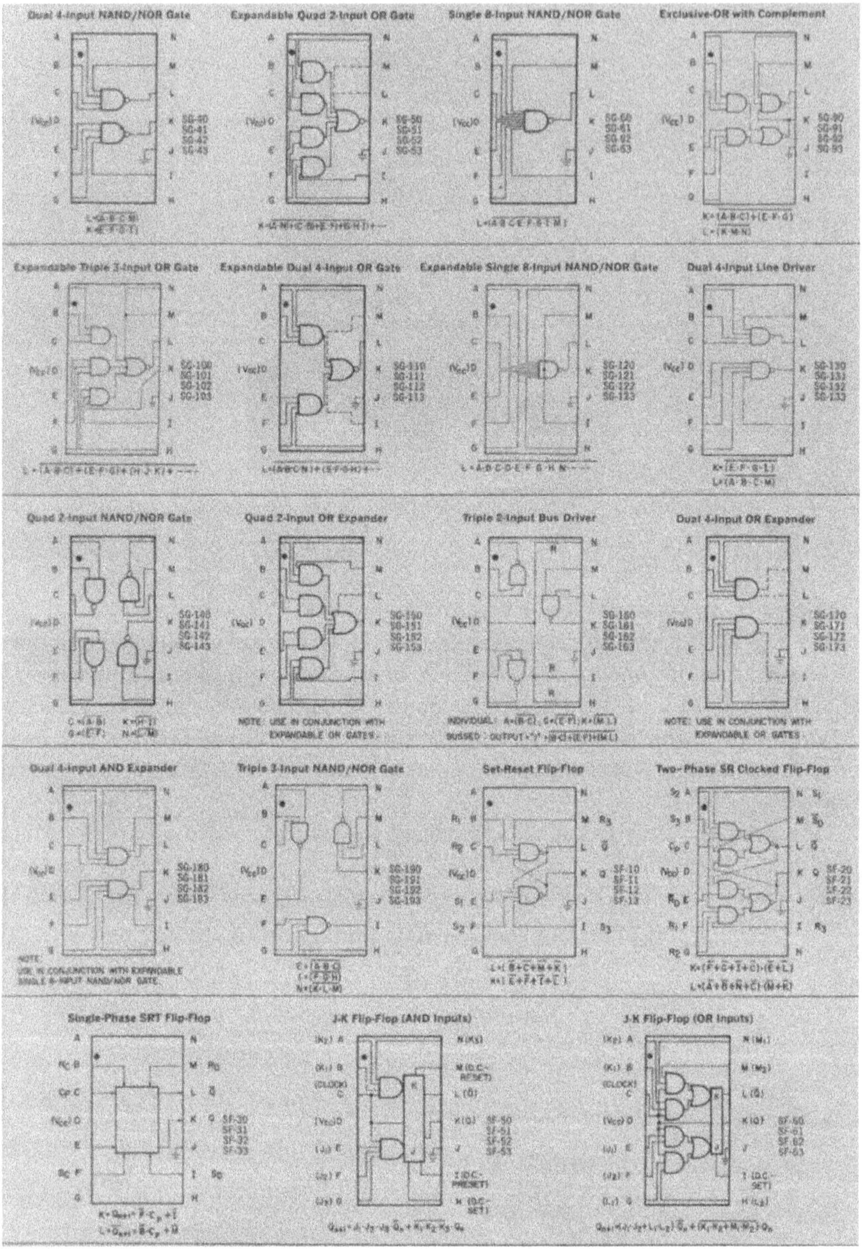

Fig. 6.32. Sylvania SUHL TTL series (May 1965)

The Sylvania integrated circuit masks were designed with registration tolerances of 0.06 μm and with minimum features of 2.5 μm. The small junction depth with low capacitance resulted in an improved frequency and switching performance with a high beta at microamperes current, and reduced falloff beyond 10 mA. Sylvania quickly introduced a family of discrete planar transistor 2N2784 (Fig. 6.31), 2N709, 2N709A and 2N2475 and Fairchild's dominance in planar transistors was for the first time seriously challenged. In comparison with Fairchild circuits, the major difference was that the Sylvania transistors were small, with very tight dimensional control.

In 1965 Tom started work on a 16-bit memory array (Fig. 6.34.) This work was "transferred" in February 1966 to Transitron where Tom Longo moved to start an IC business. Longo became the first director of ICs and transistors, and later general manager of the company.

The first Transitron16-bit 10 MHz bipolar TTL RAM consists of 16 two-transistor flip-flops arranged in a 4 × 4 matrix. Two write- and two sense-amplifiers were also built into the element. The delay time between addressing and read or write was 20 nsec. The memory cell would operate from a nominal 5 V power supply.

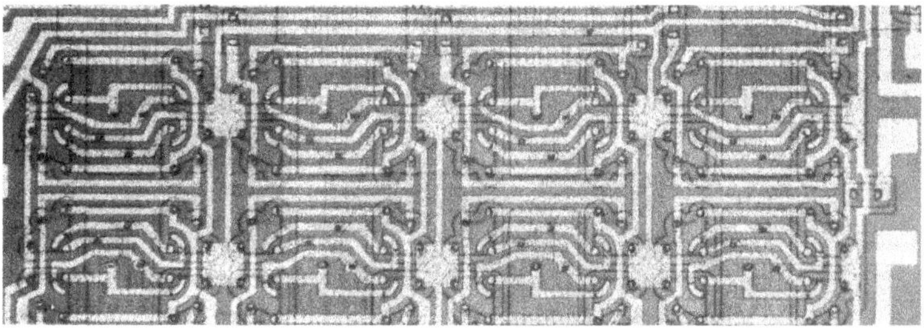

Fig. 6.33. Sylvania 16-bit memory array SM-80, introduced in September 1966

In January 1970, Lester Hogan, General manager of Fairchild Semiconductor, realized that the development of advanced technology was beyond the ability of Gene Blanchette and hired Tom Longo. Longo became Vice-President in charge of development of bipolar products and reported to Joseph Van Poppelen.

Fairchild did not have any reasonable or competitive memory products before 1968 The only memory device available was an 8 bit Micrologic element with a cost of $2 per memory bit (Fig. 6.35) and a 256-bit static RAM 4100 designed by H. T. Chua. Both part were based on the original Planar II process, and except for application of Chua's memory in an experimental computer at the University of Illinois, both parts did create any significant sale.

Under Longo's leadership, Fairchild introduced the new Isoplanar process. Isoplanar process was developed by Dougles L. Peltzer and allowed significant

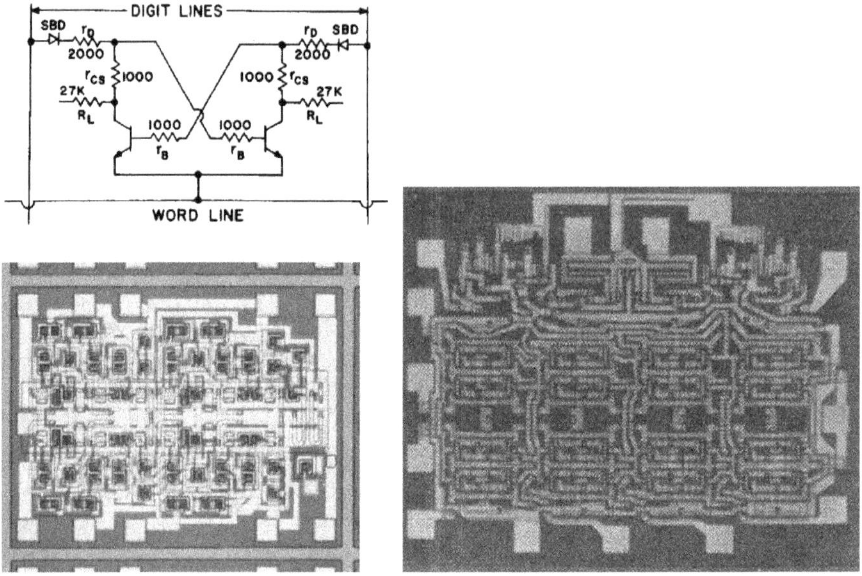

Fig. 6.34. Prototype of 2×2 bit Transitron TTL memory (a) and Transitron TMC 3162 16-bit RAM, introduced in July1966 (b)

scaling of bipolar transistors. Bill Herndon designed the the industry's first static 256-bit RAM (Fig. 6.36.) Fairchild introduced the μL 93410 at the WESCON Trade Show in Los Angeles in August 1971.

At the end of a semiconductor industry recession during 1970 and 1971, Longo's group introduced a new product line including low power Schottky logic (the 9600 series) and later ECL logic. The group started work on a 1k static RAM, which Longo offered to his old friend Seymour Cray who was developing the Cray-1 computer. Fairchild's 1k (and later 4k) static RAM was used in the Cray-1 as a main memory. The cost per bit dropped to 17 cents and other main-frame computer manufacturers such as Univac, CDC, Burroughs, Fujitsu, and Siemens also switched from core memory to semiconductor memory.

Although Longo created for Fairchild a very profitable product line, he recognized, limitations of bipolar memories and initiated the development of Fairchild's CMOS memories. The timing, however, was not right. During the recession in 1976, when Wilf Corrigan eliminated highly profitable business lines and replaced them with consumer electronics (watch ICs and video game ICs.) Fairchild's business was going down hill and got far worse in 1978 with tremendous problems. Fairchild was weakened dramatically and the company became a target for takeover. United Technology was considering purchasing Fairchild; however when they learned about Fairchild's situation they backed off from their plan.

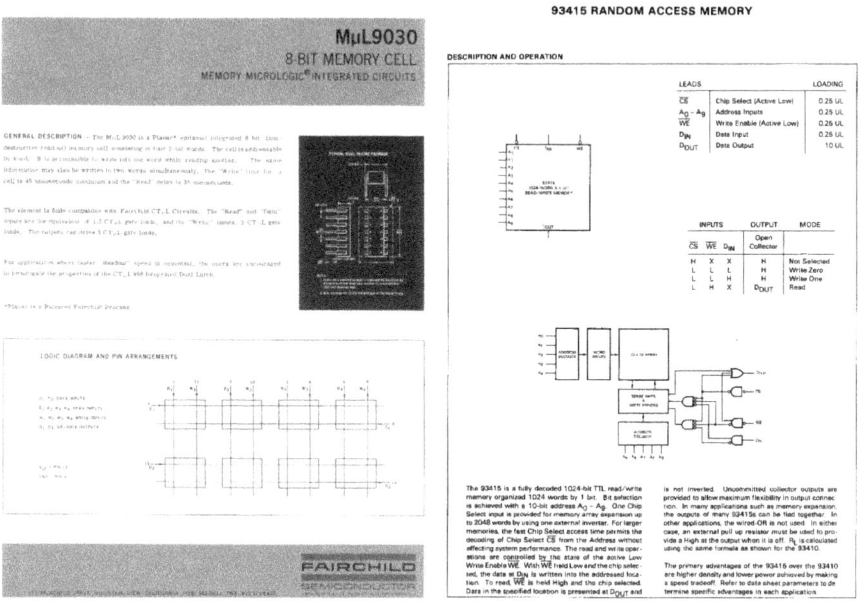

MμL9030
8-BIT MEMORY CELL
MEMORY MICROLOGIC® INTEGRATED CIRCUITS

FAIRCHILD
SEMICONDUCTOR

93415 RANDOM ACCESS MEMORY

DESCRIPTION AND OPERATION

LEADS		LOADING
$\overline{CS}$	Chip Select (Active Low)	0.25 UL
A_0 - A_9	Address Inputs	0.25 UL
$\overline{WE}$	Write Enable (Active Low)	0.25 UL
D_{IN}	Data Input	0.25 UL
D_{OUT}	Data Output	10 UL

INPUTS			OUTPUT	MODE
$\overline{CS}$	$\overline{WE}$	D_{IN}	Open Collector	
H	X	X	H	Not Selected
L	L	L	H	Write Zero
L	L	H	H	Write One
L	H	X	D_{OUT}	Read

The 93415 is a fully decoded 1024-bit TTL read/write memory organized 1024 words by 1 bit. Bit selection is achieved with a 10-bit address A_0 - A_9. One Chip Select input is provided for memory array expansion up to 2048 words by using one external inverter. For larger memories, the fast Chip Select access time permits the decoding of Chip Select $\overline{CS}$ from the Address without affecting system performance. The read and write operations are controlled by the state of the active Low Write Enable $\overline{WE}$. With $\overline{WE}$ held Low and the chip selected, the data at D_{IN} is written into the addressed location. To read, $\overline{WE}$ is held High and the chip selected. Data in the specified location is presented at D_{OUT} and is not inverted. Uncommitted collector outputs are provided to allow maximum flexibility in output connection. In many applications such as memory expansion, the outputs of many 93415s can be tied together. In other applications, the wired-OR is not used. In either case, an external pull up resistor must be used to provide a High at the output when it is off. R_L is calculated using the same formula as shown for the 93410.

The primary advantages of the 93415 over the 93410 are higher density and lower power achieved by making a speed tradeoff. Refer to data sheet parameters to determine specific advantages in each application.

Fig. 6.35. Fairchild's 8 bit memory cell M?L9030 (1966) and preliminary data sheet for 1 k static RAM 93415 (1969)

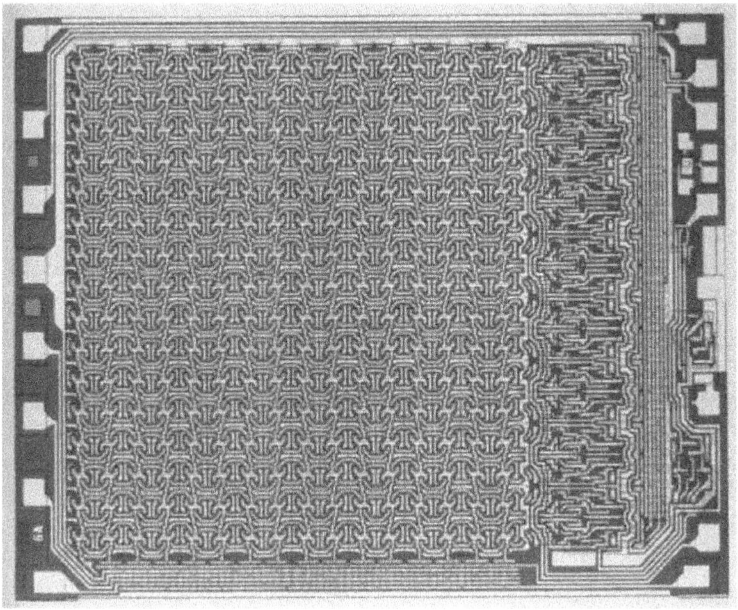

Fig. 6.36. Fairchild's 256-bit static RAM 93410 (2650 × 3400 μm)

Tom Longo left Fairchild in 1984 with a small group of engineers, and established Performance Semiconductor Corporation, the fast CMOS microprocessor and static memory manufacturer. The business was mostly oriented to military components and the most successful part was, at that time the world's fastest 16 bit 40 MHz microprocessor 1750. In the late '70s the military saw a need for standardized computing and came up with the standard USAF Mil-Std-1750 for a 16 bit airborne processor. The 1750 was used in airborn computers and weapons systems with the software which could be transferred to other applications using ADA language which was the standard programming language for the U.S. Department of Defense. The 1750 was mainly produced in radiation resistant forms for military applications.

Fig. 6.37. The P1750, in 1986 the world's fastest 40 MHz CMOS 16 bit microprocessor – Performance Semiconductor Corporation

Performance Semiconductor Corporation was a revolutionary company in Silicon Valley in the midle of the nineteen eighties. They had the first submicron (0.8 µm) CMOS process. Intel, Mostek and other were mainly running N-MOS based product lines. The others, like Motorola and TI, invested heavily into BiCMOS. Tom Longo was one of the few who argued that BiCMOS will almost double the mask count and the bipolar part will not be scalable. *"It is a waste of time"* was Tom's opinion. I remember sitting in a hotel lobby in Baltimore with Tom in the spring of 1986. At that time Tony Alvarez of Cypress and others were traveling across country with a presentations showing a slide "BiCMOS = Bye Bye CMOS." Dr. Thomas A. Longo told me "*The CMOS is all that you need.*" Unfortunately I did not promote Tom's vision enough. History may hold up with the "Moore's Law" also the "Longo's Law."

Performance Semiconductor grew up to a $200 million/year business. The problems came out with cuts in military projects after the fall of the Soviet Union. Longo sold the logic business to T. J. Rodgers' Cypress Semiconductors, and with declining government orders, the company focused only on the military products was not able to sufficiently support R&D, and closed in 2002.

Advanced Micro Devices, Inc.

When Jack Gifford become head of computer marketing at Fairchild in January 1969, it was clear that he would not stay in this position for long. Gifford did not waste any time, and inspired by Talbert and Widlar was trying to start an analog company before Hogan's Heroes took over Fairchild. His team included Jim Giles, Lawrence Stenger, and Frank Botte of Fairchild.

Jack set up a business plan and wanted to focus on analog circuits. Jack and his friend, Bruce Waterfall, who was a staff assistant of Charlie Sporck at National, and who had connections to New York's bankers, contacted about fifteen people and asked for the money with absolutely no success. Bruce recognized that because Jack was young, twenty eight years old and an unknown person, there was almost no chance to raise the capital. He told Jack to "*find somebody older with more experience than you have.*"

Fig. 6.38. Jack (John F.) Gifford in 1969

Fig. 6.39. W. Jerry Sanders, III in 1970

Almost at the same time Lester Hogan "promoted" Jerry Sanders from Director of Marketing to "assistant vice-president for marketing development" at Fairchild. Sanders, unhappy with the "promotion", quit Fairchild and for a while was dreaming and planning at Malibu Beach his entrance into the music recording business in Hollywood.

Jack was trying to engage Jerry for two days to start a new company. Nothing was working and Jerry Sanders' answer was no. Next morning Jack returned one more time and said *"Jerry, listen, I know you want to go into this Hollywood stuff, but I promise you we can get twenty million in sales in two years, we can sell the company and you can take the money to go back into the recording business."*

Walter Jeremiah (Jerry) Sanders, who always said "*I want to live well,*" looked at Jack and said "*but I get to be the president, and I want to have a digital company.*" Even with Jerry Sanders on board, to raising money was no easy job. Life, however, acts unpredictably and frequently in the most grotesque way. Jim Martin, who Jerry Sanders had fired at Fairchild, joined The Capital Group and got into fund management. Jim Martin knew who Sanders was, but he was also a very good friend of Jack's, and he convinced the Capital Group to give the money to Jerry Sanders. An additional $ 50,000 came from Jonathan B. Lovelace, Sr., an original venture capitalist for Walt Disney Productions.

In April 1969, Jerry Sanders surrounded himself with seven other ex-Fairchild employees, and announced a name he reserved with the California Commissioner of Corporation for a new company – Advanced Micro Devices. Sanders said that the company was expected to be operating in August 1969 pending incorporation and completion of its building. In the meantime the company was operating out of Jerry Sanders' home on Hillpark Lane in Sunnyvale.

AMD was officially incorporated on May 1, 1969. The company formed two groups: The Analog Operations group was headed by Lawrence Stenger. The director of engineering for Analog was James Giles. Process development belonged to Frank Botte. The second group, Complex Digital Functions, was headed by John Carey with Sven Simonsen as director of engineering. Jack Gifford became Director of Marketing and Business Development. Director of Sales and Administration was the former Fairchild sales manager for computer markets, Ed Turney.

In September 1969, AMD moved into a 15,000 square-foot facility at Thomson Place in Sunnyvale. The half-million-dollar building housed the entire company including 2-inch diameter wafer fabrication with 7 μm geometry features and 8-mask process.

By March 1970, AMD's workforce consisted of 53 employees, and was manufacturing 18 complex digital and linear integrated circuits. The first significant revenue was generated from the sale of the Am9300 4-bit shift register. Later the same year another successful device was in manufacturing – the Am2501 logic counter. The counter was the industry's first binary/hexadecimal up-down counter and became immediately a huge success. In 1971 Sven Simonsen's group designed another very successful product - the Am2505, the industry's fastest multiplier. The same year AMD entered the RAM business. At the same time the shipment of second-source linear products increased by 50% over the previous year. By the end of year 1971 sales rose to $4.6 million.

Fueled by the success of AMD's MSI parts, Jerry Sanders wanted to run a very different company than what was Jack Gifford's original plan and

Fig. 6.40. Founders of AMD (1969)

what Jack proposed to his original team. Jerry Sanders had also a different life philosophy than technical members of the founding team. Young Jerry Sanders said "*I was designing a regulated power supply around Motorola's components. So I thought I better find some more out about it. So I contacted Motorola and they sent over a sales engineer. I'll never forget that. He came over. He had beautiful clothing on. He was well groomed. He didn't know a damn thing about his product line, but he offered to take me to lunch. We went out to lunch, he drove in a new car, took me to a nice lunch, a better lunch than I certainly could afford as an Engineer at Douglas Aircraft Company. And I decided then and there that I had been in the wrong end of Engineering.*"

The differences escalated quite rapidly and, except for Ed Turney, all of AMD's co-founders were unhappy with Jerry Sanders. Jerry Sanders felt that Jack Gifford began a campaign to undermine Sanders, to get the guys to throw him out, since now Gifford's group had successfully gotten the money and they wanted Jack to be the president. Just a few months after AMD's incorporation, during a Saturday morning, Jerry came to Jack's office; they proceeded to get into an argument, and got very angry. Jack grabbed Jerry and literally threw him in a chair and said, "*Listen, you son of bitch, a month ago everybody in this company came to me to get rid of you. If you do not change your ways you are going to be out of here.*" Jerry Sanders was petrified, but not for long. He went to the Board and told them that Jack did not want to help him and wanted his job. Jack Gifford was escorted from the company the next week. Sanders had a free hand to modify the original plan for analog and digital to a digital-only company. Disappointed, Jack redirected his talents to farming and become the owner and manager of the largest California tomato farm in Yuba City.

Fig. 6.41. AMD Advertisement after passing Intel in the 4k SRAM

Despite a poor showing due to the industry downturn, AMD rose to almost 1500 employees, by the spring of 1974. AMD invested in a new 3-inch diameter facility in Sunnyvale with a new 10-mask process.

In 1975, new static- and dynamic-memory products (built with the N-channel MOS process developed by Jim Downey's group) for the first time surpassed Intel. The same year Western Electric placed the largest single sales order in the history of AMD, a \$2.5 million order for the 4k DRAM Am9050. AMD at that time grew faster than any other company. Sanders, who had kept a relatively low profile during the start-up of the company, returned to his flamboyant lifestyle he kept at Fairchild, and announced the program "Run for the Sun." "Run for the Sun" was a sales incentive program to increase annual sales to \$93 million, one dollar for every mile from the Earth to the Sun. AMD fell short, but short by only \$700,000 dollars!

In 1976 AMD signed a cross-licensing agreement with Intel. Things were working well only to the point when AMD parts became better than Intel.

For many years AMD had a reputation that they were making Intel parts better. When AMD released the Am386 which was faster and more efficient than Intel's 386, Intel responded by filing a law suit in the U.S. District Court in San Jose, alleging copyright infringement. The Am386 was reverse engineered by the Ben Oliver group in a new 6-inch diameter Fab in Austin, TX.

After eighteen months of development the first lot with 0.8 μm CMOS process technology was fabricated during just eight days. It worked!

INTERSIL, Inc.

After Dr. Jean Hoerni left Amelco he started work as a consultant for Union Carbide. He knew John H. Hall from Honeywell and he asked him if he would be interested in setting up a Union Carbide operation. John H. Hall graduated from the University of Cincinnati and he started his career with a brilliant design of an on-board aircraft computer for the YF-11 Blackbird reconnaissance plane.

Hall served as Union Carbide's director of IC Development during 1962 to 1967. Here John also started process development. Hoerni hired Bruce Kerr who worked with him at Fairchild. Jean and Bruce built the first Union Carbide Fab. John's group developed a unique thin-film technology for on-chip resistors, and the first dielectric-isolation technology[4].

At the beginning of 1966, Hall and Hoerni made a proposal to the company to develop Medium Scale Integration-TTL, asking for $10 million. The company decided to fund the project. Unfortunately, in management was John Dilinger, marketing manager, and one of his friends from Fairchild who decided to build a new company in San Diego because the land was cheap there. Union Carbide decided to move to San Diego, but Jean and John did not want to move there, nor most of the staff.

Robert Freund, General Manager of Union Carbide Electronics assumed that Hoerni left in November 1966 to attend the device show and then on to a vacation in Europe. Jean left for the device show, but also spent time on Wall Street trying to get backing for a new company. As a consultant, Hoerni gave his last advice to Union Carbide: with a very strong French accent he told to Mr. Freund *"semiconductors are not Union Carbide cup of tea."*

Hoerni, then 42 years old, started Intersil with $300,000 of his own money, but he was very quickly flooded with $4.5 million from well-known electronic-industry venture capitalists such as Arthur Rock, Frederick R. Adler, and Kenneth Thornhill, and from foreign companies: Italy's Olivetti, Omega Watch of Switzerland, and Portescap, a Swiss maker of watch movements.

[4] Union Carbide's dielectric isolation process was developed by Dr. Walt Benzing. When Hoerni and Hall left Union Carbide, Benzing left also and co-found Applied Materials in 1967.

Arthur Rock held 10% of the shares and was the largest outside shareholder; Olivetti held 5% and Swiss companies together held another 5%.

Hoerni recruited two super-stars from Union Carbide: John H. Hall and John D. Marshall, excellent device and process development engineers. Hall and Marshall were complemented by Dave Fullagar, the designer of the µA 741 op amp) who left the linear circuit group at Fairchild. Hoerni's others experimentalists were Garry Parker and Bruce Kerr. Garry was the guy who did most of the work for Hoerni, and, as usual for Hoerni, they worked mostly during nights.

The mercurial Hoerni was difficult to work with, and all of his engineers had to find a way to deal with him. Hoerni frequently would get angry and moody. John Hall had at one point a controversy with Hoerni; Hoerni never said to John if he appreciated his work or not. They argued frequently, and once John exploded and told to Jean *"Look, this is the last argument; you will change the way you are interfacing with me, otherwise I am leaving."* They never had an argument after that.

At the new venture, Hoerni did not talk much about his work at Fairchild; when he talked, he frequently gave credit to Moore and Noyce. However, he said publicly that Gordon Moore opposed his planar transistor. When Hoerni demonstrated the prototype, Gordon was still very much opposed to his idea. The only reason why the world has a planar transistor is because Howard Bobb and Don Rodgers, Fairchild's sales people, sold the planar transistor to Autonetics who placed a big order, and thereby overcame Moore's and Noyce's objections.

Hoerni's negative anger was well compensated by Hoerni's vision. Hoerni knew well that what he could read about the managing of creative people in management books, or learn at management schools, was largely irrelevant. Engineering is risky and being a good leader was about personal risk. Hoerni was not only truly creative; he was willing to take a risk well beyond where other people would take it, even if he had no proven idea how to do it. Hoerni's vision of the integrated circuit was well ahead of anybody in the world. Those who knew Jean more closely knew that he was also a shy person and he never pushed himself upfront. I have difficulty imagining Hoerni in hiking boots and mountain clothing with a backpack, trekking in the Himalayas, something he did every year.

Hoerni worked very often as Shockley did – he had the idea and somebody else would test and implement the new approach. Hoerni left for Japan, and contacted Drs. T. Ohmi and S. Torri of Seiko. Dr. Torri set up a meeting with engineers and Hoerni promised CMOS devices for watches. Jean said that he would develop a technology in a few months and for only $100,000. The Japanese thought of this as impossible and wanted to just forget it, but smart Torri said, for $100,000 there is very little to lose, and he gave Hoerni the money.

Intersil had about 300 employees and sales reached $2 million by mid 1969. Two million dollars was negligible in the 1969 semiconductor market. What was not negligible was Intersil's technology. Intersil developed the first N-MOS memory technology, low voltage metal-gate CMOS technology for the quartz clock and EPROM's, and they firmly established leadership in A/D/A converters.

Intersil's marketing manager at that time was Bob Dugan. Bob Dugan, with his baby blue eyes and innocent smile, who could lie absolutely consistently, was able to convince Ray Stata of Analog Devices to do business with Intersil. Analog Devices was selling mostly hybrids and they were looking to develop IC amplifiers and A/D/A in bipolar technology. Dave Fullagar designed in conjunction with Heinrich Krabbe of Analog Devices their first ten or so linear integrated circuits. Parts were marketed by both companies. Analog Devices learned the design business from Intersil. However, the relationship fell apart when Intersil also started working with Burr-Brown on a custom circuit. Analog Devices believed that Intersil was a captive design house for them, whereas Intersil did not want to be totally exclusive to Analog Devices.

A serious problem occurred when the largest shareholder, Arthur Rock, asked Hoerni to give up on analog. Arthur Rock had a reputation for helping many successful businesses. Unfortunately, he also ruined several of them. In the case of Intersil, Rock demonstrated that he was a typical myopic venture capitalist, who could only comprehend what was obvious.

After Jerry Sanders fired Jack Gifford, Hoerni asked Jack to help him with Intersil and to give him his opinion about the analog business which Rock wanted to terminate. Jack was working on a new start up called Zitetics at that time, and Hoerni promised to help Jack to raise the money for Zitetics if he would help him with Intersil. Hoerni actually wanted to have Jack work for him and was hoping that once he was on board, he would drop his plan for the new startup.

Jack's judgment was that Hoerni had an excellent technology and told Jean "*you can do what they were telling you and it will be fine, but you have a great business here.*" Hoerni said "*would you run it for me?*" Jack Gifford agreed. In the meantime Intersil became a company micromanaged by the Board of the Directors. Hoerni was not president any more and the Board recruited James F. Riley, the widely-respected president of Signetics Corporation to take the presidency of Intersil, Inc. Almost at the same time Marshall Cox and a team of memory experts from Raytheon Semiconductor joined the company and Marshall Cox became the executive vice-president and director of marketing. Intersil investors supplied financing for the new company – Intersil Memory, Inc. of which Cox became president. Although sales were consolidated with Intersil, Cox and Riley had no plan to merge the two.
The Intersil Memory company got a quick boost when Kenneth J. Moyle (the former product manager for MOS integrated circuits at National Semi-

Fig. 6.42. Jean Hoerni and James Riley at tour through Intersil (September 1970)

conductor), Roger Smullen (former product manager for linear integrated circuits), and six other National people joined Intersil.

Riley was quite successful at Signetics, but failed at Intersil. When the company started to sink, Riley was fired. Marshall Cox, who had no ability to run a semiconductor company become the new CEO. Intersil stock went from around $20.00 to $1.50. Marshall Cox caused a lot of new financial problems. Raytheon sued Intersil, and this took a lot of cash. It was suddenly discovered that there were a lot of distributor "sales" by Cox which were fraudulent. People believed that Intersil was a corrupt company. The Japanese called Intersil and John Hall, who was managing the watch project, and asked for an explanation. John flew to Japan to give a clarification.

Bob Freund was managing Intersil for Fred Adler, who was Chairman of the Board, until Intersil merged with Advanced Memory Systems. Ori Hoch, who was president of Advanced Memory Systems, become president of the combined company The outspoken and blunt Jack Gifford told several people what he thought. Cox became very uncomfortable with Jack and told Fred Adler that Gifford was causing a lot of problems and needed to be fired. Adler called Jack into his big room and told Jack *"I am going to take a piss and when I come back I want you tell me why I should not fire you, because I intend to."* Jack did not wait and he left the room and walked out to the parking lot. Fred was chasing Jack at the parking lot and was shouting that he wanted an explanation. "*You do not deserve any explanation*" was Jack's answer. Then Roger Smullen called Jack and convinced him to talk to Adler. Adler talked Jack into not resigning, and gave him more stock options. Cox was fired in July 1976 and Ori Hoch put the company back in fiscal shape and sold it to GE for $35 a share. The deal was finalized in February 1981.

GE also acquired the RCA facility in Somerville, NJ and tried a second time to enter the semiconductor business.

Ori Hoch stayed for a while before returning to Litton Industries. Jack Gifford was promoted to CEO of Intersil in summer 1981. Jack enjoyed a good relationship with Jack Welch, but was reporting to Sector VP Jim Butler. Their relationship was never good, and eventually Gifford was fired by Jim Dykes (who reported to Butler) in December 1982. Good relationship with Jack Welch meant nothing.

Jack Gifford realized that he was not employable by others; he only can run his own company. Jack, with Dave Fullagar, and Fred Beck, founded Maxim Integrated Systems in April 1983.

John Hall told me that his professional success was the result of his close relationship with Hoerni. Jack Gifford's respect for Hoerni goes further: "*Imagine the guy who started five companies; he did it mostly with his own money and never failed. He was one of the best managers I have ever been around. And it was just amazing that he was almost wiped out because he was not typically outgoing and a flamboyant person.*"

Shockley considered Jean Hoerni and David Allison the most capable engineers he lost during the Fairchild raid. Jean Hoerni and his own achievement demonstrated that the most efficient work done in the laboratory is always "illegal." Everything that is done "illegally" is always done with highest efficiency and the utmost concentration.

Hoerni always allowed his engineers to "steal" money. Out of these "thefts" eventually became big and profitable projects. Intersil was the most innovative company of the early seventies and Hoerni's vision was the critical ingredient of it. If you want to develop genius products, you have to put a genius in charge. Genius Jean Hoerni demonstrated it over and over, and his accomplishments overshadowed the contributions of his seven other peers from the "Traitorous Eight." Only Hoerni could go into any semiconductor factory in the world – and see that all of them are using his idea.

Hoerni died on January 12, 1997 at the age of 72, uncelebrated and remembered by only very few.

References

[1] V. Busch, Endless Horizons, Public Affair Press, Washington, D.C. 1946
[2] P. E. Haggerty, Presentation at "European/North American Conference on Research," Organization for Economic Co-Operation and Development, Monte Carlo, February 1965
[3] J. S. Kilby, "Invention of the Integrated Circuit", IEEE Trans. ED-23 (1976), p. 648
[4] Semiconductor Solid Circuitry, Electronics, April 10, 1959, p. 82
[5] TI Consolidates, Electronics News, May 22, 1961
[6] K. Lehovec et al., Physical Review, Vol. 83 (1951), pp. 603–607
[7] K. Lehovec, J. Chem. Phys. Vol. 21 (1953), pp. 1123–1128

7

Linear Integrated Circuits: Pre-Widlar Era Prior to 1963

"I had one experience which gave me some slant on the way a large organization is run. I was not allowed to take spherical trigonometry because I'd sprained my ankle. Because I'd sprained my ankle I had an incomplete phys education. And the rule was that if you had an incomplete in anything, you were not allowed to take an overload. I argued with some clerical person in the administration office, and was stopped there"

William Shockley, 1974

In mid 1959 the Air Force sponsored a research program with a goal to explore the capabilities of electronic solid state "function blocks."[1] The basic objective of the program was to achieve an eventual size and weight reduction by a factor of 1000:1 with significant gain in circuit reliability.

The only companies that responded to the proposal were Texas Instruments and The Air Arm Division of Westinghouse Electric Co. The Semiconductor Department of Westinghouse was established in the early fifties with a decision to develop high power semiconductor devices. By 1958, Westinghouse introduced the industry's first silicon power transistors, 2 and 5 amp units rated at a maximum collector current of 7.5 amps with the highest available power rating of 150 W. However, by far, the biggest Westinghouse contribution to the semiconductor industry was collaboration with the Siemens-Halske AG in Germany which resulted in a low-cost and hig-quality silicon-growth process. Westinghouse licensed U.S. chemical companies as producers of this type of silicon.

In 1959 Westinghouse Electric Co. was already heavily involved in military contracts, and with the corporation's overall sales over $2 billion, they split Pittsburgh Semiconductor Division into manufacturing and two research

[1] The term FEB (Functional Electronic Block) was introduced by J. Kilby in 1959

groups and assigned one of these groups to the Air Force proposal. The official name of the newly created group was Westinghouse Molecular Electronics Division.

In January 1960, Westinghouse VP of R&D, S. W. Herwald, demonstrated in Washington, D.C. the first working "molecular electronic functional blocks". The multivibrator shown in the center of Fig. 7.1 substituted a single monolithic structure for six resistors, two transistors, and two capacitors.

The technology used in the first Westinghouse "electronics functional blocks" was similar to the technology proposed by J. Kilby at Texas Instruments except for the isolation "mesa" pockets which separate the transistor's base from the remaining P-type layer.

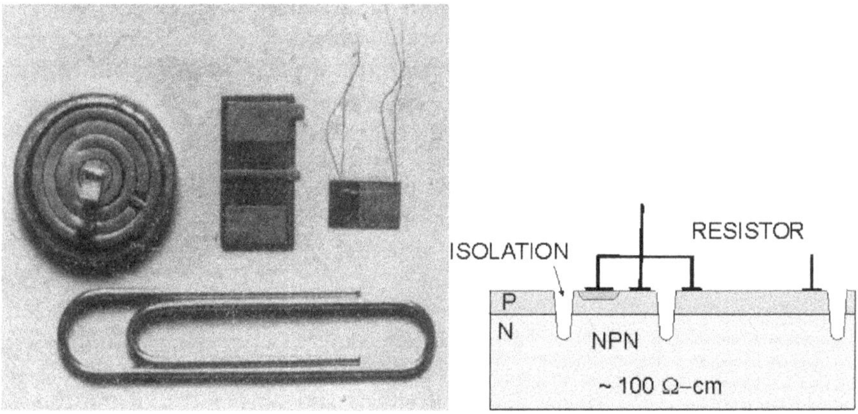

Fig. 7.1. The Westinghouse "Molecular Electronics Blocks:" Two stage 5 W Audio Amplifier, Multivibrator and Video Amplifier

The main processing steps of Westinghouse technology are shown in Fig. 7.2. The starting material was a 1 inch diameter P-type <111> silicon wafer with resistivity at least 100 Ω-cm and thickness of 100 μm. Using an oxide mask the buried collector (N-type) region with doping 10^{20}–10^{21} cm^{-3} was diffused to provide a low resistivity collector region. In the next step the hydrogen clean at 1230°C was performed, and N-type epi layer with thickness about 10 μm was grown on the top of the P-type substrate at a temperature of 1230°C for 40 minutes. The resistivity of the epi layer is determined by a trade off between the requirement of electrical isolation between individual components and transistor saturation resistance. The typical resistivity of the layer was 30 Ω-cm to 50 Ω-cm.

In the next step the wafers were placed flat into the diffusion tube for P-type base diffusion. First, the oxide was doped by Gallium at 1125°C for 75 minutes to form gallium sesquioxide at temperature 900°C in hydrogen ambient. Then the oxide mask was selectively etched to define emitter, collector, and contact to resistors. After the oxide etch, the emitter region were covered by wax and the entire wafer was exposed to silicon etch to remove about half of the thickness of the P-type base layer. After removing the wax and cleaning the phosphorus diffusion (1075°C for 20 minutes from phosphorus pentoxide at 310°C using dry oxygen as the carrier gas) was used to from the emitter region. At the same time, the phosphorus diffusing through the P-type layer (base and resistors) create that low resistivity N-type layer which extends through the P-type layer into an epi layer. In the next step the aluminum approximately 0.5 μm thick is evaporated over the entire wafer surface and selectively etched using a resist mask.

The electrical isolation of components is accomplished by a proper choice of contacts and "mesa" trenches. For example, if the electrical contact is placed at the top of the resistor formed in the P-type layer, the current flowing from the contact through the P-type layer will inject the holes into the N-type epi layer of the forward biased PN junction. By contacting the resistor with N-type diffusion (and therefore with N epi layer), injection of holes into the epi layer can be avoided. The relative variation of potential is limited by the "mesa" isolation etch (approximately 10 μm deep) when a significant thickness of the epi layer is removed. In this case the undesired paths of the N epi layer are minimized.

Although today it seems to be unbelievable that such devices could work, they somehow worked to the satisfaction of the Air Force and they were willing to fund continuation of the projects (Air Force Contract AF No. 33(600)39378).

At the end of 1961 Westinghouse and Texas Instruments were well in the lead in "molecular electronics" research and development – each having been given hefty support by the Air Force.

One of the problems common to Westinghouse and Texas Instruments' efforts was that each block (FEB) was optimized for a specific system. No standard circuit or circuit series were available for non-Air Force customers.

The first simple linear integrated circuits were developed and used in the Westinghouse "Molecular Receiver AN/ARC-63" delivered to the Air Force in December 1961 and shown in Fig. 7.4. The Westinghouse receiver is the first application with "functional blocks" used in RF, IF, and audio section with capacitors and etched tuning coils on a printed circuit board. The enormous cost of single-purpose circuits created an opinion that linear integrated circuits may be used only in those applications where the saving of weight and space was of major importance. At the time when most of the electronics equipment was still using vacuum tubes, low power consumption was even in military applications of secondary importance.

In the end of 1961 the maximum number of components per die did not exceed 10–12 components. The yield was very low, and Westinghouse just demonstrated the feasibility and function of designed circuits. None of the parts developed under the Air Force contract were manufactured in large

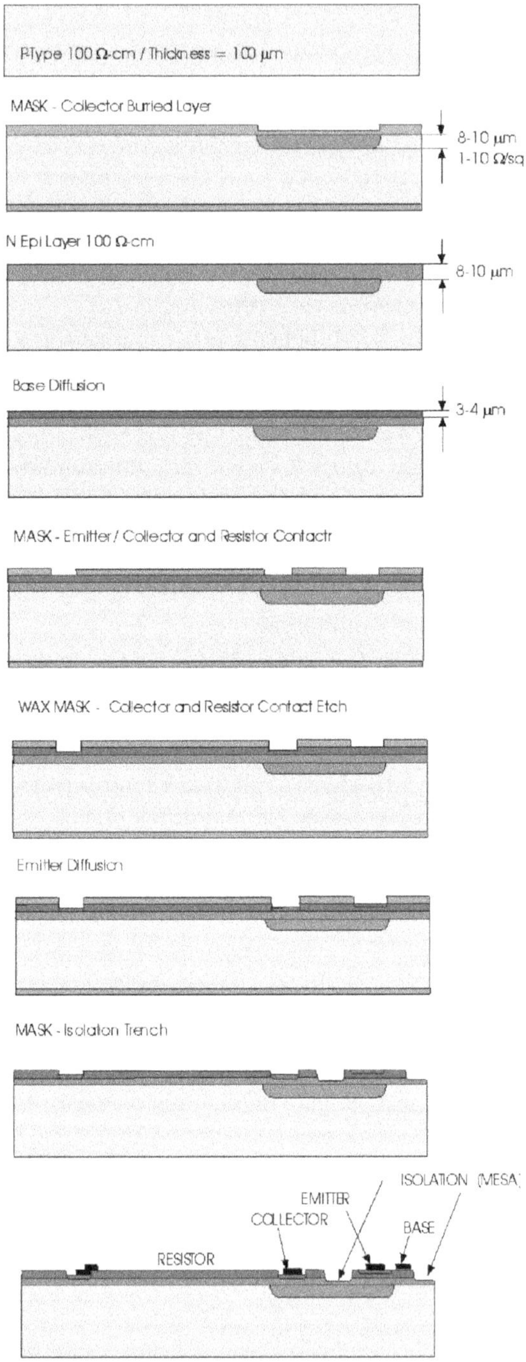

Fig. 7.2. Westinghouse process flow for "Molecular Electronics FEB" (1960–1961)

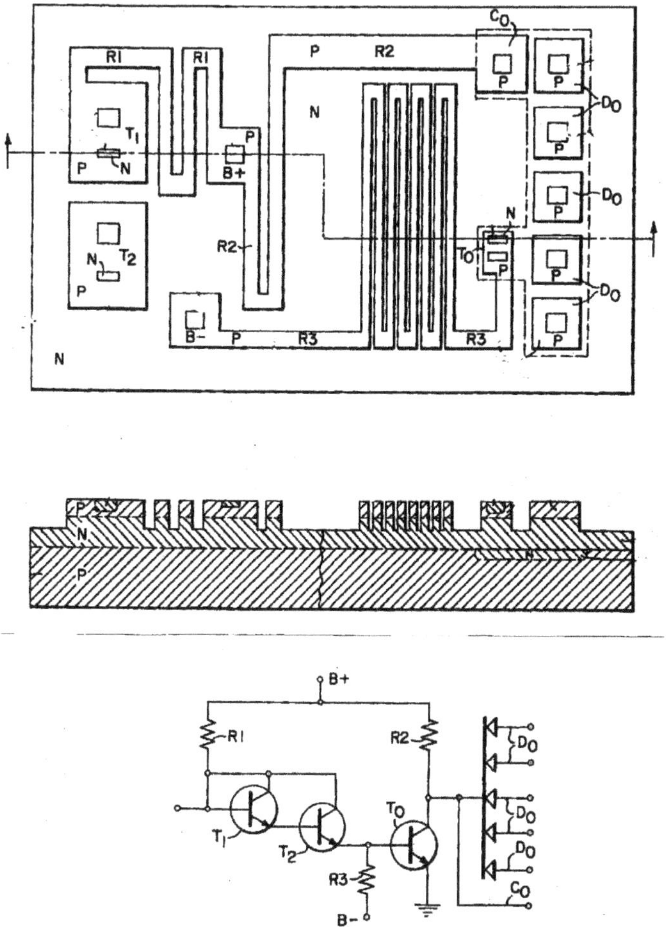

Fig. 7.3. DTL Gate Produced by Westinghouse "Molecular Electronics" process (1961)

volume. The development cost mattered only a little to Westinghouse. In 1961 there were very limited or no competitors to Westinghouse, and Westinghouse could charge military accounts a premium.

The capabilities of the integrated circuit technology at Westinghouse are depicted in Fig. 7.5. Both engineering and production costs were rising as the circuit which need to be integrated increases in equivalent number of components. The fact that the area of silicon required to integrate a circuit increases with increasing number of components suggested a good criterion of the cost of integration.

From the chart we know now that Westinghouse was overly optimistic when they predicted a hundred components on a die in 1963.

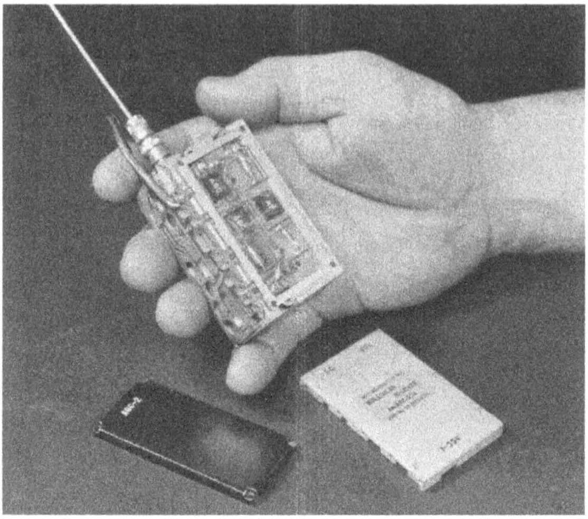

Fig. 7.4. Westinghouse Molecular Receiver AN/ARC-63 (1961)

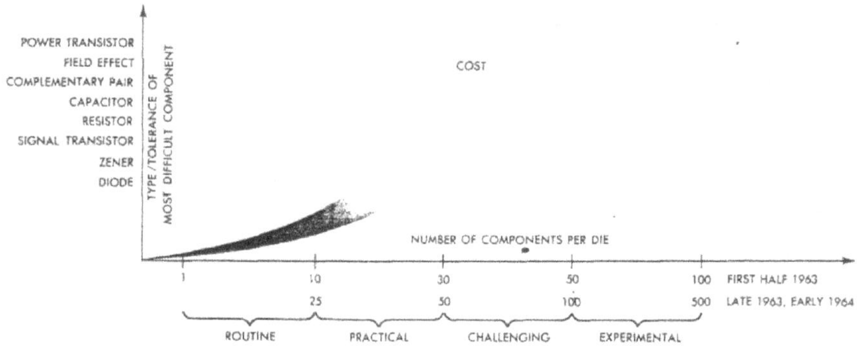

Fig. 7.5. Cost of ICs as a function of the number of components per die (Westinghouse 1961)

At the time when Westinghouse and Texas Instruments responded to the Air Force proposal there was very little known about the Fairchild planar process. Fairchild integrated technology itself was in limbo after key personnel of Jay Last's group left Fairchild. Fairchild continued in a marketing campaign and announced a move to III-V compounds and work on "integrated solid-state circuitry," however, no new products were introduced. By 1961 Fairchild Semiconductor sales were over $92 million, operations expanded to several locations with almost 200,000 square feet with some 7,400 employees. The times were good for Fairchild and there was no reason to be worried. In 1961 there was nobody who envisioned and pushed for planar integrated circuits. The situation changed when a very capable David Allison started Signetics.

David's group did not bother with discrete components, they focused only on planar integrated circuits.

In the meantime Westinghouse and especially Texas Instruments did not miss any opportunity to downplay the importance of the planar process. They consistently avoided the term integrated circuits and for a very long time they did not initiate any alternative methods to their approach of using point-to-point connections of wire between individual components. Instead they were arguing that FEB's may be extended almost infinitely. For example, TI at a presentation to the Air Force in May 1961 used the following example:

> *"The packaging density of molecular electronics is almost beyond belief. An audio amplifier with a gain of 50 dB can be produced in an FEB measuring* $1/8 \times 1/4 \times 2.1$ *inches or 230,000 these parts per cubic foot.*
>
> *Translating this into the number of the transistors, capacitors, etc., that would make an equivalent circuit with over two million of conventional parts."*

The discussion of the labor intensive part of interconnects which needs to be completed manually on each part were left out from all of Westinghouse and Texas Instruments over optimistic announcements.

Autonetics, encouraged by results which Westinghouse achieved in the military receiver design, asked at the beginning of 1960 for a very simple integrated differential amplifier which could be used in Minuteman flight control. The requirements for significant volume and weight reductions of electronics systems concurrent with improved reliability were the main driver.

In this project, the Westinghouse group was competing with Jack Kilby, Art Evans and Lee Evans of Texas Instruments who were designing simple circuits they called "feebs" (FEB – Functional Electronics Block). They used technology previously developed for digital circuits (Fig. 7.3) with addition of a PNP structure which was required and which had not been successfully integrated before.

One of the versions of a differential amplifier designed by Texas Instruments for Autonetics' flight control system in 1962 is shown in Fig. 7.6.

Texas Instruments was using a conventional vertical structure for PNP transistors and surprisingly the circuits worked. The problem was that Autonetics required a certain minimum radiation resistance. The TI vertical PNP device had a very poor radiation resistance and parts were failing.

The same technology with lose wire interconnects which Texas Instruments used for Autonetics differential amplifier was used for the first Solid State circuit available off the shelf – the 502 flip-flop. TI advertisement from April 1960 stated:

> *"This multivibrator, the TI type 502, is so real it carries a price tag; $450 per circuit in quantities less than 100, $300 each for larger quantities."*

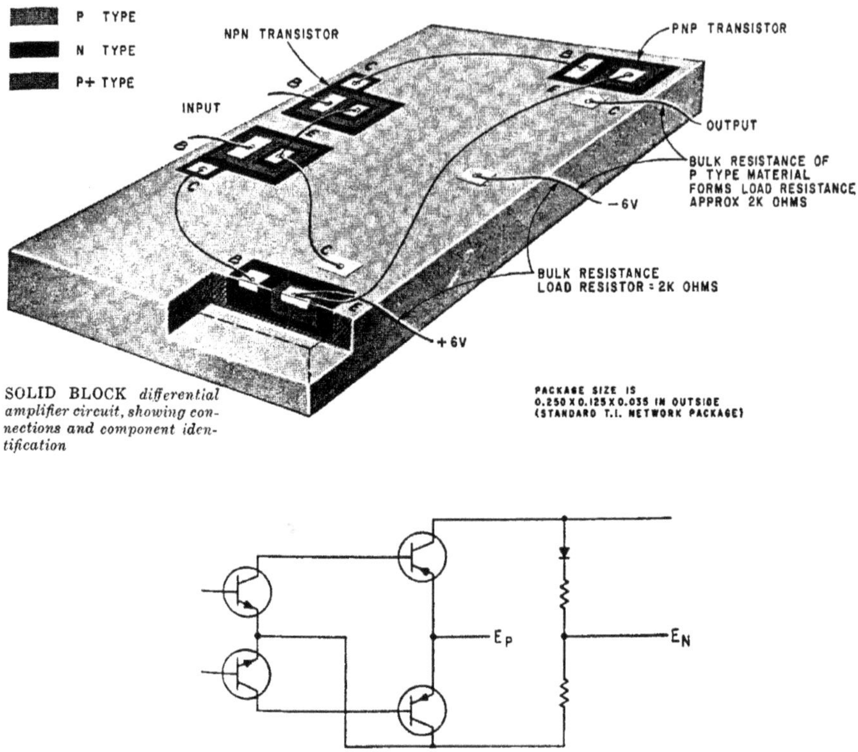

Fig. 7.6. Jack Kilby's Differential Amplifier [July 1962]

The circuits become available in the middle of 1961 and the price was about $50 higher than originally announced.

One of the Westinghouse newly hired engineers was Dr. Hung Chang Lin. H. C. Lin was a research engineer at RCA Laboratories engaged mostly in early transistor circuit development. In 1956 he followed his friend Dr. Chu and joined the Hytron Division of Columbia Broadcasting System where he worked until 1959 as Manager of their Semiconductor Applications Laboratory.

H. C. Lin brought to Westinghouse a significant knowledge of transistor circuit design from his previous experience at RCA, He was one of the first who recognized the importance of temperature effects in circuits using junction transistors. The effects of varying temperature upon the performance of transistor circuits have been evident since the beginning of the transistor art. However, the nature of these effects, their causes, and the techniques which may minimize the temperature dependence, had not generally been developed. The behavior of the transistor with respect to temperature is most adversely affected by changes in the saturation current and the d-c input

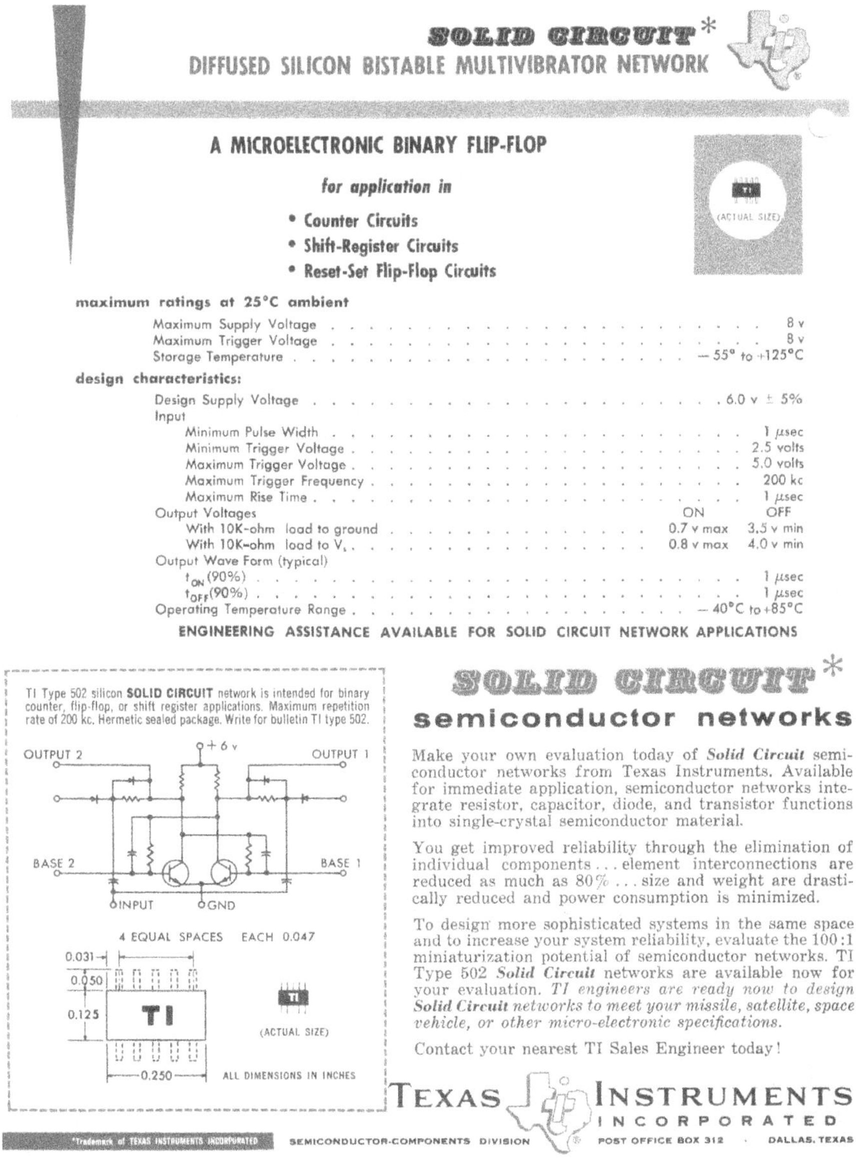

Fig. 7.7. Texas Instruments "Solid Circuit Networks" 1960

conductance. The effects of temperature on the d.c. characteristics of bipolar transistors are that V_{BE} decreases and saturation current increases with increasing temperature. If the transistor should operate at constant collector current within a range of temperatures, the emitter current must be held constant while V_{BE} is forced to change approximately by -2.5 mV/°C.

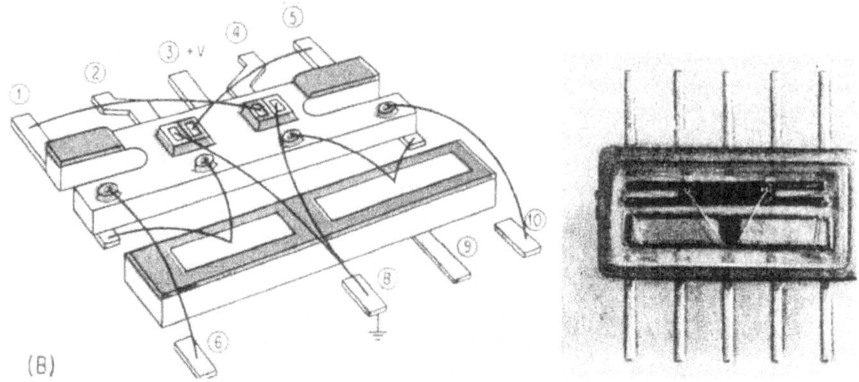

Fig. 7.8. Texas Instruments Type 502 multivibrator [1960]

H. C. Lin developed in the early fifties "single diode compensation" where base-to-emitter bias is derived from a junction diode which is constructed from the same materials as the transistor to be compensated. If the transistor base-emitter diode is identical with the compensating diode, the rate of change of voltage across the diode, as a function of temperature for constant current operation, is the same as the change of base-to-emitter bias required for constant emitter current.

While still with RCA, H. C. Lin filed a patent application "Semiconductor Device and Stabilization Thereof," in December 1954 describing a device based on this idea. Fig. 7.9 is showing a junction alloy transistor with an alloyed diode formed on the same semiconductor material. The device was never manufactured beyond laboratory samples. Ten years later almost everybody recognized the importance of Lin's solution.

In August 1962 at the Western Electronic Show and Conference (WESCON) in Los Angeles, H. C. Lin presented another very important contribution for the future design of Linear Integrated circuits. In a paper titled "Diode Operation of a Transistor in Functional Blocks" he argued that for the reason of matching of components it is more convenient to use the transistor structures rather than a diode. Lin's idea of matching devices changed the design methodology of linear integrated circuits.

In theory such brilliant ideas should be recognized by company management and used immediately in products which could easily dominate the market and leave the competitors stunned. At Westinghouse there was no such visionary manager.

In 1962 Autonetics (A Division of North American Aviation, Inc.,) established the Molecular Electronics Team (MET) under leadership of T. Mitsutomi. This change forced Westinghouse and later Texas Instruments to abandon their "mesa integrated technology" and develop the planar process similar to Fairchild's processing technology.

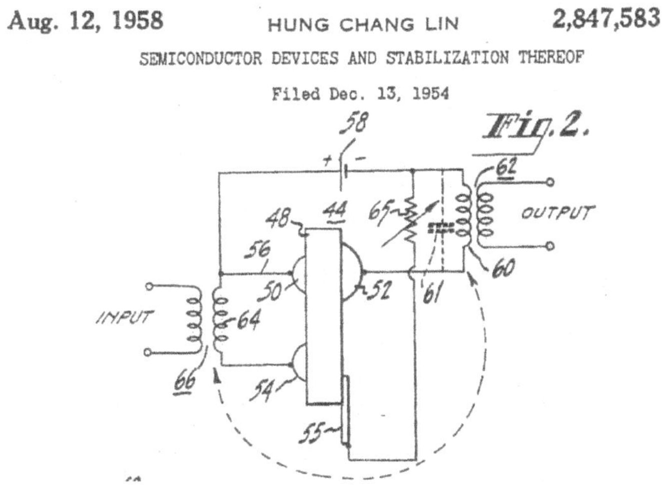

Fig. 7.9. Diode compensation of d-c operating point of junction transistor

Fig. 7.10. Dr. Hung Chang Lin, inventor of matching of integrated devices and the PNP lateral transistor

At the beginning of 1963 when working on Autonetics' differential amplifier, H. C. Lin came up with the idea of the PNP transistor which could be produced without any change in processing and did not require any additional masking steps. In July 1962, C. Harry Knowles, was allured from Motorola by Westinghouse, and become head of Westinghouse R&D and Lin's boss. Knowles made a bet that such a device could not work. Lateral PNP action due to the wide base makes this not a great device, but it worked and solved the problem.

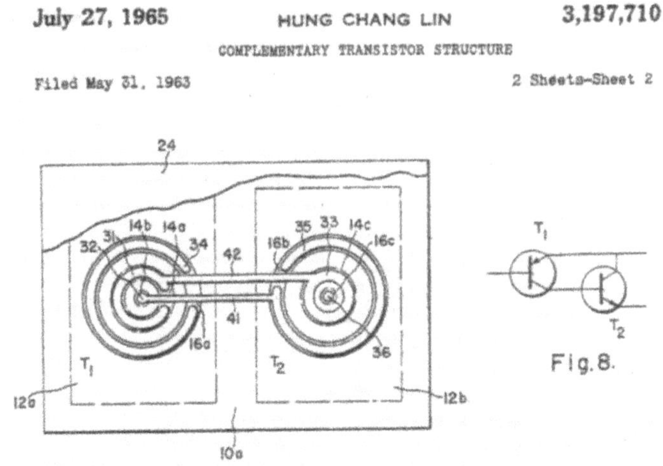

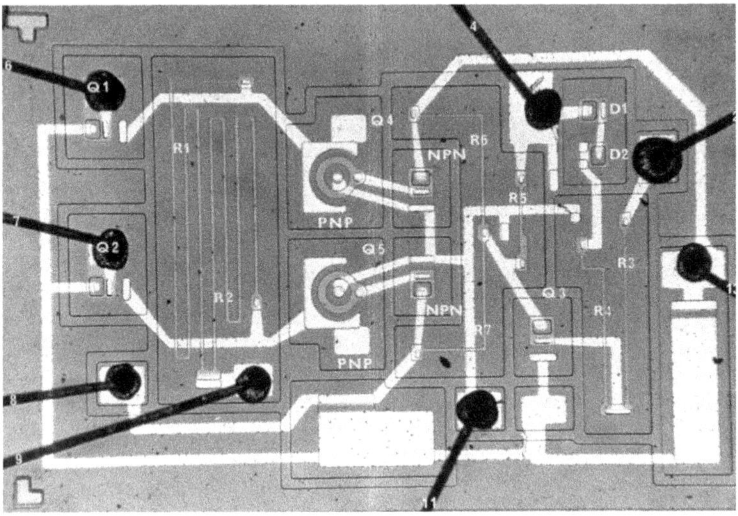

Fig. 7.11. Westinghouse differential amplifier using Lateral PNP transistors (1963)

Harry lost his bet, but was compensated by a million dollar Autonetics contract. Texas Instruments responded in a way which was to become the pattern of semiconductor cowboys – they sued Westinghouse. The court litigation was later settled. The pattern and precedents set up by the "Texas legal firm" remains until today.

In 1963 nobody believed that general purpose linear integrated circuits could be a serious business. Common opinion was that the cost of integration is high for all circuits except for digital circuits where available "off shelf" standard logic circuits can be produced in large quantities. The standard line of linear circuits may not give the system engineer the flexibility he desires.

The cost of masks required for new circuit design varies from $3,000 to $20,000 and delivery schedule may range from three to eighteen weeks. Back in 1963, that was a lot of money. To minimize these costs and accelerate circuit design Westinghouse and Autonetics developed a technique whereby analog integrated circuits could be breadboarded using "off the shelf" circuits. This approach was basically a more sophisticated version of technology than was originally developed by Molectro Corporation. The "off the shelf" circuit was a chip with separate output pads for each circuit element. An example of a components menu on a Westinghouse breadboard block is shown in Fig. 7.12.

The individual components are connected by thermo-compression bond wires during circuit development. Once the developed circuit is thoroughly tested only the specialized metal mask is required. The cost of a fully specialized linear integrated circuit ranges from $20 to $200, depending on quantity. If all processing steps are recycled and only the metal mask is dedicated to the particular circuit, the expected cost reduction per circuit will be $12 in quantities of 50 or more.

An advantage of the breadboard approach is that all devices have isolation and parasitic effects comparable to what might be expected in an integrated circuit. The disadvantage was that the prototype device could not be easily hermetically sealed for lifetime testing. Secondly, it is more difficult to achieve close-tolerance resistance by a series-parallel combination of the various resistor taps.

An example of a linear circuit manufactured by the bread-boarded technique is shown for a case of Autonetics' differential amplifier used in the controller of electromagnetic accelerometer. Because of the need of PNP transistors this circuit troubled not only designers at Autonetics but also Westinghouse, and Texas Instruments. The circuit schematic diagram is shown in Fig. 7.13. In Fig. 7.14 is shown a diagram of the prototype wiring.

Westinghouse was pushing the microcircuit breadboard design up to 1966. The most advanced version was called "Insta-Circuit." The circuits were sold to The Norden Division of United Aircraft Corporation and Westinghouse Aerospace Division. Dr. Gene Strull who was in charge of the operation with four other engineers developed over 100 circuits. The average cost of an "Insta-Circuit" was $70 to $200 in small quantities.

It is not well known that Westinghouse used this technology for the TV camera that was used by Apollo astronauts sending TV signals from the moon back to earth.

Despite the large investment into a new Molecular Electronic facility in Elkridge, MD, things were not going well for Westinghouse. Donald W. Gunter, Westinghouse Semiconductor General Manager, lectured at a board meeting in May 1961: *"Part of the business philosophy of the Semiconductor Department is that the operation should have the flexibility of a small company to adopt new research findings and a rapidly expanding market, while*

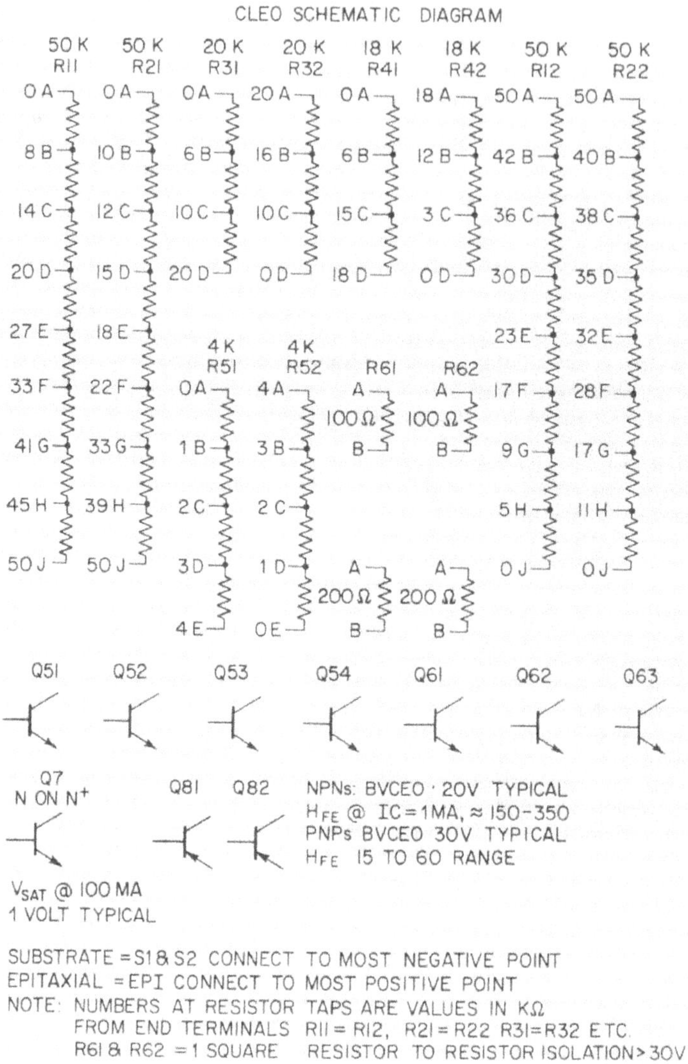

Fig. 7.12. Components menu of a Westinghouse breadboard custom chip

tapping the full strengths of a major corporation." He concluded his presentation with the phrase which became the buzzword of every single meeting between engineers and top level management: *"People are an investment with us, too."*

Unfortunately, Gunter's philosophy never happened and his phrase was never sincere. Gunter was the type of manager who can manage business if there is no competition and his customer is the Department of Defense.

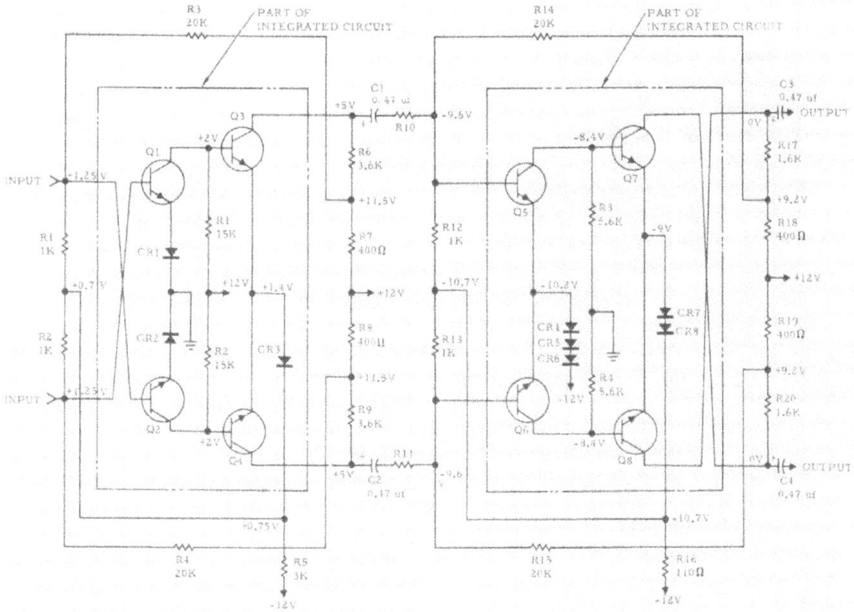

Fig. 7.13. Four stage Differential Amplifier Schematic Diagram [Autonetics 1961]

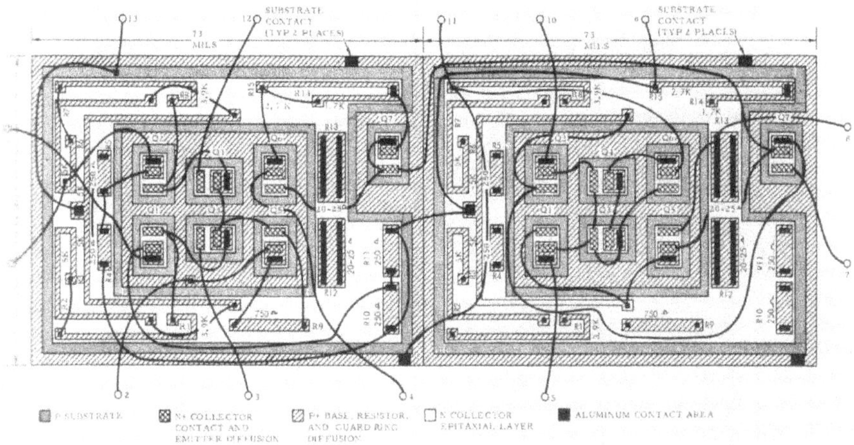

Fig. 7.14. Prototype of differential amplifier with wire interconnect

Size is the enemy of innovations and people in the semiconductor industry always come last. History proved that it is impossible to get effective innovation in an environment of more than a few hundred people. During 1966 and 1967 Westinghouse was no longer able to dictate prices and was forced

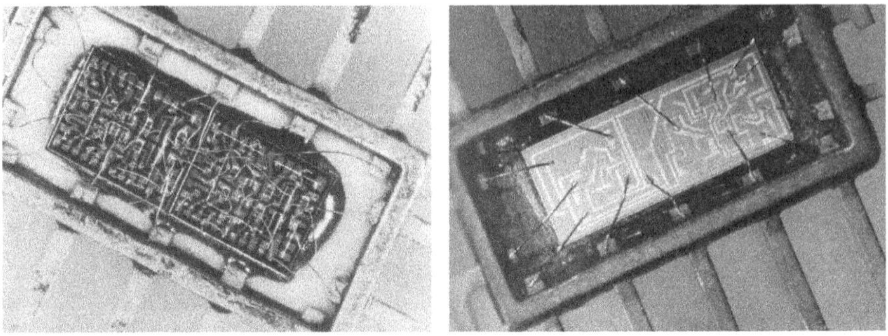

Fig. 7.15. The four stage differential amplifier designed as breadboard custom device a) and with custom made metal mask b)

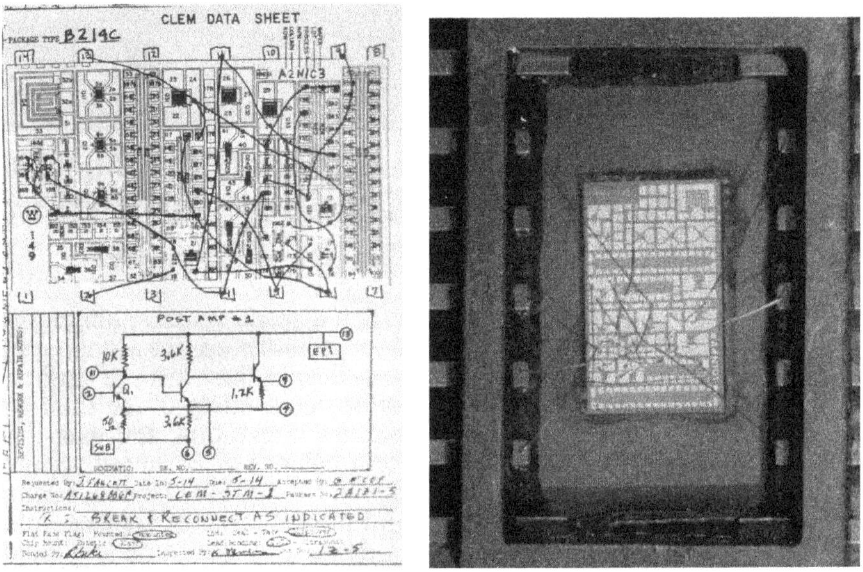

Fig. 7.16. Westinghouse "Insta-Circuit" (1966). Circuit breadboard design using data sheet (left) and fabricated linear circuit (right)

to compete with the leaders: first Fairchild, second Texas Instruments, third Motorola, and fourth Signetics.

In 1967 the Westinghouse Molecular Division in Elkridge, MD had about 1200 employees in 170,000 square feet of manufacturing space and was under Gunther leadership losing some \$3 million a year on sales of \$15 million. Although the facilities were considered excellent, and despite some excellent technical work (for example 2 μm geometry feature demonstrated in 1965 (Fig. 7.17)) the company was still using 1-1/2-inch wafers when the competitors switched to 2-inch wafers.

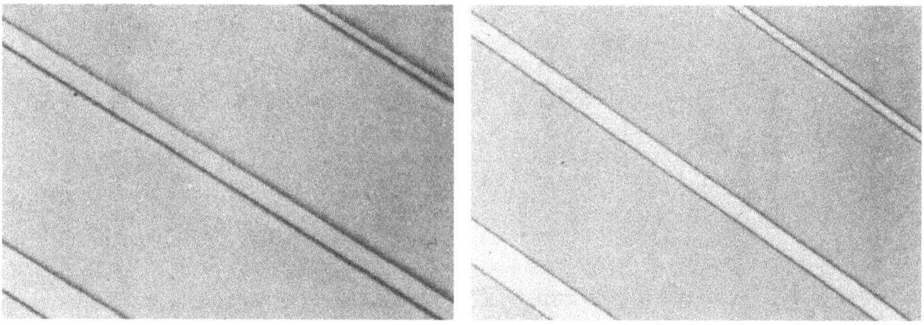

Fig. 7.17. Westinghouse lithography with 2 mm features (1965). Resist (left), oxide (right)

Fig. 7.18. State-of-the-art clean room – Westinghouse 1965

Except for the early integrated circuit days, when Westinghouse tried to establish the W200 line of DTL but lost out to the Fairchild 930 line, Westinghouse had never been known as real innovator. Fairchild Camera & Instrument announced in December 1968 interest to buy the entire division. The deal, however, did not go through, and at the end of 1968 Westinghouse Molecular Electronics disappeared forever from the semiconductor business.

8

Robert J. Widlar – The Genius, The Legend, The Bohemian

"A Fairchild researcher trained a frog to jump at the sound of a bell. The researcher measured the distance the frog would jump, then removed the frog's legs and rang the bell again. The frog did not move, thus proving the Fairchild R&D group hypothesis that removing a frog's legs deafens the animal."

Robert J. Widlar,
describing Fairchild's R&D group in 1967

The achievements of Bob Widlar demonstrated that an engineer could have a major impact on the business of a large company. Bob Widlar created new products, new applications, and a new market – linear integrated circuits. Robert John Widlar is a typical example of how a company benefited from somebody who consistently remained an inventive genius over an entire career if they let him be himself.

At the same time the personality of Bob Widlar is a clear manifestation of unhealthy changes which occurs in our society. In the current semiconductor business environment, with Human Resources Departments defining what engineers can and cannot comment about, it is very unlikely that we will see his kind again.

Bob was a fiercely independent individual, very happy to be by himself, and he did not care a lick about how others saw him. He did almost everything in a stunning way, which was absolutely natural to him, but completely weird to so-called "normal people."

Bob Widlar never talked about his early years, or anything personal. He was very charming but always a bit mysterious. He seemed to be more artist than engineer; he cherished creativity more than the concept of general intelligence. Widlar's boss should be able to know the subject better than him; otherwise Bob is unmanageable. To those who never knew Bob, he may appear simultaneously crazier yet saner, more naïve and knowledgeable, more destructive and more constructive. What is the result of such originality? In the same way as great painters, great writers, great poets, each of Widlar's circuit creations had at least one feature which was far ahead of the crowd. It is natural that such accomplishments may irritate less creative people who would like to live "Widlar's Way." It is the human nature of engineers working in semiconductor research to down-play the accomplishments of others while embellishing the accomplishments of their own. Perhaps for this reason the personality of Bob Widlar is surrounded by many Silicon Valley legends and myths. Some of them are ranging from fantasies and lies, to uncritical imitation of Widlar's behavior.

To move a product from idea to market requires a significant effort. History and good business textbooks teach us that if a company becomes large, such efforts often fail if conducted only by management without a strong and self-appointed leader. Such a leader is called a "Product Champion" in modern business textbooks [1,2].

Between the discovery or invention, and the efforts to commercialize it, is a large gap, see Fig 8.1 [3]. Engineers and scientists who are on the discovery side often do not understand the concerns of commercialization personnel who are on other side of the gap. Both sides have different objectives and rewards stimuli; technical people find value in discovery and pushing the frontiers of knowledge, while commercialization people need a product to sell and often consider the value of discovery as theoretical and useless.

Theoretically, to overcome these antagonistic differences, company management should reconcile the differences of both groups. Unfortunately, the managerial position is frequently also a "political" position related to the power and financial compensation. The best way to keep the managerial position is to keep status quo with no disturbance. To accept even a small level of risk associated with new discovery could be risky and it is always safer to reject any new idea.

Politically correct managers operate through the command-and-control system: they provide direction, plans, and rules that define the work. They make sure engineers and scientists follow directions and comply with these plans and rules. A shift from leadership to the command and control role is the major change which the "politically correct" establishment infused into the semiconductor industry in the end of the seventies.

To ensure that research is focused along corporate strategic interests, politically correct companies recently reduced their efforts to only one factor – to satisfy Wall Street. Unfortunately, purely market-driven project initiation can easily fail, especially when the project is not accompanied by previous research. The main problem of a company which is only market-driven is that any surprising discovery events which may somehow randomly occur in the course of a market-driven project are not welcome. Due to the conflict with expectations, there are neither resources nor time for the unexpected discovery.

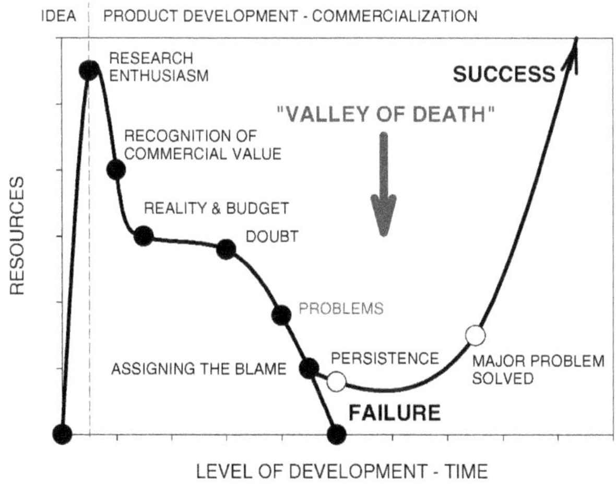

Fig. 8.1. Success/Failure Principle [3]

To overcome all of these problems the Product Champion must successfully accomplish a series of discrete activities ranging from influence tactics and bargaining techniques to coercive actions with persistence in doing things with or without permission.

There could not be a better example of a Product Champion than twenty six year old Bob Widlar at Fairchild. Two years after graduation, Robert J. Widlar was not questioning if Fairchild Semiconductor had the resources or interest to get his brilliant idea of Linear Integrated Circuits to market. Widlar in the position of a newly hired low level engineer managed to get it designed, produced, and introduced to the sales force without the knowledge or approval of Fairchild's management.

To understand Bob Widlar's personality one must go back many years and trace the roots of Widlar's family. In 1902 bold, proud and Bohemian Frantisek Vithous realized that a person needs to decide whether to be stupid and content or clever and full of renunciation. We are either a "slave of life" or we "live life by our way." Empty-handed Frantisek left his small village in south Bohemia, not too far from the famous Pilsner Urquell brewery, and ended up on other side of the Atlantic in Cleveland. There he met a beautiful Bohemian girl, Marie Zakova, who became his wife. Frantisek's rules went as far as changing his name to Frank. This was the only change he was willing to make in his new country. If somebody wanted to talk to him he had to learn the Czech language first. There is a word in the Czech language which is untranslatable to English: Frank was "*sekac*"[1] and he would not allow anyone to tell him what to do. The "*sekac*" man trusts and believes only in

[1] "*sekac*" could be expressed as combination of valiant, intransigent and consistent with ethics

himself, and he is able to go alone against all. The genes of "*sekac Franta*" and the genes of gentle, loving and caring Marie defined the initial conditions of the highly nonlinear Widlar's equation of life.

Frank's daughter Mary Vithous married Walter L. Widlar Jr., on August 24, 1935. Widlar's German ancestors settled in the Cleveland area at the end of the eighteen century.

Fig. 8.2. Frantisek Vithous, Bohemian grandfather of Robert J. Widlar

Walter L. Widlar, graduate of Lakewood High School and a self-taught engineer, worked at the Cleveland WGAR Broadcasting Company where he was in charge of the mobile short wave AM and FM transmitters' design. He was a pioneer in the early work of mobile broadcast pickups and relays and he developed several unconventional circuits for frequency modulation. He established himself quickly as an expert and the Widlar name become a hot commodity in the broadcasting industry. The Gammatron vacuum tubes advertisement (Fig. 8.4.) used Widlar's endorsement for their vacuum tubes.

The second son of Mary and Walter, Robert J. Widlar was born on November 30, 1937. Walter Widlar revolutionized not only the mobile broadcasting business, but he was also very ingenious in his home. Robert J. Widlar was the first baby monitored by wireless radio. As illustrated in Fig. 8.3 the local Cleveland newspaper informed the public about Bob's father's latest invention: the first "step-saving" device designed for his wife.

In early childhood young Bob very soon developed a strong interest in electronics. Supported and guided by his father who frequently took his son to Bird Electronic laboratories, Bob at age fifteen was not only able to repair any radio or TV, but he developed a strong bond with his dad.

In 1952 the local Cleveland newspaper printed a short story about "Scientist Bob Widlar" (Fig. 8.4.) The career of a passionate reader and a user of "The Devil's Dictionary" by Ambrose Bierce did not start by usual way. The first job of young Bob Widlar was "Floor Sweeper" in a drug store (Fig. 8.5). The second job, however, put Bob back on track and he proudly put into his resume the title "Technician" in Edbin Furniture Company in Cleveland where he repaired radios and televisions. At that time radios and TVs were sold in furniture stores. Edward Bentowski's furniture store played an important role in Widlar's life not only as Bob's second employer, but also as a place where a young sales person, Mary Vithous, met Walter Widlar.

• • Walter Widlar of the WJAY staff has installed a microphone at the cradle of the new baby boy at his home. Thus, when Mrs. Widlar is bossing the laundress in the basement, a loudspeaker there informs her when the baby is crying.

Inventive—Walter Widler, former WJAY engineer now merchandising for that station, has worked out a new "step-saving" device for his wife, according to the station. Widler has planted a mike on his baby's crib in the ursery. A loud speaker can be placed in the home wherever Mrs. Widler happens to be. If the infant cries, she'll know it.

Fig. 8.3. Press announcement of the world first baby monitoring radio

What is not well known is that Bob Widlar's father was working on projects managed by William Shockley. Vannevar Bush, of MIT, and others persuaded Roosevelt that the technical brains should gear-up for action. The President established The Office of Scientific Research and Development (OSRD). When Shockley become research director of the Antisubmarine Warfare Operations Research Group in 1942, the Navy Department at Columbia University was reporting to him. In 1942, Walter Widlar resigned his position at WGAR Broadcasting Co., and become a member of the scientific staff of the Division of War Research at Columbia University where the U.S. Navy Underwater Sound Laboratory established a classified sonar project as part of anti-submarine war effort.

At the request of the U.S. Army's Aircraft Radio Laboratory at Wright Field, Dayton, Ohio, Walter Widlar, Lewis Southworth, and Henry N. Jasper developed an underwater sound pickup-radio transmitting device which could pick up the sound of moving submarines and transmitted it by a short-wave radio transmitter in the buoy to the patrolling plane. Walter Widlar's experience with designing the FM modulated transmitter was crucial to this project.

In 1943 Walter Widlar joined Bird Electronic Corp. in Cleveland and continued research on the sono-bouy. Almost half a million of these devices

Electronics, Machines, TV and Wrist Radios Are Leading Interests for Scientist Bob Widlar'

"Oh, where did I put that J9 vacuum tube?" mumbles a curly-haired Ignatian, in the midst of a pile of radio equipment.

Frantically searching for the tube that will complete his latest radio discovery is Bob Widlar of 3B. Here, in his basement workship, he is often found, making advances and experiments in the field of electrical engineering.

This future engineer is preparing for his life's work by taking a Latin-Scientific course at Ignatius.

Bob already has had experience in designing electrical equipment. He is now designing and building radio equipment for one of Mr. E. H. Konerman's "Radio Plays."

As a member of the Science Club, under Fr. J. W. Stejskal, S.J., he also gains experience in the electrical department.

In his spare time, Bob works on equipment such as crystal sets and radios. His present attention is on perfecting a wrist-radio.

To round out his radio experience, Bob also repairs radio and television sets. His workshop, stocked with many parts, can easily supply his needs for any repair that might be necessary.

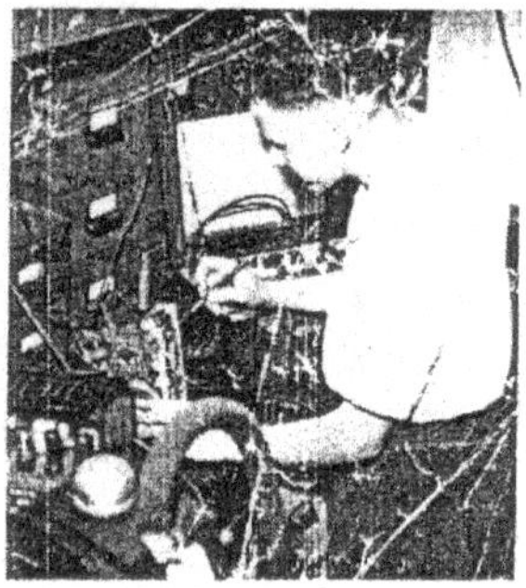

Bob Widlar

His only other preoccupation, besides radio and electronical machines, is the study of internal combustion engines. A good example of this type of engine is the one used in automobiles. Bob, not quite 16, does not have his own car, but hopes to have one in the future.

After leaving Ignatius, Bob plans to continue his scientific training in preparation for a career in science.

Fig. 8.4. The future engineer Bob Widlar

		Ohio		
Technician	Bird Electronic Corp. H.E. Stevens	1800 E.38th St. Cleveland, Ohio	6/55	9/56
Technician	Edbin Furniture Co. Edward Bentowski	E.65th & Foreman Cleveland, Ohio	2/55	6/55
Unemployed	(School)		9/54	2/55
Floor Sweeper	Marshalls Drug Stores R. Jeske	Snow & Ridge Rds. Cleveland, Ohio	6/54	9/54

Shirley A. Cameron
SIGNATURE OF WITNESS

Robert J. Widlar
SIGNATURE OF EMPLOYEE

Fig. 8.5. List of early jobs in Widlar's resume submitted to Fairchild Semiconductor

were dropped into the oceans to protect America's coast from Nazi U-boats during World War II.

In September 1, 1945 all three engineers were decorated by the Navy for *"development of the vital equipment to the Navy and others"*.

Bob saw in his father a role model, and he always wished his father could be proud of him. Unfortunately, in 1953 Walter L. Widlar died. Walter Widlar's heart was scarred by frequent rheumatic fevers and he collapsed under a massive heart attack. Bob's father was only 45 years old.

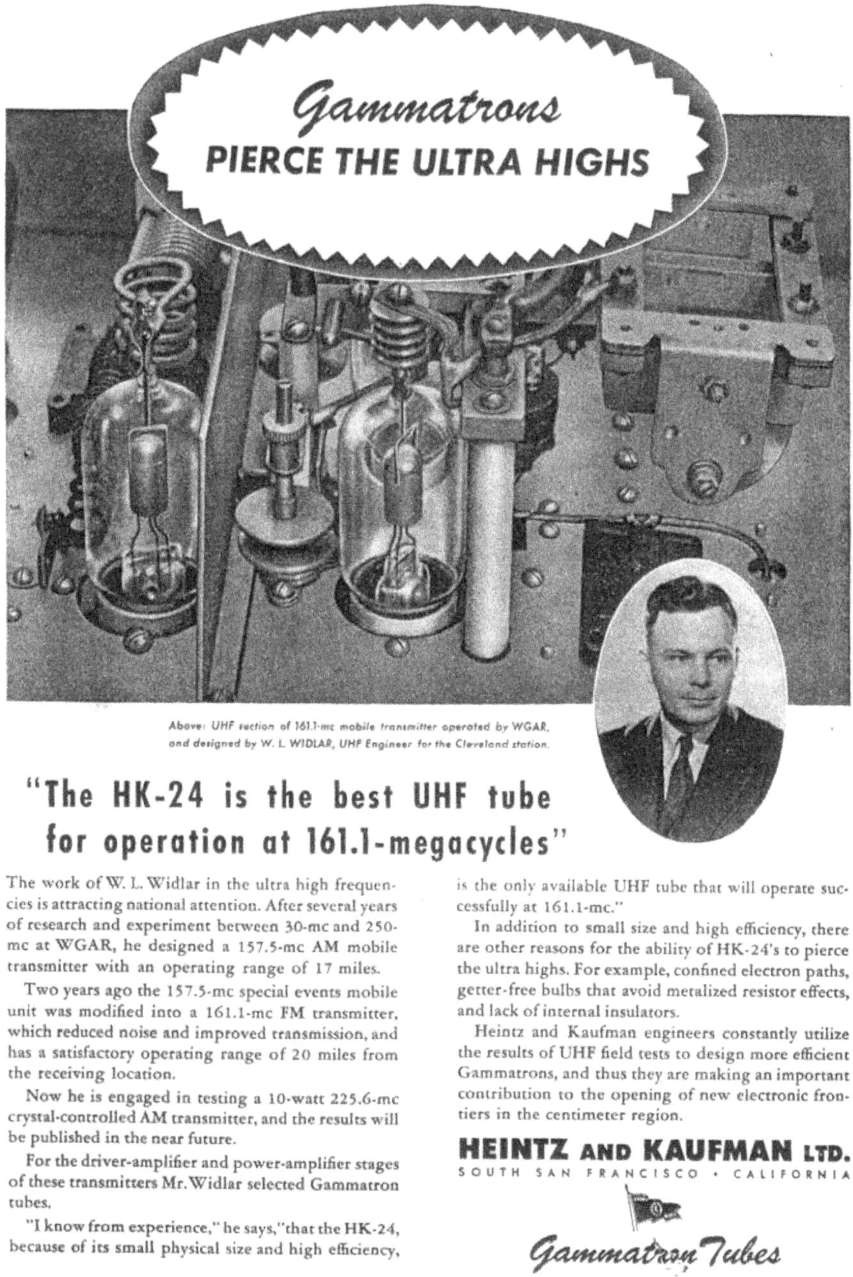

Fig. 8.6. Gammatron Tubes advertisement [Electronics Magazine, February, 1944] exploiting "UHF Engineer" Walter L. Widlar

Young Bob Widlar joined the United States Air Force in February 1958. One of his duties was teaching classes on electronic equipment and devices. The very first Widlar publication was a crispy clear textbook "Introduction to Semiconductor Devices" (Fig. 8.9.). When I was reading this text, I realized why Bob Widlar was so successful in his future work. He had an extraordinary capability to simplify complex problems.

Fig. 8.7. Robert J. Widlar at age 16

After enrolling into the University of Colorado in Boulder and tasting student life enhanced by free spirit genes from his Bohemian grandfather, it become clear very quickly, that Bob Widlar's military career could not flourish. In 1961 Bob separated from the Air Force and joined Ball Brothers Research Corporation in Boulder where he designed analog and digital circuits for NASA's orbital station.

Widlar, due to early training by his phenomenal father and due to his own hard and diligent work, had more knowledge of transistor circuits than most of his professors at school. Bob graduated as a straight A student in the summer of 1963 while still working for Ball Brothers.

In the course of this work he worked with Jean Hoerni and Shelton Roberts of Amelco who produced a 2N930 radiation hardened transistor. This encounter resulted in Widlar's desire to work for a semiconductor manufacturing company (Fig. 8.8.)

ATTN C S ROBERTS AND J A HOERNI

REF [redacted] TO YOUR TECHNICAL BULLETIN NO. 1 /MARCH 1963/. THE RADIATION TOLERANCE OF CURRENT GAIN FOR THE 2N930 TRANSISTORS INDICATED IN FIGURE 1 IS ONE OR TWO ORDERS OF MAGNITUDE LOWER THAN THAT EXPECTED ON THE BASIS OF BULK DAMAGE ALONE / SEE S C ROGERS, "RADIATION DAMAGE TO SATELLITE ELECTRONICS SYSTEMS", I EEE TRANS., VOL. NS-10, JAN. 1963/. SINCE IRRADIATION WAS BY CHARGED PARTICLES, IT IS POSSIBLE THAT THE EFFECTS OF SURFACE CHANNEL DOMINATED /SEE D S PECK, RR BLAIR, W L BROWN AND F M SMITS," SURFACE EFFECTS OF RADIATION ON TRANSISTORS", BELL SYS. TECH. J., JAN. 1963/. SINCE I AM ENGAGED IN A RADIATION EFFECTS STUDY FOR NASA, I WOULD BE INTERESTED IN RESOLVING THIS QUESTION. THE [redacted] FOLLOWING INFORMATION WOULD BE HELPFUL... THE BIAS APPLIED TO THE TRANSISTORS DURING IRRADIATION, THE TIME ELAPSED BETWEEN IRRADIATION AND PARAMETER MEASUREMENT, THE UNIFORMITY OF DEGRADATION AMONG INDIVIDUALS ANY OBSERVED RECOVERY AFTER REMOVAL FROM THE RADIATION FIELD. I SHALL PHONE YOU THIS AFTERNOON /10 MAY/ IN HOPE THAT YOU CAN PROVIDE FURTHER INFORMATION.

R J WIDLAR

Fig. 8.8. Correspondence between Bob Widlar and Jean Hoerni and Sheldon Roberts (1963)

Young Widlar was firmly in command of transistor physics. Driven to excel, he was also well aware that success seldom comes for free. The untold story about Widlar's success is his hard work. Widlar spent countless hours in libraries studying the Bell transistors papers and scribbling numbers and ideas on receipt or napkins. Widlar worked constantly. The thrill of finding solutions of technical problems was the only thing for which Widlar could leave a bar in middle of a drinking spree, find a quiet place and organize his mind.

As almost everything else in Widlar's life, his start at Fairchild was remarkable. Jerry Sanders, the Fairchild marketing guru, conducted in the summer of 1963 a series of marketing presentations with sample demonstrations. One of these presentations took place in Denver and Bob Widlar attended the meeting. At Ball Brothers Bob was developing a transistorized DC voltage regulator and the reference source and was very interested in Sanders' briefcase with samples. Widlar was especially interested in a small box with about hundred NPN's 2N1613 – Fairchild's first transistor with high beta that held up pretty good down to 1 μA. List price was about $ 100.00 each. After the presentations, Jerry and Bob ended up in a bar, where after a few hours of technical discussions about Fairchild ICs Jerry was not able to control the situation and the transistors with a value higher than the cost of the most popular car at that time – Volkswagen Beetle, ended up in Bob's possession. Jerry regained control of his facilities the next morning – but he did not regain possession of those lost transistors! Fortunately for Widlar, smart Jerry Sanders was more concerned about Widlar's talent than the lost transistors and the integrated circuits samples. Sanders referred Widlar to a Fairchild recruiter. A subsequent series of phone calls resulted in Widlar's interview at Fairchild. Widlar took a few days vacation in August 1963 and shortly after,

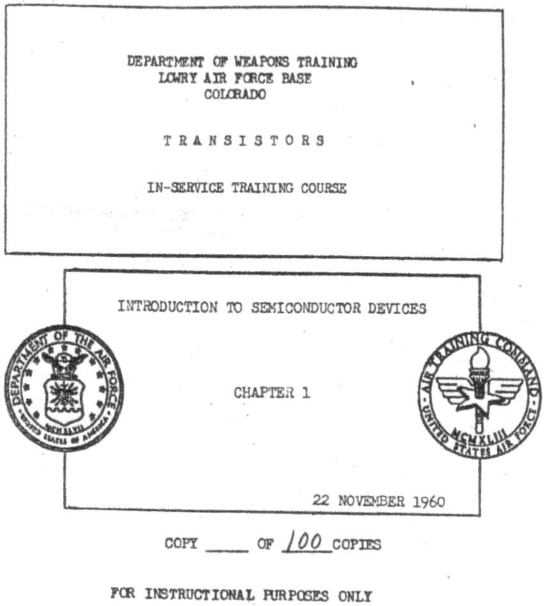
DEPARTMENT OF WEAPONS TRAINING
LOWRY AIR FORCE BASE
COLORADO

T R A N S I S T O R S

IN-SERVICE TRAINING COURSE

INTRODUCTION TO SEMICONDUCTOR DEVICES

CHAPTER 1

22 NOVEMBER 1960

COPY ____ OF 100 COPIES

FOR INSTRUCTIONAL PURPOSES ONLY

Fig. 8.9. The very first Bob Widlar publication – Air Force Training Course (1960)

when he returned from California, he mailed to his colleague at NASA, W. Gallagher, a letter announcing that for "*seventeen or eighty-five reasons*" he will be working on "*integrated circuit things*" for "*Circle-F*" (Fig. 8.11.)

Widlar joined Fairchild in September 1963 as an Electronics Engineer, reported to John C. Barrett who was heading a Linear IC Design and Application Department. Widlar should work on the evaluation of digital integrated components with application in analog circuits. Initially, development of integrated circuits was concentrated primarily on digital applications. Digital circuits were more tolerant to the limitations of monolithic integration, and they were less sensitive to the temperature dependence of transistor parameters. They could be constructed around elementary logical functions into basic building blocks and could be produced in high volume quantities.

Widlar, however, considered digital design and digital parts as low class work which did not need much talent. His argument was *"every idiot can count up to one."* The real challenge for him was analog design. At the same time he realized that integration of linear systems was limited by capabilities of current technology and no progress could be made unless better process control was achieved. Bob Widlar also recognized perhaps the most important fact of engineering work – information is not knowledge. Contrary to Fairchild's R&D organization directed by Gordon Moore and remote from production, Widlar put his hands on every aspect of the design, layout, fabrication, test, and product data sheet. Widlar, in a similar way to Shockley,

Fig. 8.10. Robert J. Widlar showed up at Fairchild Semiconductor in his new red Pontiac convertible with the top permanently down (1963)

Mr. W. Gallagher

I just came back from vacation in California. I kinda like it out there for seventeen or eighty-five reasons. In fact, I should be punching cattle at the Circle-F ranch out there in about a month - workin' on them integrated circuit things.

I'll tell you a few war stories next time I see you.

R. J. Widlar
Electronics Engineer

mbm

Fig. 8.11. Part of Bob Widlar's letter from August 1963 to W. Gallagher at the Solar Physics Branch of NASA in Greenbelt, MD

did not trust information from people who were not members of his innermost circle. Widlar learned very well that there are plenty of people who are able to talk convincingly about things they do not understand.

Fortunately, Fairchild had a person such as David V. Talbert. Talbert was a very capable engineer, but he did not like to talk. He never talked when he worked, and he was not terribly excited when other people talked. Regis McKenna described an anecdotal story about Dave Talbert: *once I went over to see him at National. I went into his office and we sat back and chatted for about a half an hour or so. I was late for a meeting and Don Valentine said to me, "Where were you," and I said, "Well, I was over chatting with Dave Talbert." He said, "For the last half hour?" I said, "Yeah, for the last half hour." And he says, "You've probably talked to him more than his wife does..."*

Dave Talbert graduated from Nebraska State University in 1957 with a Bachelor degree in Electrical Engineering and joined Fairchild Manufacturing Group in January 1962. Before Fairchild, Talbert worked in vacuum tubes division of Westinghouse. Not surprisingly, Bob and Dave bonded very easily, became the best friends, and would spend hours discussing everything.

Up to end of 1961, Fairchild R&D Device Development Group was headed by Phil Ferguson. Working for him was Heinz Ruegg, head of linear circuit group, and Donald Farina, head of logic circuit group. G. Parker, H. Wolf, B. Jones, and C. Craig worked on transistors; P. Almond and Randy Parker were developing the power transistor. When Phil Ferguson and Don Farina left Fairchild with Howard Bobb (Military Marketing Manager) Pierre Lamond replaced P. Ferguson.

Pierre LaMond who studied physics in his native France, was sent by the French Army to U.S. to learn about semiconductor industry. LaMond changed in U.S. name to Lamond and joined Fairchild in January 1962 when he was 31 years old, after three years at Transitron, Inc., where he was manager of the device development group. At Fairchild he become manager of the Device Development Section reporting to Gordon Moore. Because Gordon Moore had no knowledge of electronic circuits, in reality the design groups were managed by Victor Grinich, the only electrical engineer from the "Eight Traitors."

P. Lamond was not a popular boss and he was ridiculed frequently and served as a subject of many jokes and other climatic humorous twists. A fair note needs to be said about Pierre: in the time when he took over management of the Device Development Group, there was no coordination of work. In Moore's loosely managed R&D organization without assigned responsibility basically, every engineer followed his own path to set himself defined goals. Pierre, in a very short period of time set up order, priorities, responsibility and deadlines for all projects the group was working on.

An important day for the Linear Integrated Circuit was April 1, 1962, when preliminary circuit work was done with the Micrologic kit block to determine probable performance of a differential amplifier designed by John Barrett (Fig. 8.17). Dave Talbert worked on development of a process which included an epitaxial layer. His work was very successful and in May 1962, Gordon Moore reported *"the epitaxial isolated device have produced interesting results that have a considerable potential and the information should be considered "Company Private" for a long time."* Talbert's second major contribution was the pinched resistor 4 to 18 kΩ/sq resistor using the full emitter depth to pinch across the base stripe.

Project 151 was the most important for the future of the linear integrated circuits. Swiss-born Heinz Ruegg and G. Bechtel worked on a variety of analog circuits including operational amplifiers. Helmut Wolf worked on general purpose amplifiers. Other members of the linear group Hans Jaket and Bohumil (Bob) Polata worked on an amplifier for Autonetics contracted by Radiation, Inc., and later on the MOS differential amplifier suggested by Frank Wanlass. Remaining members of the group were Alan Grebene, Bill Lehrer, H. Nichols and G. Tripp. All linear designs were based on using Micrologic epi transistors and NiCr resistors; the NiChrome resistor deposition was developed by Bill Lehrer.

Effectively January 1963 the Device Development Section split into two parts. Heinz E. Ruegg became head of the Integrated Linear Circuit section and G. Bechtel, and H. Wolf moved to Micrologic group. Robert Seeds became manager of Logic Circuit Development section. P. Lamond re-organized the Linear Circuit development one more time, and limited work to only:

- µA001 Differential Amplifier (Terminated)
- µA002 General Purpose Amplifier (H. Jaket, B. Polata)
- µA003 Application defined Op Amps (A. Grebene)
- µA004 Video Amplifier (G. Bechtel, B. Frescura)

N. Gault, E. Porter, and D. Talbert were responsible for production line in Mountain View facility. Talbert was a very secretive and closed boss. He did his work, but he was a very tall, very strict, very stern sort of person who did his work superbly but didn't talk a lot about it. Talbert was very skeptical about the remote Fairchild R&D group. His philosophy was to do research by doing things. He refused to write a weekly report. When asked to report progress Talbert stood up and said *"If my boss does not know what I am doing and needs to learn it from weekly report he should not be my boss,"* and without another word left the room. Since this moment (July 1962) Talbert's name vanished from monthly Progress Reports.

Fig. 8.12. David V. Talbert (This picture was taken just a few months before the tragic car accident in October 16, 1989 when Dave Talbert was killed)

Development of linear integrated circuits at Fairchild before Widlar's era was dictated mostly by defense contracts. Fairchild was using an approach which was industry standard at that time – the circuit was designed in the same way as a circuit with discrete components. In discrete components design, any resistor value is available and the cost of the component is the same. In the integrated version the only appropriate diffusion for the diffused resistors was the base diffusion. This, however, does not allow enough flexibility to design circuits equivalent to the discrete counterparts. Fairchild R&D, as many others design organizations at that time, believed that the only solution was using thin films resistors.

A typical example of circuitry designed in early 1963 at Fairchild is shown in Fig. 8.13, which shows the layout and schematic diagram of the µA001 Differential Amplifier designed by Czech-born engineer Bohumil Polata. The circuit used Nichrome resistors deposited in vacuum over oxide protective coating. Although the 001 circuit was very simple, the die size was 1725×1725 microns and yield was very low.

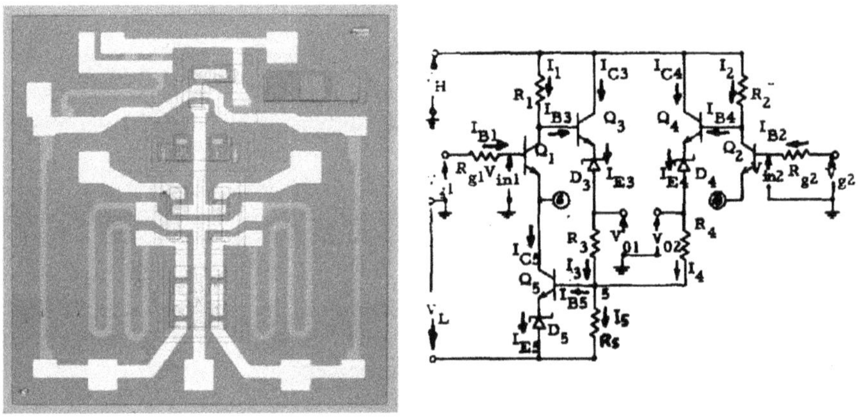

Fig. 8.13. Fairchild Differential Amplifier µA001 designed by Bohumil Polata in 1963

Bob Widlar came onto the Fairchild scene in the style of his Bohemian roots. Heinz Ruegg invited Widlar for an interview. Widlar boosted his appearance with a few drinks. When he looked on some analog circuit designs that Ruegg's boys were doing, he told him *"what they are doing is bullshit."* Ruegg did not appreciate Widlar's comments, especially from somebody who never did any work on integrated circuits. David Hilbiber, who was developing processing in Ruegg's group was also not impressed with Widlar because he had a feeling that Widlar was deficient in knowledge of semiconductor processing. Ruegg and Hilbiber sent him for an interview in Mountain View, to the Application Engineering Department headed by John Hulme. John Hulme hired

Widlar on the spot over the objections of most other who had interviewed him. This experience formed Widlar's attitude about Fairchild's R&D group and he never changed his opinion.

John Hulme hold degrees in Electrical Engineering from University of Utah and University of Southern California in Los Angeles. Before joining Fairchild, he worked at Lockheed as circuit designer and Fairchild customer using Fairchild discrete transistors. John was hired by Fairchild in 1961 by Bob Norman. He worked on the development of Micrologic products which were a relatively small effort in the company at the time, compared to discrete devices. He eventually became the manager of the Application Engineering Department. Fairchild's Application Engineering Department consisted of approximately 50 engineers divided into five groups: Digital IC Design & Application group was headed by Maurice O'Shea, Linear IC Design & Application group was headed by John Barrett, the Product Characterization group was managed by Henry Blume, Darek Bray headed the Consumer Application Engineering and John Reinhard was head of Test and RF Development Group. The Application Engineering Department was a link between R&D in Palo Alto, marketing and manufacturing.

John Hulme assigned Widlar as a junior engineer working for John Barrett. Widlar, then 27 years old, proved to Ruegg that he was deadly serious with his criticism one month after he joined the company. He offered, along with his criticism, also, better solutions. Everybody recognized merit of Bob's ideas and Pierre Lamond wrote in the Monthly Report dated November 1, 1963:

> *"Robert Widlar from Application has come up with a circuit which we feel has some merit. A simplified version of his circuit is presently being breadboard by us. The two candidates will be compared with each other and decision made on the basis of relative performance and circuit complexity."*

Bob Widlar was eventually applied so much pressure on John Barrett's group that Barrett resigned and went to work for Hewlett-Packard. The merits of Widlar's circuit was so important that Pierre Lamond concluded in the monthly report: *"problems have appeared in practically every one of the project area, and the differential amplifier made with NiCr resistors has run into processing problems and it has also been found that it was conditionally stable for capacitive load in excess of 80 pF. Furthermore, all the linear circuit seem to be suffering from channeling"* Lamond once again terminated several original projects in R&D group and future Fairchild linear circuits were designed by the Bob Widlar way.

Widlar had a similar attitude as Talbert; once they realized that a person is not on the level of the problem they wrote them off. Widlar started work closely with Dave Talbert and very quickly learn "processing tricks." Talbert and Widlar conspired; they keep their data under a lid. They were moonlighting and running one experiment after another. Talbert was running

Fig. 8.14. **a**) John Hulme in Fairchild office area (1966); **b**) Fairchild's John Hulme, Ben Anixter, and Mike Markkula (1969)

literally two jobs, process engineer for digital Micrologic as day job, and a night job as process engineer for Linear Integrated Circuits. The cycle time in Mountain View line was approximately six weeks, Talbert pushed Widlar's lot in two weeks. They both worked harder than hard.

Bob Widlar met another person at Fairchild who in a significant way contributed to Widlar's success – Mineo Yamatake. Mineo's family after being released from a U.S. concentration camp returned to their native Hiroshima after WWII. But there was no Hiroshima; there was nothing. Twelve years old Mineo, with the help of his uncle, returned to America. He applied for a job at Fairchild in 1959 but was not hired. Young Mineo started working for Link Aviation where he met John C. Barrett. Barrett, although he was himself a very laid-back person, recognized Mineo's quality, and when he joined Fairchild, he brought Mineo on board. Many claim that this was Barrett's biggest contribution to Fairchild. Mineo become Widlar's most reliable and trusted technician and their collaboration lasted until Widlar's death. Another technician who worked easily with Widlar was Ken Kraft.

Not everybody could work with Widlar. Bob Sleeth in the summer of 1962 walked in off the street into Fairchild personnel office and asked for a job as a technician. *"They gave me a test on transistors which I failed miserably, having not a clew what one was, as my electronics background was with the military in vacuum tubes,"* Bob recalled. Fairchild's response was *"Can you start on Monday?"* Bob worked on the first analog circuit which was contracted by Radiation Inc., for Autonetics. Radiation paid for an NPN differential pair with collector resistor load $20,000 apiece. The circuit was designed by John Dell, and it took 9 lots[2] to yield one operating part. By mistake Bob Sleeth "fried" the first good part in one second. After about another 9 more lots one more goof part was working. When Widlar joined Fairchild, Bob Sleeth was assigned to Widlar. Bob Sleeth's characterization says all: *"The match made in hell, but we managed not to kill each other"*

Talbert's function at Fairchild was what is called today a "Process Integration Engineer" and at that time was called an Operations Engineer. David Talbert was involved in additional highly secret and important project. Dolores Brasch, as a diffusion technician, was responsible for evaluation of sheet resistance of diffused layers. There was a dedicated measurement set up that included an oscilloscope. No adjustment of the instrumentation was ever needed because over and over the sheet resistance was always in a certain range. Before deciding if Dolores should be asked out, David, as a pure engineer, wanted to have the data first. David turned the oscilloscope knob and changed the range. He wanted to see if Dolores will be able to figure out where the problem was. Dolores not only passed the IQ test which "Eight Traitors" implemented as a vehicle for hiring of the first wave of Fairchild employees at the end of 1959, she also passed David's tests. Several years later, David and Dolores married.

Talbert was a master in organizing key operations into his "sphere of influence." After Dolores Brasch became a mask designer he moved her and her drafting table into his office. With Widlar designing, Mineo measuring, Dolores working on masks and Talbert pushing lots through production line, they have all under control. Widlar and Talbert ignored the rest of the world and they did not pay attention to anybody or anything except their work.

Talbert and his future wife Dolores, called "Big D", became Widlar's life' long friends. "Big D" was able to manage both cohorts and keep them straight. They both needed a keeper. Bob Widlar shaved, wore ties, and kept regular hours and they were on a roll.

One of Widlar's very early works at Fairchild was his involvement in Fairchild's reliability program. During 1964 Fairchild Semiconductor proved a reliability failure rate of 0.0032%/1000 hours at 120°C with planar epitaxial transistors. Bob and Dave started implementing improved manufacturing methods and process control which reduced the price of the planar process and prompted the development of linear integrated circuits.

In the course of this work, Widlar learned about the parasitics of integrated components and their limitations, both in type and in the range of values. Linear integrated circuits differ from their discrete counterparts by the fact that:

a) no large-value resistor may be used because of area limitation
b) no large value capacitor may be used
c) no inductors can be used

[2] At that time lot consists of 12 wafers

Fig. 8.15. Dolores Brasch-Talbert (left) and Bob Noyce (right) on April 18, 1962 cut red ribbon to open Mountain View plant's patio

d) the integrated transistor has parasitic components which are not present in discrete parts
e) transistor characteristics and circuit resistance can be better matched at all temperatures

In the spring of 1963, just before Widlar joined Fairchild, Howard Bogert developed a design technique called "Circuit Breadboarding", which became the only technique used to design analog integrated circuits for the next thirty years. Bogert's approach uses a set of integrated circuit elements – resistors, diodes, and transistors produced by integrated circuit process and individually packaged. The breadboarding device contained at least partially the parasitic components of device isolation which would appear in the final integrated version of circuits.

The Fairchild experience was based mostly on the design of digital integrated circuits and led to the so-called "Point System" where each component item was rated with certain points. If the total number of points was higher than 150, the circuit was considered not able to be integrated. The point systems and weight for each component is shown in Table 8.1.

Dave Talbert established at that time ($\sim$ 1963) another rule: the die which will yield must be smaller than 900 $\times$ 900 microns.
Widlar very quickly acquired knowledge of Fairchild's state-of-the-art technology, and equipped with the latest results of integrated transistor characterization from Fairchild RTL and DTL Micrologic, Widlar formulated one of his favorite theorems: do not attempt to match discrete counterparts by

Fig. 8.16. Mineo Yamatake

Table 8.1. Fairchild's integration "Point System" criterion

COMPONENT	# POINTS
Resistor of value $R[\Omega]$, $R < 300\Omega$	$4 + 65/R$
Resistor of value $K[k\Omega]$, $R > 300\Omega$	$K + 4$
Each External Connection	4
Group of transistor with common Collector ($n < 5$)	$2(n - 1) + 5$

IC's design. He started work on what became later the μA702: the industry's first linear integrated circuit, which departed from conventional circuit techniques as used in discrete components networks. Widlar put H. C. Lin's theory of matched and compensated devices into practice. Reference to Lin's publications is in each Widlar paper that he published at that time.

The concept of integrated circuits whose operation is dependent upon matched active and passive components is more obvious today than then. Actually today it is difficult to find an engineer who is able to design circuits with discrete parts.

Fairchild's process was designed to optimize the performance of NPN transistors in DTμL digital applications. Those transistors were using a 2 to 5 Ω-cm epitaxial layer which restricted the current handling capability of the transistors. The small transistor (25.4 × 25.4 μm emitter) would exhibit its current-gain peak at 1 mA due to a limited velocity of majority carriers

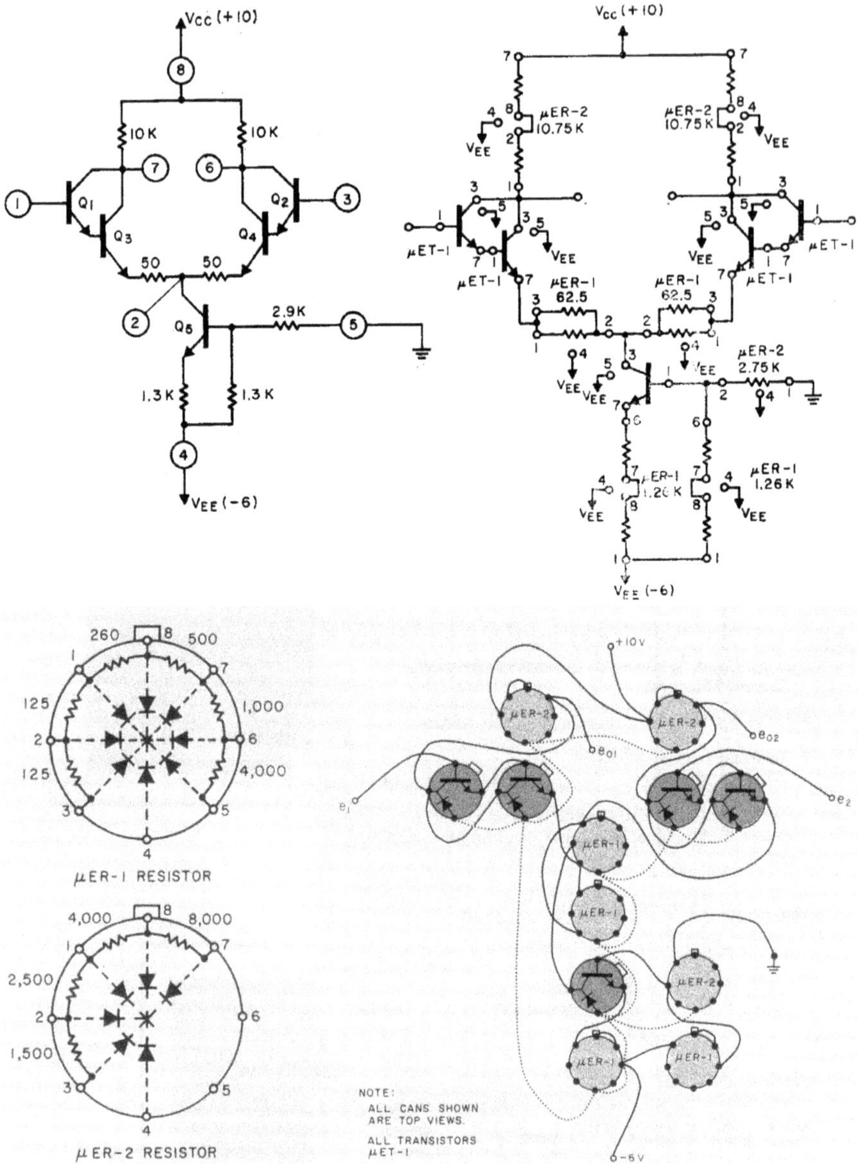

Fig. 8.17. An example of a breadboarded circuits of a differential amplifier. All isolation diodes of "transistor cans" are connected to −6 V

in the high resistivity collector bulk. Therefore, when operating at moderate currents, the transistor had to be made considerably larger than minimum size geometry.

The typical characteristics of NPN transistors produced by the Fairchild Planar I process after Talbert implemented improved the process control and changed several technological operations are summarized in Table 8.2 and in Fig. 8.18, and Fig. 8.19.

Table 8.2. Electrical parameters of large geometry integrated transistor at 25°C (Fairchild "Planar I" process, 1963)

PARAMETER	CONDITIONS	VALUE
Collector-emitter sustaining voltage LV_{CEO}	$I_{\mathrm{C}} = 10\ \mu\mathrm{A}$	60 V
Collector-emitter breakdown voltage BV_{CEO}	$I_{\mathrm{C}} = 10\ \mu\mathrm{A}$	50 V
Collector-isolation breakdown voltage BV_{CIO}	$I_{\mathrm{C}} = 10\ \mu\mathrm{A}$	110 V
Emitter-base breakdown voltage BV_{EBO}	$I_{\mathrm{E}} = 10\ \mu\mathrm{A}$	7 V
DC Current gain $h_{\mathrm{FE}}@V_{\mathrm{CE}} = 10$ V	$50\ \mu\mathrm{A} < I_{\mathrm{C}} < 100$ mA	~ 60
Saturation resistance R_{sat}	$I_{\mathrm{C}} = 10\ I_{\mathrm{B}}$	800 Ω
Transconductance g_{m}	$I_{\mathrm{C}} = 0.5$ mA	15–20 mA/V
Collector-base capacitance C_{CB}	$V_{\mathrm{CB}} = 20$ V	18 pF
Collector-isolation capacitance C_{CI}	$V_{\mathrm{CI}} = 20$ V	8 pF
Emitter-base capacitance C_{EB}	$V_{\mathrm{EB}} = 20$ V	10 pF

Except for the higher saturation resistance, resulting from the top-side collector contact, the integrated NPN transistors were nearly identical to discrete planar transistors made with similar doping levels. The 2N1613 transistor which Bob Widlar knew, thanks to a binge with Jerry Sanders, was used as a good approximation to estimate circuit's behavior.

In Fig. 8.19 is shown the collector-base leakage current which is negligible at room temperature; however, it can become significant at elevated temperatures. The major problem in integrating linear integrated circuits was leakage current of the collector-isolation diode because of the large junction

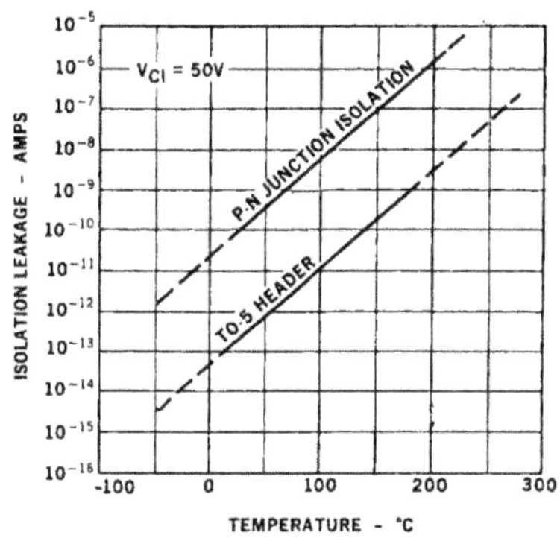

Fig. 8.18. Collector-isolation diode leakage of 35 μm × 28 μm "Kit Transistor" (Fairchild "Planar I" process 1963)

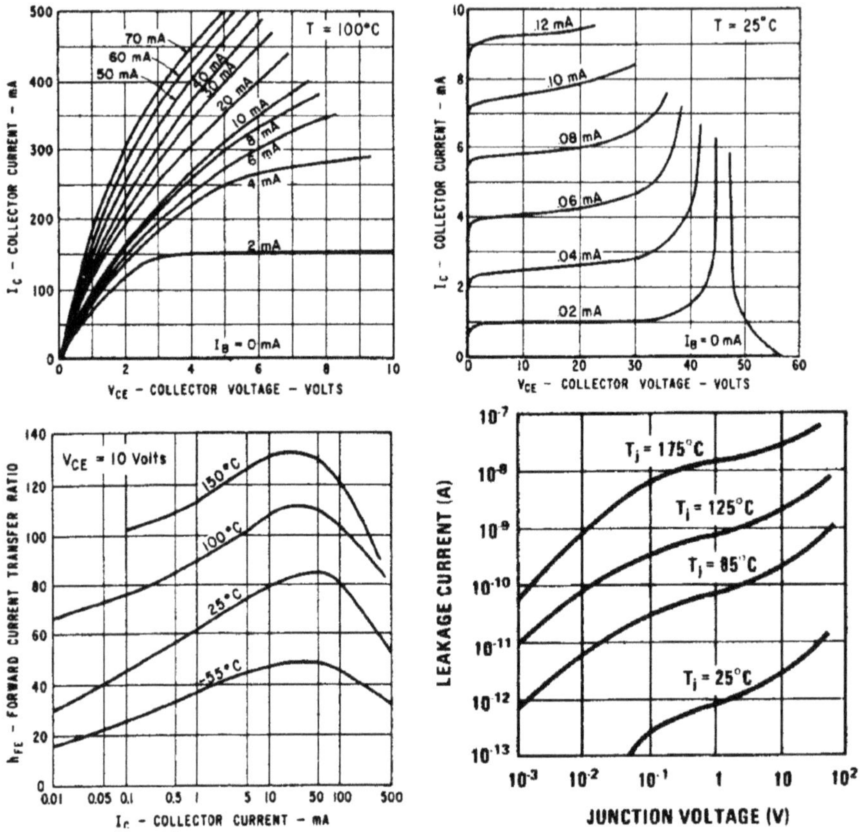

Fig. 8.19. I-V and h_{FE} characteristics of large geometry integrated transistor (Fairchild "Planar I" process, 1963)

area. The leakage current as a function of temperature is plotted in Fig. 8.19. The package leakage could contribute to the total leakage quite significantly. In the 1960's the TO-96 and TO-99 with 8 or 10 leads was a standard package for almost all IC's, digital as well as linear.

Bob Widlar's "702 notebook" contains a schematic of a discrete component operational amplifier (Fig. 8.20) which he used as a starting point in 702 design[3]. The 702 was the first linear integrated circuit which used only diffused resistors, and where possible, transistors were substituted for large value resistors. The "µA702", as shown in Fig. 8.21, contains only nine NPN transistors and Widlar used two key innovations: conversion of differential signal to single-ended output without losing half of the gain; and the DC level shifter using only NPN devices. Transistors Q4 and Q5 form a simple common emitter amplifier. If the gain of these devices is large enough, the

[3] This amplifier was very similar to the PhilbrickP65 or Nexus DA1

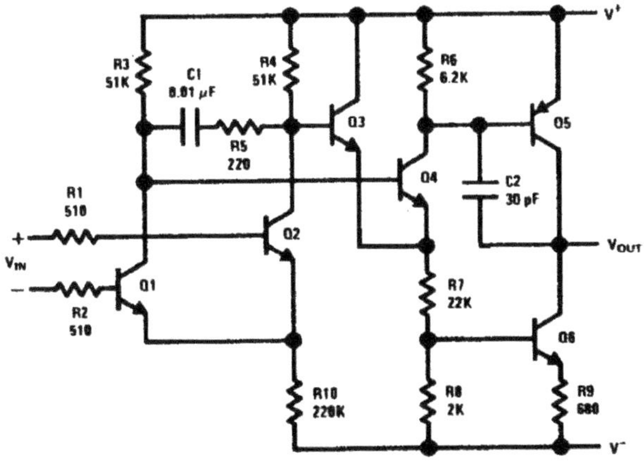

Fig. 8.20. Discrete components operational amplifier [R. J. Widlar Notebook 1964]

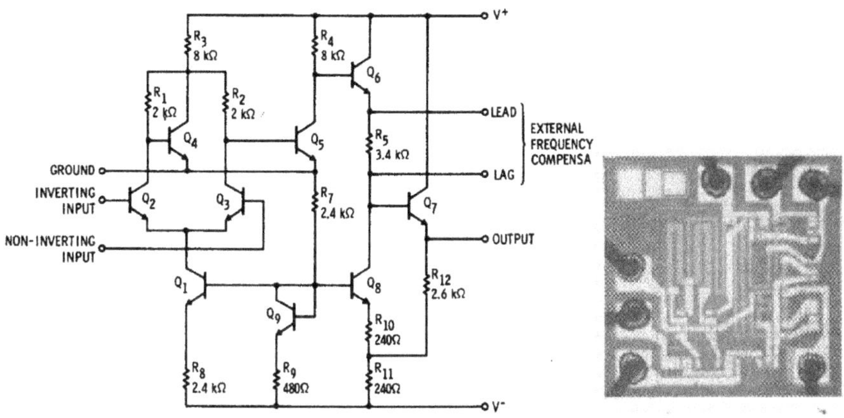

Fig. 8.21. Circuit diagram of μA702

voltage at the base of Q4 changes much less than collector voltage. If the collector current of Q2 increases the voltage drop on R1 also increases. In the differential stage, a decrease of collector current of Q3 is the same as an increase of collector current of Q2; therefore, the voltage on R2 decreases. Both currents in the differential stage Q2 and Q3 contribute to the voltage which is driving Q5. Emitter follower Q7 with R12 and Q8 with R10 and R11 form a positive feedback loop.

The '702 circuit had several limitations: low gain, and (due to grounded emitters Q4 and Q5) almost no CMR, and very limited output drive capability. However, there was a feature which was unheard of at that time for discrete circuits: connecting a small capacitor to the external frequency compensation pad (Parallel to R5) the bandwidth can be extended to a range of 25–30 MHz. Considering that the 702 was an NPN-only circuit, Widlar set precedents which have held for more than a decade. Low DC offset, low drift, and capability of operation over a wide range of power supply was another nice feature of the 702.

Up to this point, Widlar and Talbert did all this work semi-illegally with John Hulme sponsorship. When the '702 performance showed signs of progress, Widlar found another ally – Floyd Kvamme, Floyd was a technical guy, but was working in marketing. He knew potential customers for the 702 and he contacted them. Reaction of several customers which Floyd contacted ranged from positive to phenomenal. Floyd passed the results of his marketing survey to Tom Bay. Tom Bay and Bob Noyce decided that Fairchild will be selling linear parts. Problem was that either of them did not knew who Bob Widlar was. One day in May 1964 they both went to see Widlar, and introduced themselves. Widlar knew who they were, but had no previous contact with them. This was the very first encounter when Widlar met Tom and Bob, Fairchild's Pinnacles. Tom said that Fairchild will start a marketing campaign to sell the µA702, and listed to Widlar items he needed from him to start marketing of the µA702. Widlar who knew about low yield, felt that he still need some time to finish the part said, *"what do you mean?"* to which Tom responded "w*e are going to announce the µA702 and we are going to sell it,"* said Tom. The genes passed on to Widlar from "sekac Franta" activated instantly and Widlar blurted *"Fuck You!*

Although many engineers feel the same way as Widlar did, they were forced today to be "yes man" and agreed with their boss even in the situation when they absolutely know that decision will hurt the company. In the sixties no Human Resources Department defined what we may or may not say. We were allowed to be sincere and we did not need to pretend to be always happy and positive. The problem with happiness and positive hypocrisy is that somebody ultimately pays for it. I am frequently witnessing that people do not know what they are doing, when a part or experiment is ruined, we just take a new part or run the experiment again. Robert Widlar could not stand this attitude. He always explained to his technician what he was expecting and always emphasized *"if you do not know what are you doing, don't do it."*

In the sixties we were allowed to tell a girl that she had beautiful legs or nails, (and beautiful girls enjoyed it immensely.) Young successful men (like Jerry Sanders) or older bachelors (like Sherman Fairchild) took full advantage of it. At that time Electronics News was advising to readers that "fun headquarters for the electronic field" is in Gaslight Club of Beverley Hilton and Electronics Magazine was suggesting to marketing departments and advertising agencies the best selling approach as one shown, for example, in Fig. 8.22. This is something that is difficult to imagine today, when politically correct empty style replaced substance.

Widlar completely seized the moment and continue: *"You are not going to sell these things, because you do not know how to sell them. You guys do not know anything about this part. Nobody inside this company knows anything about it. If I give this to you, you will just screw up my work."* Widlar refuse to cooperate.

Although Bob Noyce told me later that he took Widlar less seriously and liked the Widlar enthusiasm, Tom Bay was offended, and his comment was "*this arrogant nut is driving me crazy.*" Noyce knew about Texas Instrument's morass with linear parts for Minuteman II. Widlar's part was a precious stone and a similar masterpiece like Hoerni's planar process, and after all, Fairchild got it for nothing.

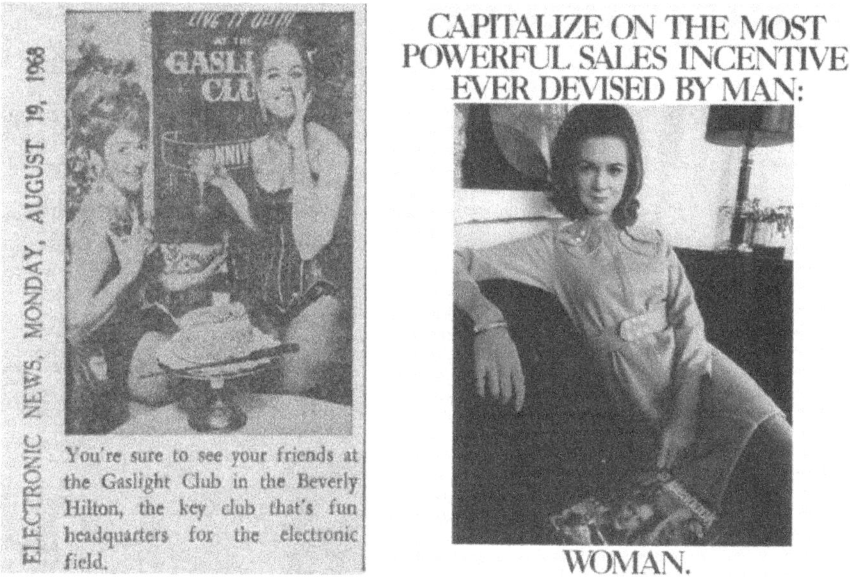

Fig. 8.22. The Electronic News and McGraw Hill Electronics Magazine ad prior to politically correct era in which hypocrisy substitutes for common sense

At the same time Jack Gifford was transferred from Los Angeles field office to Mountain View. Jack was hired by Don Valentine, who ran the Fairchild field sales in 1963. Jack was formerly working as a design engineer for a small company in Los Angeles and to be a salesman *"was the furthest thing from his mind."* Soon twenty four years old Gifford was selling some $3 millions worth of semiconductors to Fairchild's biggest sales territory. Hughes, who built the Phoenix missile systems, was the biggest customer.

John Gifford reported in his new position to Floyd Kvamme and Floyd asked Jack to put together plan for Fairchild linear circuits' roadmap. Analog technology was and still is Jack's forte. Jack with his background in control theory and his understanding of amplifiers enthusiastically started to strategize linear product development. Widlar strenuously objected to Gifford's

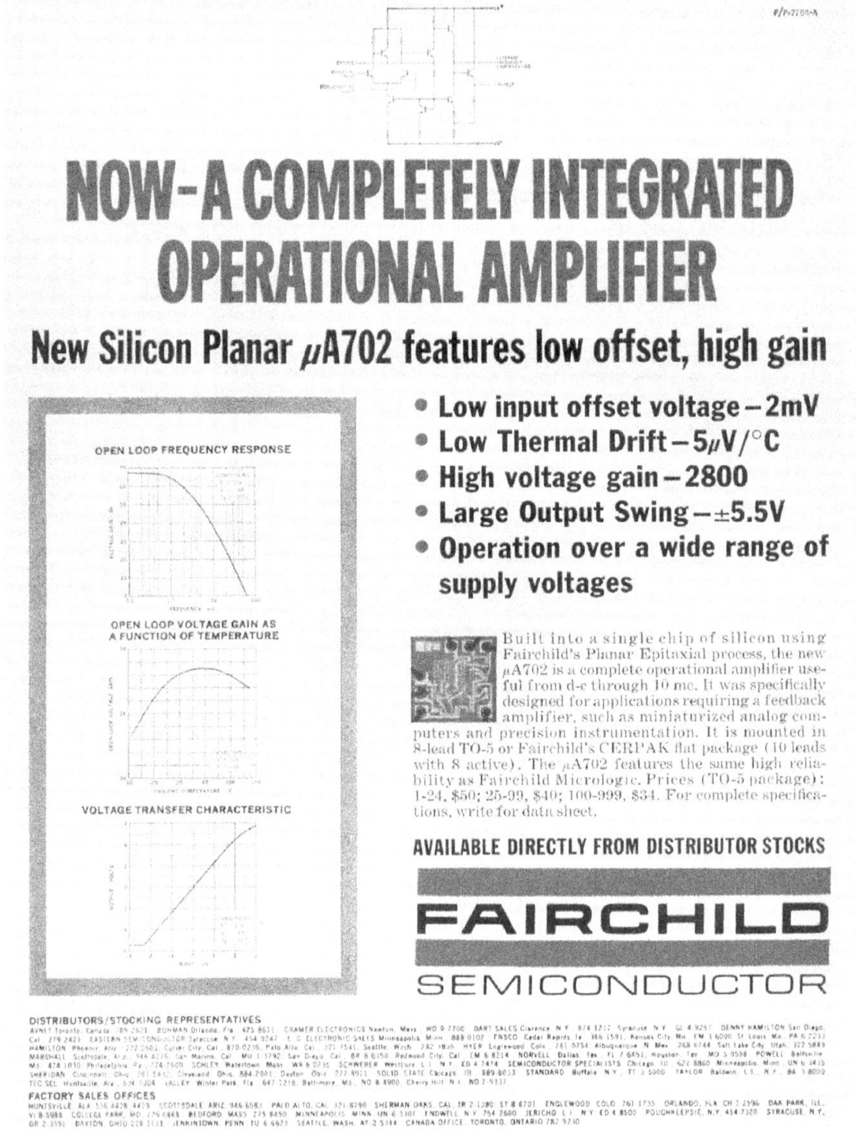

Fig. 8.23. Fairchild introduced the µA702 in October 1964. Note the price and the text "Silicon Planar µA702"

product direction and ignored it. Widlar was sufficiently established to ignore Jack, but he took Jack's quite grandiose plan including PCM, A/D and D/A products and op amps, corrected English and posted plan on his bulletin board. Bob and Jack were very strong, independent personality and although Jack's plans make a sense Bob opposed it because it was not his idea.

Jack willingness to share an occasional drink or two with Bob gradually smoothed edges. Jack Gifford become at age twenty five, the first Fairchild Linear Circuit Product Manager and big supporter of Widlar's circuits

Jack started building his marketing group, and he hired Mike Scott, Gene Carter and Mike Markkula. Jack learned how to go along with Bob. Bob loved talk and solved engineering problems, especially in the bar. After one of these technical sessions, Jack and Bob ended up in Bob's tree house, which Bob maintained near Big Basin. Bob had there various cans marked with names like, for example Pierre Lamond, and they served as shooting targets for his middle-of-the-night shooting. He did not live there permanently, but it was very convenient hideaway in early morning hours after bars closed.

Fairchild released the first μA702 advertisement in October 1964 (Fig. 8.23.) The advertised price of the '702 was $50.00, but you could not get the part for this price. The initial retail price was around $300.00 and even with this high price you could not get them easily. I remember that I placed an order to Schweber for 5 pieces. When I was questioning the status of my order a couple weeks later, I learned that I needed to order other parts from that distributor. The distributor was using the 702 as a bonus if customers ordered a large volume of discrete transistors or diodes. I destroyed the first unit almost immediately after connecting to the power supply. After this experience, I spent hours checking to see if there was any reason why a part should burn up before I turned the power supply switch on.

Bob wrote me *"The product and its design aren't the only important things with ICs. Even the best product can get in trouble if the manufacturer does*

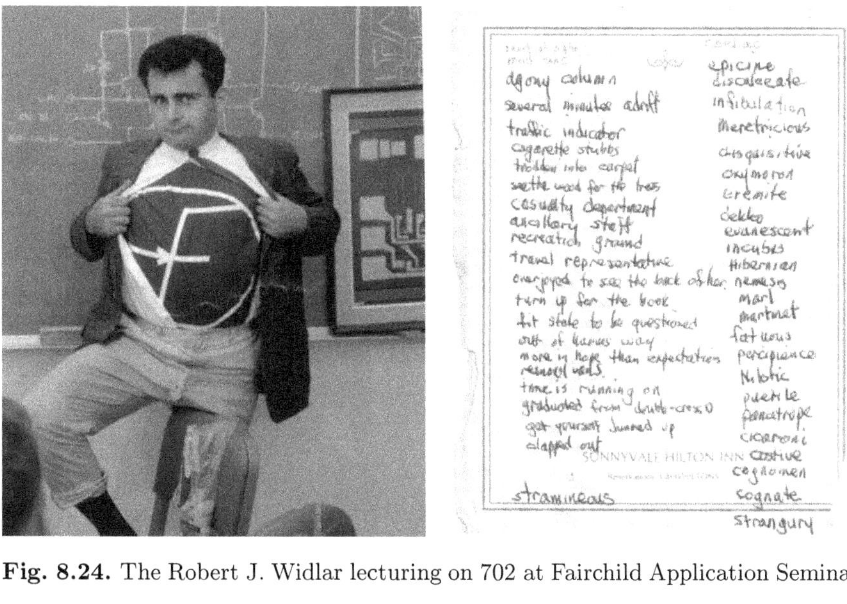

Fig. 8.24. The Robert J. Widlar lecturing on 702 at Fairchild Application Seminar

not supply application back-up for it." Widlar knew that new and unproven technology would need a great deal of support. He prepared Application Notes covering the main '702 applications. He was touring with Jack Gifford across the country, lecturing and giving advice to engineers at numerous Fairchild seminars. Bob developed quite a professional level of showmanship and with additional help from the Ambrose Bierce's "Devil's Dictionary," and with Widlar's humor the seminars were attended in masses in places such Madison Square Garden in New York.

The µA 702 gave rise to one of the greatest commercial and technical successes of the young semiconductor industry, the masterpiece – the µA 709 operational amplifier. The 709 was far more powerful than the earlier chip, and it quickly became the most widely used operational amplifier in electronics.

The '709 is faster and far more efficient than the '702 and had ten times the amplification: whereas the 702 can boost an incoming signal by some seven thousand times, the 709 can raise it an astounding seventy-thousand-fold. Until Widlar created the '709, such a chip was thought to be impossible.

In the '709 design Widlar brilliantly traded the restrictions imposed by the limited types of components, poor tolerances and limited range of integrated component values against using a large number of active devices with free choice of device geometry, and close matching of active and passive devices over wide range of temperatures.

The first stage of the '709 is similar to the input stage of the '702. Resistively loaded differential stage Q1 and Q2 is biased by current source Q11 / Q10. The differential to single-ended converter Q3, Q5 and Q7 form a simple amplifier. Emitter follower Q3 is biased by low current by the voltage across resistance R3 which depends on the difference of voltage between Q5 and Q15. The common-mode input voltage range is extended by negative feedback loop to bias the front end of the amplifier. The common-emitter stage Q6 and Q15 increases the current through source Q10 / Q11 in presence of the common mode input voltage. Current mirror Q10 / Q11 regulates and maintains constant current to Q1 and Q2

The transistor driven by the second stage, Q8, is resistively coupled to Q6 and a lateral PNP transistor Q9 which is working as a level shifter. The third stage of the 709 consists of common emitter Q12 and with complementary Q13 and Q14 transistors, where Q13 is a vertical PNP transistor. Because complementary followers are off for base voltage ~ 0.7 V, there is a range of input voltages when neither transistor is conducting. Widlar used a negative feedback loop R15 to reduce distortion of output signal which is working unless the signal frequency is too high. The output stage had no protection against output short-circuits.

The lateral PNP transistor was a quite challenge in 709 design. D. Hilbiber in R&D group was working on circular structure of transistor similar to H. C. Lin design in Westinghouse. Hilbiber did not want to be stuck in production with device which was not properly working yet and he refused to release

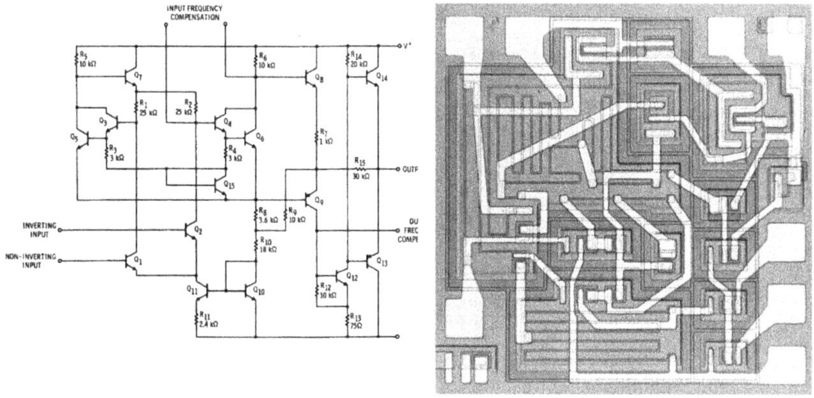

Fig. 8.25. The Fairchild Operational Amplifier µA 709 has fourteen bipolar transistors and fifteen resistors. Actual die size: 1880 × 1880 µm

his device for Mountain View production line. Bob Widlar, as was his habit in such situation, demonstrated his superiority. He worked continually for about 170 hours and developed device which was using two resistor diffusion regions as a transistor region. The gain of the transistor was not higher than 0.1, but this was enough for level shifter.

Widlar introduced the µA709 at the National Electronic Conference in October 1965 [4]. The µA709 hit the market in November, 1965. In December the Bendix Corporation placed an order for 10,000 units. Widlar and Talbert accomplished the impossible – customer demand for parts designed by Widlar exceeded the estimated demand by a factor of 10 and Fairchild was sold out in Linear Circuits for almost two years (1964–1966).

In the beginning of 1964 there were the first indications of managerial problems in the Fairchild organization. The skyrocketing phenomenon and success of Widlar's part delayed bigger visibility of company difficulties because declining sale of digital parts was offset by linear parts. In September 1964, non-military products accounted for well over a third of the Fairchild sales.

Widlar's 702's and especially 709's did not have any challenging competitors on the market. In 1966 the only integrated operational amplifiers available on the market were:

Amelco	A13-250
General Electric	4JPA107/135
Motorola	MC1530/31
Philco	PA702/PA 712
RCA	CA3015/16 (available 1967)
Texas Instruments	SN512/522/524/526/(724 available 1967)
Westinghouse	WS161Q
General Instruments	PL210/212/250/251

Philco's PA702 was a replica of the Fairchild μA702; all other parts could not really compete with the Fairchild line. Texas Instruments announced in October 1964 the SN524A (military) and in beginning of 1966 the SN 525, 526, and 724 general purpose op amp (Fig. 8.27 and 8.28) But the 724 was not available until 1967.

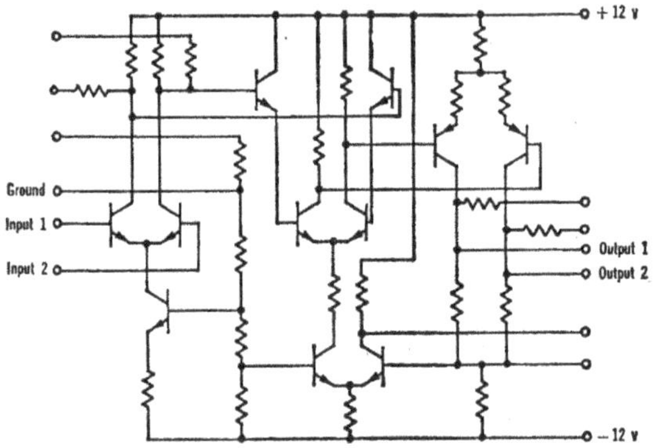

Characteristic	Typ. Discrete-Component Amplifier	1965	
		SN525A	SN526A
Gain, Open-loop, db	94	88	60
Input-voltage Offset, mv	1	1	3
Temp. Coefficient, Input-voltage Offset, μv/°c	5	5	10
Input-current Offset, μa	0.2	0.3	0.03
Common-mode Rejection, db	92	100	80
Output-voltage Swing, v	± 11	± 8	± 5
Output-current Peak, ma	2	1	10
Input Impedance, megohm	0.3	0.07	2

Fig. 8.26. Texas Instrument's op amps (1966)

The parameters listed in Fig. 8.27 and 8.28 look pretty good, but these were only "typical characteristics" and the guaranteed specification were much poorer.

The only other somewhat reasonable device on the market was Westinghouse's 161Q and Amelco's D13-001 and RCA CA3015/16. It took Motorola and RCA another two years to come up with any reasonable operational amplifier. Bob Dobkin told me recently that Motorola's first amplifiers where designed by someone from Burr-Brown.

Although the μA709 was far better than the monolithic amplifiers on the market at that time, there were problems. The very first version was afflicted by leakage current to the substrate through isolation diffusion. Fairchild had a sudden need for 10,000 a month but was producing less than 300 amplifiers which met the specifications per week. Widlar, Talbert, and Yamatake were working feverously. Talbert improved the original Fairchild Planar I process

SN724 General-purpose Operational Amplifier

TYPICAL CHARACTERISTICS

Voltage gain	1200
Input impedance	1 MΩ
Input offset voltage	12 mV
D-c drift referred to input	20 μV/°C
Output signal swing, single-ended	±7.5 V
Common-mode rejection	55 dB
Temperature range	0° to +70°C

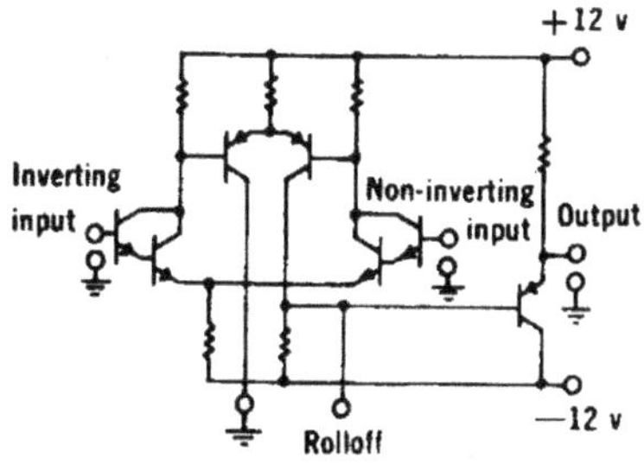

Fig. 8.27. Texas Instrument SN724 general-purpose op amp (March 1966)

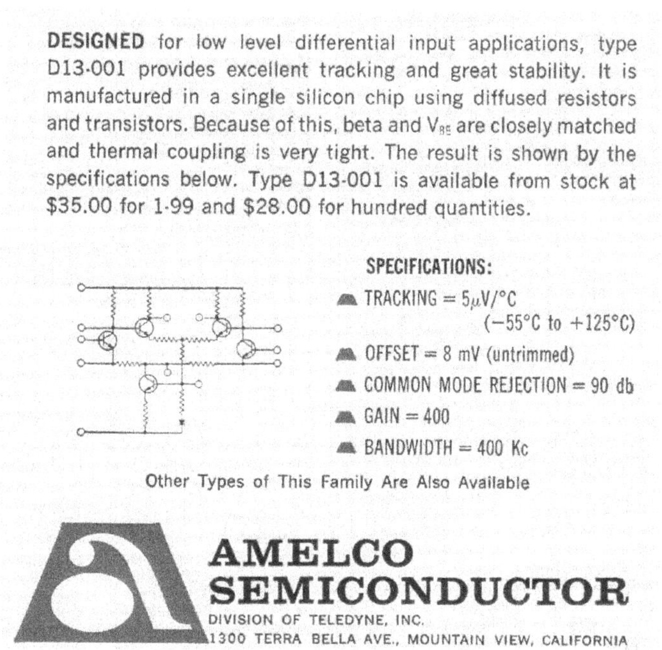

Fig. 8.28. Amelco D13-001 Operational Amplifier (This was rather differential amplifier, because output swing was very small)

and reduced leakage through isolation with "base diffusion overlay" over P isolation diffusion and did all masks revisions.

Widlar generated several circuit revisions which resulted in the μA709A. The '709A was the first Fairchild circuit which was using the more durable chromium masks in production. Yamatake evaluated countless circuits and measured everything that could be measured. Talbert was especially under big pressure due to the fact that the production line was in 1965 switching

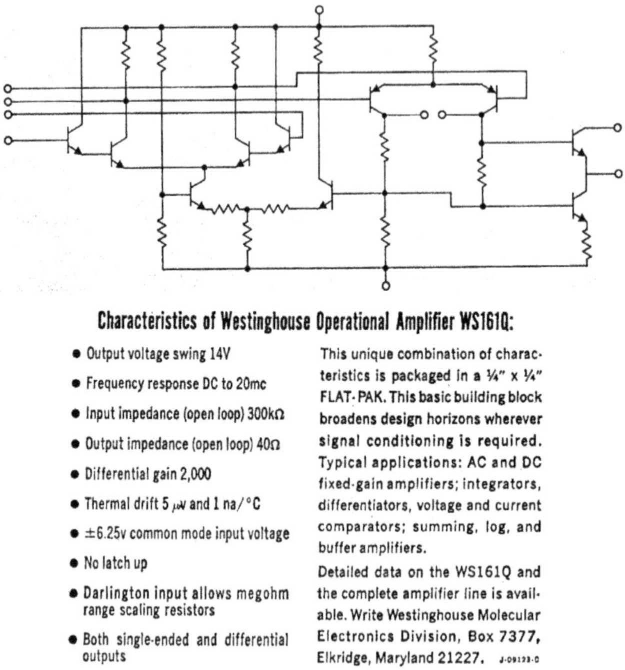

Fig. 8.29. Westinghouse Operational Amplifier WS161Q

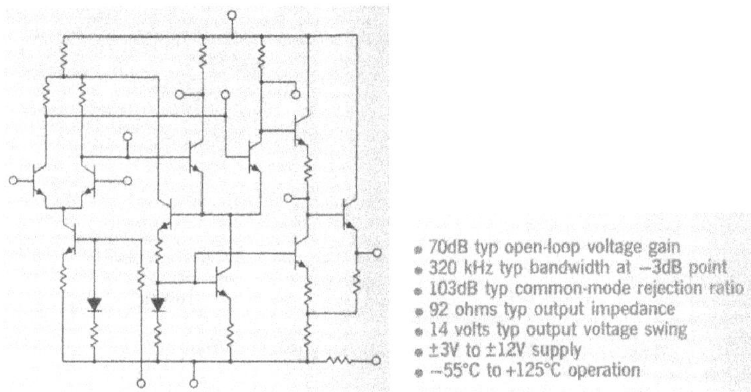

Fig. 8.30. The RCA CA3015/16 (September 1966)

from 1-1/2-inch wafers to 2 inches wafers, and Fairchild was introducing ultrasonic bonding.

In 1966 Bendix only used 80,000 of Fairchild's μA709 op-amps. Fairchild was making '709s around the clock. By August 1966 production stabilized and Fairchild was selling 5000 pieces of 709s each week.

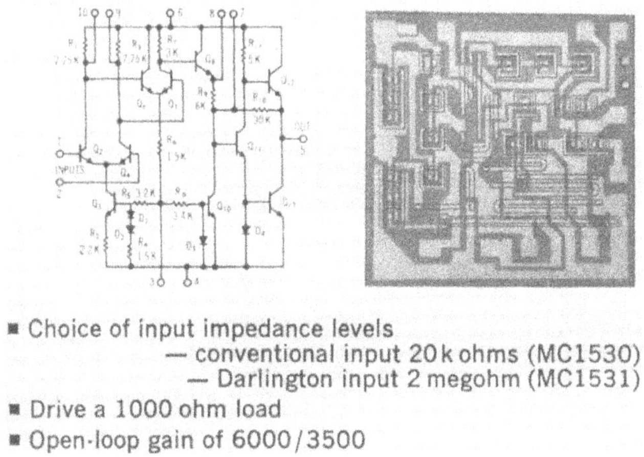

Fig. 8.31. Motorola 1530/31

When Talbert and Widlar realized how much money Fairchild was making on their work, they asked for a pay raise. Widlar's compensation was growing rapidly, but not as fast as he wanted in view of his established notoriety in the linear circuit industry. Widlar's grandfather's Bohemian genes did not miss a cue: *"Pushing a wheelbarrow would be a suitable challenge if it paid enough."*

They argued that all work on Fairchild linear integrated circuits was done by production engineers without any major investment, and without the contribution of Fairchild's R&D group. Most of Fairchild's linear circuits innovations originated from the production group. The "Weekly Report" style of R&D organization was a subject of numerous jokes and created a significant tension between the Palo Alto and Mountain View "linear groups." Talbert especially had no respect for Palo Alto R&D group and its leader Gordon Moore. He argued that Moore never did any work himself and all information and knowledge he used was information passed on to him by others.

In April 1964, a major reorganization had been made in the electronics operations of Fairchild Camera & Instrument Corporation. The move includes the long-expected formation of a new Fairchild Instrumentation Division, combining the electronics equipment capabilities of Fairchild Semiconductor division in Mountain View and Dumont Laboratories division in Clifton, NJ. Robert N. Noyce, vice-president and general manager of the Semiconductor division had been named to a new post of group vice-president, with responsibility for both the Semiconductor and Instrumentation division. The new division was head-quartered in Clifton with John S. Auld as general manager. The Dumont tube activity became the Dumont Tube division under general manager Fred Walzer. Charles E. Sporck, operations manager for Fairchild Semiconductor, had been named to succeed Robert Noyce as gen-

The Fairchild Camera & Instrument Corp., in its first step toward "making the company as significant in instrumentation as it is in semiconductors," recently named **Robert N. Noyce** group vice president of the reorganized semiconductor and instrumentation divisions.

The aim, explains Noyce, a 37-year-old physicist who holds a doctorate from the Massachusetts Institute of Technology, is to make specialized semiconductors available for higher-performance, lower-cost instruments. The best way to do that, he adds, is to tie the two divisions together.

With the change in organization, he says, "I hope to create an atmosphere in which any product idea will be well nurtured and have a chance to grow. Noyce adds that work being done now in the semiconductor labs is expected to result in products other than semiconductors.

Fig. 8.32. Press announcement about Fairchild reorganization [6]

eral manager. Gordon Ness was marketing manager of Dumont Laboratories and Thomas H. Bay continued as marketing manager of the Semiconductor division. Fairchild Camera and Instrument also consolidated its Winston Research and its Industrial Products division into one unit under Ray G. Hennessey.

After Jay Last and Jean Hoerni left Fairchild, for a while, James Nall, Bob Norman, Donald Farina, and Bob Anderson kept working on integrated circuits. Bob Norman was a very innovative designer and he is largely responsible for the early Micrologic concept. He was also one of few who forecast the emergence of solid state memories. Norman who headed Fairchild's circuit design group wanted to apply for a patent for his idea of semiconductor memory where flip-flops could be used as memory elements. Gordon Moore decided that such an idea was economically so ridiculous that it did not make sense to spend money on patent application. A few years later such semiconductor memory became a major commodity semiconductor product. A few months later, Robert Norman, followed another spin-off and left Fairchild. The lack of foresight of the remaining Fairchild "four" resulted in growing impatience between Fairchild engineers and the not yet determined Fairchild management.

Gordon Moore quickly forgot the reasons of his criticism of Shockley. Gradually, a series of small controversies resulted in growing frustration and frequent spin-offs from Fairchild Semiconductor. Jack Gifford wanted to be an analog salesman par-excellence; he wanted to learn more about what he was selling and was also under pressure from Widlar. When he enrolled into an internal Fairchild engineering class which was taught by about a dozen Fairchild engineers, Gordon Moore opposed his attendance because the "class was reserved for Ph.D's only"[4]

[4] The class material become later the contents of the book "*Physics and Technology of Semiconductor Devices*" by sole author A. Grove.

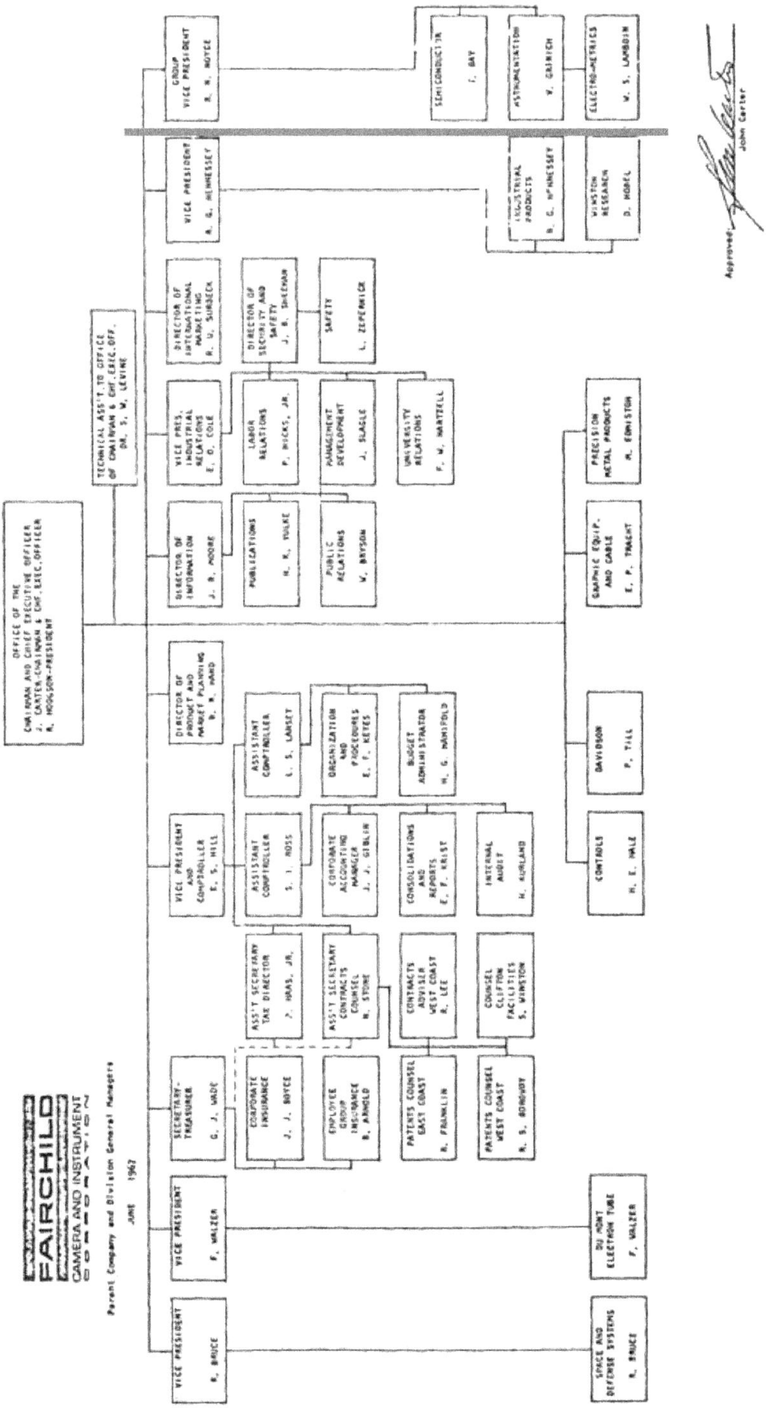

Fig. 8.33. Fairchild organization (1964)

Several very close co-workers felt somewhat disappointed when Sherman Fairchild granted "Eight Traitors" with bonuses in 1961. There was no reward for hard work and accomplishments of employees with badge number higher than eight. Ex-Shockley David F. Allison, formerly head of the Fairchild's Device Development section, David James, formerly head of Physics section, Mark Weisenstein, and Lionel Kattner gave up on Fairchild in August 1961 and formed Signetics. Victor Grinich became acting head of the Device section while C. T. Sah became head of the Physics section. "Tom" Sah was also affiliated with University of Illionois, he returned to Fairchild and presented often very impressive presentation, however, when daily problems need to be solved, Sah was nowhere to find.

Both Widlar and Talbert were very critical of Fairchild's R&D group in Palo Alto. They had enough experience with Fairchild managerial style and they did not believe Robert Noyce's announcement that at Fairchild "any product idea will be nurtured" (Fig. 8.32.) They started plan a new adventure. Peter Sprague knew that Danbury based National had different troubles and one way that he can get out of the hole was to get access to technology which National did not have. Peter Sprague hired David V. Talbert as VP and General Manager of NSC reorganized Microcircuit division. At the time of merger, Molectro had about 30 employees; the National plant in Danbury had about 600 workers. Widlar left Fairchild with Talbert.

When Widlar walked out of Fairchild in December 1965, Fairchild, per Bob Noyce instruction, was paying Widlar until April 1966. *"Maybe they did not believe that I was actually leaving, some people are really a little slow"* was Widlar's comment. Fairchild never tried to get back what they paid to Widlar.

Robert J. Widlar who designed at Fairchild the μA702, '709, '710, '711, and '726 submitted to John Hulme a one-line resignation letter written in the red pencil "I resign effective December 31, 1965." Widlar refused to sign a Fairchild Exit Interview Form, but he scribbled in the line for "Reason of Separation" the sentence "*I want to be RICH!*" The word rich was in capital letters.

When asked by John Hulme what it would take to keep him, Bob said "*one million tax free by whatever way you choose.*" In the discussion Widlar said that his goal was to be worth $1 million by age 30, and that was about a year away. John comments was *"I think you might make it, let me know if you succeed."* A couple years later when National stock became valuable, John received a call about 2:00 AM one morning and was advised that the goal had been achieved.

Molectro history sterted in 1962 when James R. Nall, expert in masking techniques at Fairchild, and by D. P. Spittlehouse formed the Electro Radiation, Inc. which was later renamed to Molecular Science Corporation in Santa Clara, and later became National Semiconductor.

Molecular Science Corporation changed in August 1963 one more time its name to Molectro Corporation and moved to 2950 San Ysidro Way in Santa Clara. The company had a contract with Jet Propulsion Laboratory to develop solid state IR and UV detectors. The new company was financed by Electronics Capital Corporation in San Diego with total commitment of $2 million, and the right to 68% of the total common stock. Jim Nall, president of the company, developed the first tape-programmed computerized step-and-repeat camera for precision photomasks. This new approach allowed them to align two masks within ±250 Å and minimum spacing of lines on different masks 2.5 μm. Originally the work on the stepping camera was expected to take from 6 month to 1 year, but the work took most of the first two years.

Molectro came up with the idea of the first Application Specific Integrated Circuit. The first family of custom RTL circuits was announced at 1963 Western Electronic Show and Convention (WESCON.) The line failed to get into production at the time, however, and was "re-introduced" in the 1964 IEEE Convention in New York. Molectro was producing universal silicon, and customers would define the unique metal interconnect mask (Fig. 8.32). A good idea was born too early. Companies working on circuit design had no engineers with knowledge of integrated circuit technology. The cost of integration was much higher than the cost of similar circuits designed with discrete devices, and as always, there was skepticism about a new technology.

In two years Molectro was running into unmanageable financial troubles. Jim Nall had originally a commitment from Electronic Capital Corporation which includes $600,000 in five years, 8 per cent debentures (convertible into 68% of the common stock) and a term loan of $1.4 million. The Electronic Capital Corporation in June 1964 reduced its financial support of Molectro to "a minimum" and put all or part of its investment in the firm for sale. Jim Nall resigned as president of company and in July 1965 Molectro's plan in Chapter XI was ratified. As proposed by Molectro, $92,000 was deposited in the Golden Gate National Bank in Los Altos. The plan provides for Molectro to remain in possession of all manufacturing equipment and to pay 20 per cent to general unsecured creditors of more than $500. Those claiming from $100 to $500, and smaller, should be paid in full. The plan was based on the ability to complete the sale of the stock of debtor to National Semiconductor Corporation.

With time passing, it became clear that manager with sense for time schedule and deadlines with delivery need to be put in the charge of Fairchild engineering. Pierre Lamond was installed as an integrated circuit operations manager on June 1966.

Naturally, Jack Gifford was mad and disappointed that Widlar and Talbert did not involve him into their plans. Jim Giles who was hired by Widlar continued to work on linear at Fairchild. But Jim was rather application engineer, they need new design engineer.

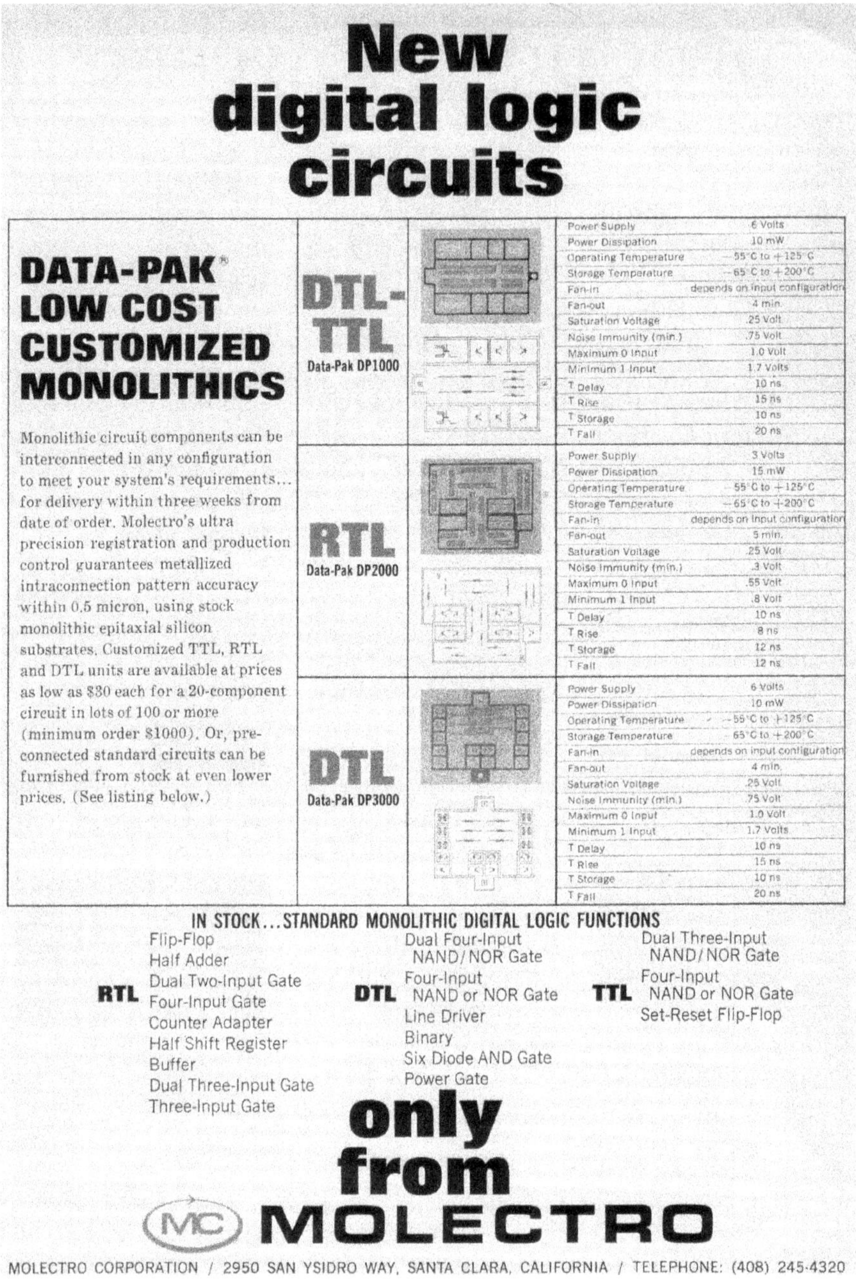

Fig. 8.34. Molectro Corporation's advertisement released in March 1963

6—PALO ALTO TIMES, TUESDAY, AUG. 16, 1966

Molectro Corp. appoints two as part of reorganization

SANTA CLARA — Molectro Corp., one of the early manufacturers of integrated circuits in Santa Clara County, has been dissolved as a corporation and reorganized as the Microcircuits Division of National Semiconductor Corp., an NSC spokesman said today.

National Semiconductor, located in Danbury, Conn., acquired Molectro in July, 1965.

In a management reorganization aimed at increasing production, David V. Talbert has been appointed vice president and general manager of the Santa Clara operation.

Talbert is the former product manager at Fairchild Semiconductor of Mountain View.

Robert J. Widlar, another former Fairchild man, has been appointed manager of electronics at Santa Clara. He was section head for applications engineering at Fairchild.

Francis G. Walker, vice president and general manager of the former Molectro, continues as manager of administration.

DEVELOPMENT

According to F. Edmund Naumann, merchandising manager for NSC, the Microcircuits Division will continue development work on the monolithic integrated circuit for introduction into new markets.

"We manufacture a hybrid integrated circuit in Connecticut and now hope to combine this technology with that of the monolithic circuit," Naumann said.

He explained that each of the circuits has a separate function. A small user of circuits finds the hybrids more practical, he noted.

"A large user purchasing circuits, however, finds the monolithic variety more practical because they combine transistors, resistors, and capacitors on a single silicon chip," he noted.

Fig. 8.35. Molectro re-organization as refered in Palo Alto Times

Fig. 8.36. David V. Talbert, VP and General Manager of NSC Microcircuit division with Gerald Schneider, and J. F. Hegarty (from L to R)

In Britain young engineer David Fullagar was looking for bigger adventure than work for Radar Establishment. David's friend, Wadie Khadder, who was already working for Transistron in U.S. painted a rosy picture of beautiful American girls which were not obesite at that time. David contacted Transistron and received Transitron letterhead with interview notification. The

NEWS RELEASE

FAIRCHILD SEMICONDUCTOR
A DIVISION OF FAIRCHILD CAMERA AND INSTRUMENT CORPORATION

FOR USE: UPON RECEIPT

313 FAIRCHILD DRIVE
MOUNTAIN VIEW, CALIFORNIA
NEWS BUREAU: (415) 962-2530

Contact: R.L. Molay

New responsibilities assigned to veteran Fairchild managers include these important changes:

Pierre Lamond was named integrated circuit operations manager, with the new staff positions of manufacturing manager and engineering manager reporting to him.

Fig. 8.37. Fairchild Press Releases from June 9, 1966

Transistron stationary had on the left side of sheet a drawing of beautiful Transitron corporate building. Tom Longo, general manager of Transistron, was in summer 1965 in London and he interview and hired David.

In September 1965, David with letterhead in his hand was looking for the building pictured on the stationary. Instead he found old textile mill building and his first job at Transitron was work on 16 bit memory. The die size was large and David needs to solve first problem how to glue together two pieces of Rubylith which was used for mask design. In less than one month David realized that he was screwed when he was hired for about half of salary typical to his status in comparison with his American colleagues. It took another three months to find a new job.

David Fullagar joined Fairchild in January 1966, one weekend after Talbert and Widlar left. David joined Marvin Rudin group at Fairchild R&D organization in Palo Alto as an office roommate of George Erdi. The first David's assignment was to improve everything on the 709. David was considering two stage amplifier with compensation rather than three stage amplifiers, and he presented the first version of the 741 to Marvin Rudin. Rudin joined Fairchild from Pacific Semiconductors and was not very technical, but the guru of the group, Gart Wilson was excellent. They recommended to David to go to Mountain View if he wants to build such amplifier. *"It will be very difficult to do such work here"* Rudin said.

Jack Gifford hired from the R&D group Dave Fullagar and brought him to Mountain View. At the time when Widlar and Talbert struggled to build an epi reactor in Molectro, and had no time to work on new circuits. the Fairchild group was still on the move. D. Fullagar completed the μA741, the first internally compensated, low cost, high gain operational amplifier in 1968 and started the μA725 which was completed by George Erdi. Fullagar, as one of the first, proposed three terminal regulator. Unfortunately, Mike Markkula of Fairchild marketing decided that there is no market for such regulator and idea was killed (especially if idea originated from somebody as young as Fullagar.)

Fig. 8.38. µA741 – Product which will be always associated with Dave Fullagar (left)

For a while, linear circuit development continued with Marvin Rudin and Garth Wilson from the R&D group, but Fairchild never got further than '741 and '725. The end of Fairchild's Linear business was obvious.

When Rudin and Wilson formed Precision Monolithic Corporation, Fullagar did not joined them because he thought Rudin will not make it. Instead he joined Hoerni's Intersil in February 1969, as a manager of linear integrated circuits.

Despite the huge success of the µA702 and '709 there were a lot of open questions about future trends of linear integrated circuits. For example, Jim Solomon of Motorola was as late as 1966 not convinced that linear circuits would be a huge business. He met Widlar for the first time February 9, during the 1966 International Solid State Circuits Conference in Philadelphia. R. S. Pepper of Sprague Electric Co., moderated a panel session about present and future status of linear integrated circuits.

Somebody in the audience asked a question about the standard amplifier which could be produced in quantities as digital counterparts. James Solomon of Motorola (Fig. 8.40), (later Widlar's rival at National) replied that *"it seems doubtful that a universal amplifier would be feasible."* Circuit demands vary greatly and would not be satisfied by one or two standard amplifiers, he said. Solomon elaborated also difficulties with process control which must be solved if linear ICs should be successful.

Bob Widlar grabbed a microphone and said *"The argument that there are problems in manufacturing linear integrated circuits is rubbish."* Widlar pointed out that linear integrated circuits could be made as easily and as well as digital ICs are now being made. He warned, however, that their advantage in circuit design would be realized only by fully employing their unique properties of good device matching and tight temperature coupling.

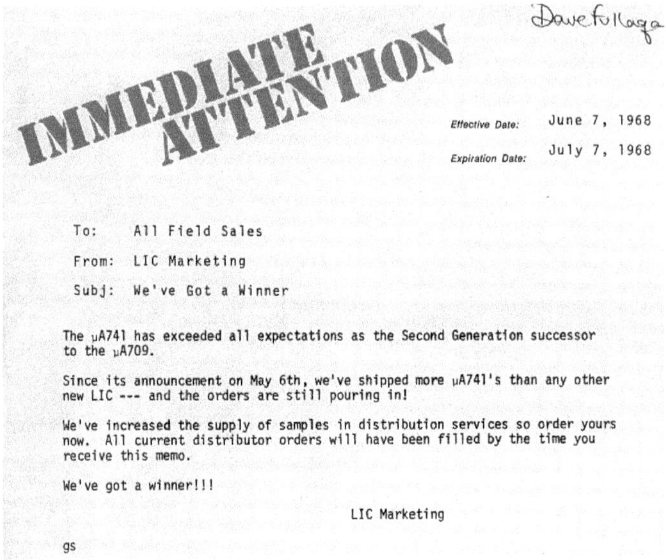

Dave Fullaga

IMMEDIATE ATTENTION

Effective Date: June 7, 1968

Expiration Date: July 7, 1968

To: All Field Sales

From: LIC Marketing

Subj: We've Got a Winner

The μA741 has exceeded all expectations as the Second Generation successor to the μA709.

Since its announcement on May 6th, we've shipped more μA741's than any other new LIC --- and the orders are still pouring in!

We've increased the supply of samples in distribution services so order yours now. All current distributor orders will have been filled by the time you receive this memo.

We've got a winner!!!

LIC Marketing

gs

Fig. 8.39. Fairchild's Internal Memo from June 7, 1968

Fig. 8.40. James E. Solomon, Motorola 1966

At that time Linear Integrated Circuits had only two heavy weight believers – Robert J. Widlar and H. C. Lin who was a member of the same conference panel and who defended Widlar. Solomon changed his opinion a few years later and wrote the best technical paper about an Integrated Operational Amplifier. The paper did not lose anything on its merit during the last thirty years.

Fairchild Semiconductor patented Widlar's current source and the biasing of the input stage and the conversion of differential input to a single-ended output; however, they never seriously enforced their patent rights.

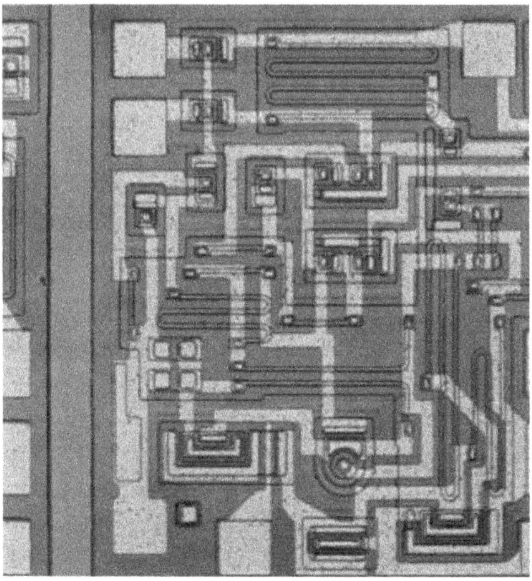

Fig. 8.41. Motorola's equivalent of the '709 – MC1709G. Transistor with circular configuration is the lateral pnp transistor

Fairchild did not license the manufacturing rights of the μA709 to any manufacturer (except to ITT Semiconductor, which bought a blanket license for all of Fairchild's semiconductor products and technology). Twelve months after Fairchild introduced the μA709, Motorola, Raytheon and Texas Instruments introduced their versions of 709 with vaguely familiar names: MC1709G, RM709, SN52709L. Philco-Ford, ITT Semiconductor and Westinghouse Molecular introduced the equivalent of 709 a little bit later.

The original '709 was much imitated. The copies were not an exact equivalent of μA709 and characteristics of actual performance varied from manufacturer to manufacturer and from batch to batch. Texas Instruments' equivalent exhibited oscillations in circuit configuration where Widlar's product worked perfectly fine. Raytheon's part had an excessive noise. Only Philco with their PD7709 solved problems with the 709 and they also got a very good yield, better than Fairchild.

Nothing similar to the story of Widlar and the 709 ever happened in the semiconductor business again. Millions of 709s have been sold, and 709's were made until recently – a unique longevity record in an industry whose products usually become obsolete within only a few years. Widlar's products rapidly become million-selling industry standards. Widlar's impact on design was so obvious, that the company management had to recognize its dependence on a single engineer.

References

[1] V. Jolly, "Commercializing New Technologies: Getting from mind to market," Harvard Business School Press, Boston, MA, 1997
[2] M. Maidique, "Entrepreneurs, champions, and technological innovations, Sloan Management Review, Winter 1980, pp. 59–76
[3] B. Lojek, "Lojek's Success/Failure Principle", RTP'96, p. 456
[4] R. J. Widlar, A monolithic high gain DC amplifier, Proc N.E.C. 1964, pp.169–174
[5] R. J. Widlar, US Patent 3,364,434, Filed April 19, 1965
[6] People Column, Electronics, May 31, 1965, p. 8
[7] Industry Report, Solid State Design, September 1964, p. 43

9

National Semiconductor – A New Type of Semiconductor Company

"The building process at National and our plan to compete was straightforward. We knew exactly what we wanted to do, and who can do it."

Charles E. Sporck

Sperry Semiconductor was a division of computer giant Sperry Rand Corporation in South Norwalk, Connecticut. The division developed the equivalent of Fairchild's mesa transistors which were announced in mid 1959.

In May of 1959, eight former Sperry scientists led by Dr. Bernard J. Rothlein separated from Sperry and established in Danbury, Connecticut,

National Semiconductor Corporation. Dr. Rotlhein, who started his career in Signal Corps where he worked with Paperclip scientist, became president of the new company residing in an 11,000 square foot plant. The investors, Don Weeden and Don Lucas, put enough money into the start up to launch a new company. In mid 1961 the National plant was employing some 250 employees in two buildings totaling 40,000 square feet, and manufacturing over 60 types of silicon alloy and mesa transistors.

In March 1963 National Semiconductor merged with Clark Semiconductor Corporation. Clark Semiconductor Corporation pursued the relatively small VHF power amplifier market and they were quite successful with their MT-16 and MT-24 8 watt amplifiers working at 135 MHz using Clark's proprietary triple-diffused silicon power transistors. At the time of the merger Clark Semiconductor employed about 20 persons and had sales in the $300,000 range.

In 1963 National Semiconductor directed its efforts toward integrated circuits. The company decided to focus mainly on the analog type of circuit applications where performance at very low current levels is required. Because of the need for precision resistors and high ohmic values, the company chose a hybrid circuit rather then a monolithic circuit approach. The plan was not to offer a general line of circuit functions for sale but to continue working with specific customers in developing circuits to fulfill a specific function. National Semiconductor hybrids were using silicon active devices and Cermet resistors with a range of values of 30 Ω to 3 MΩ and tolerances ±1%. The idea of CHIC (trademark for Cermet Hybrid Integrated Circuit, Fig. 9.1) should have enabled production of relatively complex customer-designed circuits without the compromise of function or performance, particularly in analog circuits where the limitations of monolithic circuits are most severe. The CHIC circuits would supposedly have been appropriate technology for small quantities because of the low tooling cost of each new design.

Under Rothlein's leadership, National grew from $3 million in 1961 to $5.3 million in 1965. But profit was decreasing. The same year, Fairchild Semiconductor did more than $80 million.

Meanwhile, a suit brought against National by Sperry Rand reached litigation in 1965. The big plans of the young company started falling apart when Sperry Rand won. The court enforced an order to destroy National's entire inventory and freeze company assets. The company was forbidden to use any product based on Sperry technology.

Peter J. Sprague, the 27-year-old nephew of Robert C. Sprague, chairman of Sprague Electric Company, who previously set up a successful food processing operation in Iran and Turkey, became chairman of National at the end of 1966. With 150,000 shares of his own, Sprague was supported by two other major investors. They "recommended" to Rothlein that he leave the company (Rothlein become head of another semiconductor company, Norden Division of United Aircraft Corporation) and Sprague started to reorganize the company.

Fig. 9.1. Sperry Rand mesa silicon transistors (1959)

Peter Sprague recognized that in integrated circuits National's history was bleak, primarily due to start up costs and problems with the RTL digital series of circuits, and manufacturing problems with a new voltage regulator designed by Bob Widlar.

At the end of 1966 a group of mid-level executives started to plan a defection from Fairchild. Charles E. Sporck, as many others, recognized that Bob Noyce was a very charismatic leader but not a good manager. Fairchild Research and Development was closer to Academia than to manufacturing facilities. There were constant problems with technology-to-production transfer. Each side blamed the other side even for ridiculous problems. Charles

Sporck tried to solve these problems, but a satisfactory solution was never implemented. It became obvious that as long as Gordon Moore was in charge of R&D and Bob Noyce was in charge of the Division, no major changes would happen. This could be now somehow surprising to many. We know Moore and Noyce mainly from the later years, and they are very courteous, reverential, and even acknowledged mistakes they made when they climbed up. This is how they both matured.

In the mid sixties the competition in the semiconductor industry increased significantly, and mismanagement become much more obvious. There were causalities. Molectro was one of them. Philco Corporation announced in August 1963 that they were dropping transistors and would focus only on integrated circuits. Signetics, who never even started with discrete devices, challenged Fairchild and others. The Japanese competition of Nippon Electric Co., Hitachi Ltd., Tokyo Shibaura Electric Co. Ltd., Fujitsu and Matsushita in 1965 expanded production of silicon transistors and they were serious about integrated circuits. In January 1967 Hitachi, Ltd., and Tokyo Shibaura (Toshiba) Electronic Co. applied for Government permission to negotiate patent licenses with Fairchild Semiconductor for manufacturing planar integrated circuits. Both companies already produced integrated circuits under a sub-license agreement with Nippon Electric Co. which had a license from Fairchild.

Today it is very difficult to raise money and start a semiconductor business. In the sixties, the only credential you needed to raise money was that you were previously associated with somebody who was in the semiconductor business.

Sidney L. Siegel, VP of marketing at Pacific Semiconductors wrote in May 1961 an excellent essay about the status of the semiconductor business and "financial operators" who fueled money into the "wonderful drug – silicon". He wrote: *"The semiconductor industry is loaded with immature managerial talent. There may be good salesmen or good technical people, but there are few good businessmen. In some cases the good businessmen are not close enough to the business and are at the mercy of their inexperienced managers."* Charles Sporck recognized the price erosion of semiconductor parts and realized that a smarter way to come up with $10 million to get into the game, and not be obligated with 70% to investors, would be to take over an already established business. "The days when two or three brainy guys can start out in the business are gone," was Charlie's comment.

In order to understand why $10 million is needed, one may compare the budget set up by a consulting firm Integrated Circuit Engineering, Inc., and released in 1965. This report provides an estimate for a facility capable of producing about 200 integrated circuits per week. The report was prepared for a customer who was considering whether or not an in-house operation could be economically profitable. This report is also interesting because it lists the type of manufacturing equipment we were then using.

The breakdown of the figures is the following:

Design and Evaluation	Personnel:	1 Engineer	$1,000/month
		1 Technician	$540/month
	Equipment:	1 Curve Tracer	$1,300
		1 Oscilloscope camera	$500
		1 LC meter	$200
		1 Oscilloscope	$1,700
		1 Oscilloscope Plug-in	$200
		2 Power supplies	$300
		1 Voltmeter	$400
		Miscellaneous	$3,400
1. Photomasking Operation	Personnel:	1 Technician	$540/month
	Equipment:	1 Camera facility	$10,000
		1 Plate developing facility	$1,000
		1 Optical microscope	$2,000
		Miscellaneous	$2,000
2. Photoresist operation	Personnel:	1 Technician	$540/month
		1 Operator	$350/month
	Equipment:	1 Chemical hood	$1,500
		1 Skinner station	$500
		1 Alignment and Exposure station	$2,000
		3 Dust hoods	$900
		1 Microscope	$880
		1 Camera	$400
		1 Ultrasonic Cleaner	$500
		1 Probe station	$1,500
		1 Curve tracer	$1,325
		Miscellaneous	$1,495
3. Diffusion laboratory	Personnel:	1 Engineer	$900/month
		2 Operators	$700/month
	Equipment:	1 Chemical hood	$1,500
		4 Furnaces	$6,800
		4 Sets of glassware	$2,000
		1 Dust hood	$300
		1 Probe	$1,500
		1 Curve tracer	$1,075
		1 Microscope	$880
		Miscellaneous	$2,045
4. Thin Films	Personnel:	1 Operators	$550/month
	Equipment:	1 Vacuum system	$7,477
		1 Alloying furnace	$2000
		1 Chemical hood	$1,500
		Miscellaneous	$2,023
5. Assembly and test	Personnel:	1 Technician	$540/month
		2 Operators	$700/month
	Equipment:	1 Die bonder	$2,000
		1 Wire bonder	$2,000
		1 Microscope	$880
		1 Probe	$1,500
		1 Curve tracer	$1,075
		1 Scriber	$1,500
		2 Dust hoods	$ 600
		Miscellaneous	$2,445
	Summary:	Equipment	$75,000
		Personnel cost	$6,000/month
		Overhead	$7,000/month
		Depreciation	$2,000/month
		Materials	$1,000/month
		Operating cost	$16,000/month

Although $10 million was a lot of money in the sixties, the future was painted in rosy colors and "financial operators" encouraged new ventures. In 1960 an Electronics Industries Association forecasted that factory sales of the electronics industry would more than double in 1970 (Fig. 9.2.) Sales in 1959 were $9.2 billion. Most of the $11 billion in added sales would be registered in the military market. Expenditures for military electronics equipment were expected to rise from 1959's $5 billion to $9 billion by 1965 and to $12 billion by 1970. By then, 60% of the total dollar value of equipment produced by the domestic electronic industry would be going to the armed services. Expenditures represented 20–25% of the total estimated military budget of $60 billion.

The reader should note that the population of the United States in 1970 was expected to reach 220,000,000. The 1970 Gross National Product had been forecasted at $725 billion, up from about $475 in 1959.

Charles Sporck, motivated by a desire to run his own show, explored opportunities. Sporck remembered well Talbert and Widlar's success at Fairchild, and he met with Dave Talbert at the end of November 1966. Talbert referred Sporck to National's board of directors and serious negotiation started in January 1967.

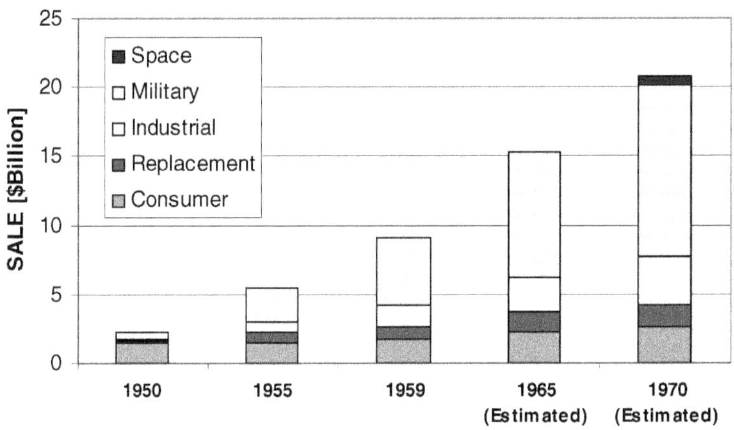

Fig. 9.2. Electronic Industry Sale as reported by EIA in 1960

In February 1967 National Semiconductor Corporation announced that they reached an agreement with a team of executives to join a company. The core of the team was a group of five from Fairchild Semiconductor (Charles Sporck, Fred Bialek, Pierre Lamond, Floyd Kvamme and Roger Smullen); also joining were a former employees of Texas Instruments (R. Kenneth Davis), Parkin-Elmer Corporation (John F. Hughes), and Hewlett-Packard Company (Kenneth Moyle). The group was headed by Charles E. Sporck, who was general manager of Fairchild Semiconductor. Friday, March 1, 1967, the new managerial team started to turn National Semiconductors upside down.

Fig. 9.3. Charles E. Sporck (1967)

Fig. 9.4. Peter J. Sprague (1967)

Almost at the same time Fairchild announced their adjustment: Thomas H. Bay had been moved from the Instrument Division of Fairchild back to the Semiconductor Division as general manager to replace Charles Sporck. Victor Grinich took over Tom's position in the Instrument Division. Floyd Kvamme, marketing manager for integrated circuits was replaced by Ben Anixter. Roger Smullen and Pierre Lamond's positions were combined into a single post and John Sentous was named for this function. Fred Bialek's post of over-seas operations manager was transferred to Don Yost.

The second day on the job, Charles Sporck, Fred Bialek, and Roger Smullen flew to Danbury. Despite the fact that the Danbury division was profitable and that the Santa Clara operations sustained a loss of almost $300,000, they laid off about 300 employees of the Danbury division and Fred Bialek was put in charge of the Danbury plant. Sporck said that this was his strategy to get the capital they needed to expand the operation of Molectro – alias the Microcircuit Division of National Semiconductor in Santa Clara.

Bialek was a mechanical engineer with a degree from MIT who before joining Fairchild worked as a design engineer for GM. This was not enough to be successful. Bialek, as all his peers, worked long hours each day; many paid a big penalty; often it cost them their family. They frequently struggled with the question of whether the sacrifice was worth it, but they could not stop. Fred Bialek held also one primary achievement in the semiconductor industry: when the labor union started to organize a work contract in the

Fig. 9.5. The National Semiconductor advertisement for its mesa transistor, which the company was not able to sell after successful legal action by Sperry Semiconductor

Fairchild diode facility in San Rafael, Fred, who knew that he would never be an operator or technician, crushed the union movement so unshakably that technicians and operators are up to this day at the mercy of company management. Since the sixties, when a semiconductor company runs into trouble, the easiest and the first solution is to fire people.

Charles Sporck knew that many of National's parts had no future. National dropped Rothlein's CHIC technology. He brought his staff from Fairchild and they wanted to push farther Fairchild's technology. One brilliant part was already there. Bob Widlar, besides developing an epi reactor with Dave Talbert, developed at Molectro the industry's a first series voltage regulator – the LM100. Robert J. Widlar, "Member of Establishment" (Fig. 9.7.) developed a circuit that was revolutionary in the same way as Widlar's op amps.

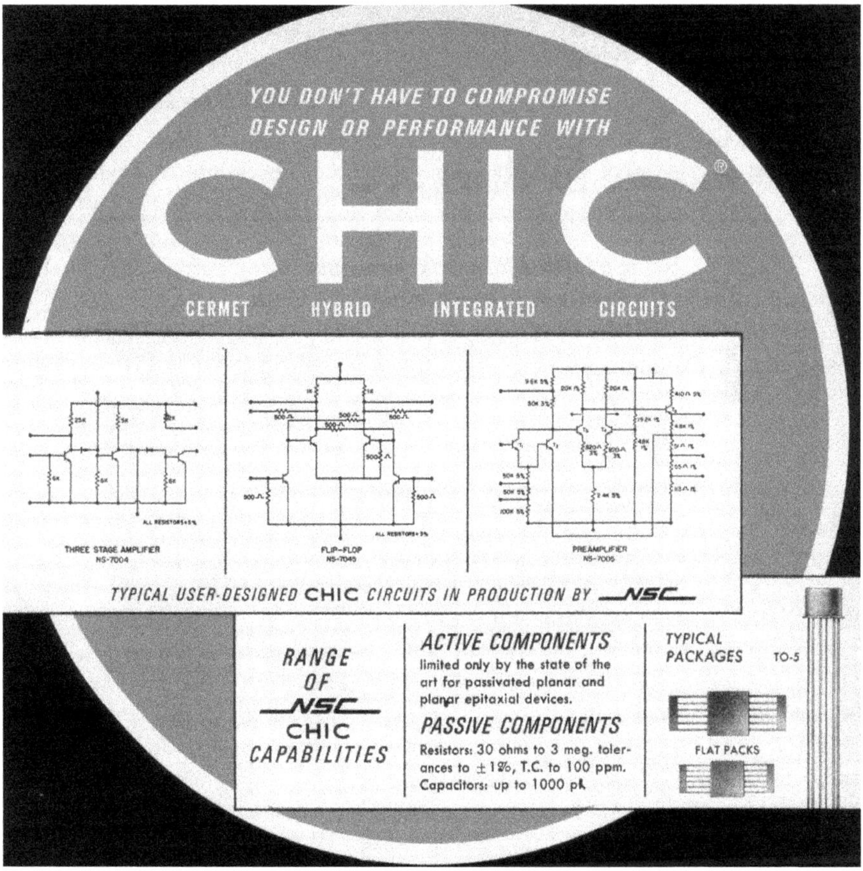

Fig. 9.6. National Semiconductor CHIC hybrid integrated circuits

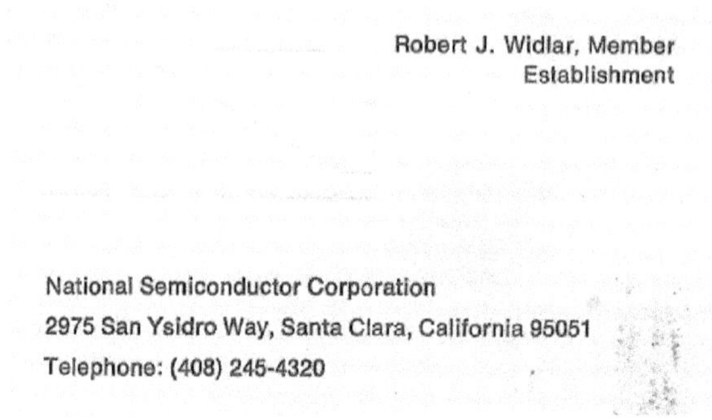
Robert J. Widlar, Member
Establishment

National Semiconductor Corporation
2975 San Ysidro Way, Santa Clara, California 95051
Telephone: (408) 245-4320

Fig. 9.7. Bob Widlar's business card

The LM100 was in 1966 a unique circuit with outstanding characteristics:

- Output voltage adjustable from 2 V to 30 V
- 1% load line regulation
- 1% stability over full military temperature range
- adjustable current limiting

There was no similar component on the market and demands for LM100 exceed by far the most optimistic expectation. The circuit was redesign in several versions and became the cornerstone of linear integrated circuits business.

The cost of mask plates for new linear circuits or circuit revision was approximately two or three thousand dollars (Fig. 9.10). The floor-planning, placement and routing was done manually. The drawings were digitized and then converted to plates. The manual drawing typically required three to four weeks of work.

The next Widlar design at National was new operational amplifier LM101. When Bob Widlar began design of the LM101 in early 1967 he had three goals: to develop an op amp with 15:1 improvement in bias current over the µA709, to increase the open-loop gain to over 100,000, and to protect the output stage against output short circuits.

Bob, as was his habit, designed a unique input stage where he connected the PNP (Q3 and Q4) devices in a differential common-base configuration and combined them with a pair of NPN based emitter followers (Q1 and Q2) to overcome the input current problem. This NPN-PNP combination is equivalent to a common emitter PNP pair with high current gain, except that the effect of collector-base capacitance is significantly reduced. Because the current load Q1' of second stage amplifier Q9, the two stages of amplifier provide a nominal gain well over 150,000. The output is protected against short circuits; the voltage drop across resistor R7 is monitored by transistor Q15; when voltage drop is too big the output sinking current is limited to a safe value.

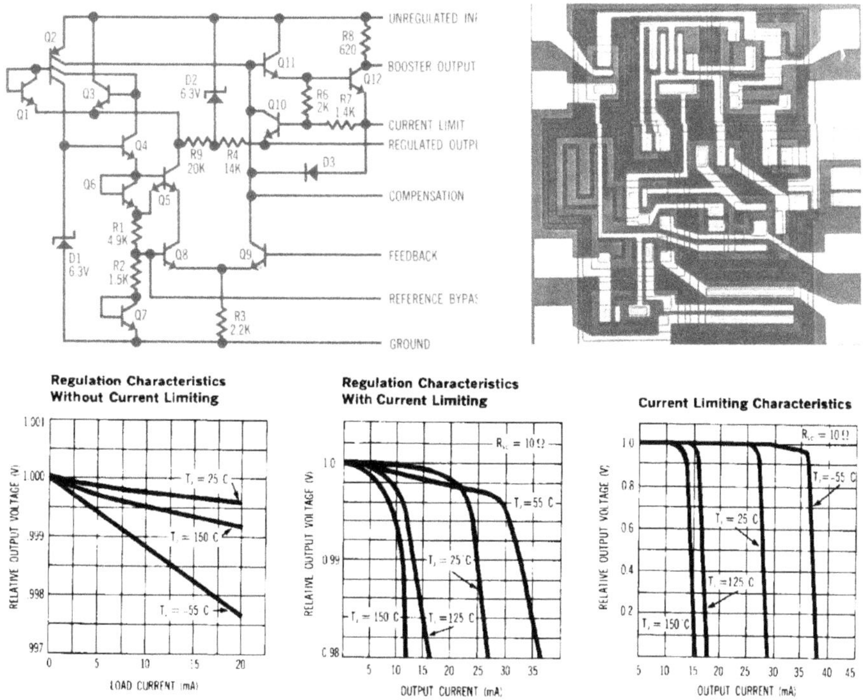

Fig. 9.8. The world's first integrated voltage regulator LM100

The LM101 had essentially the same electrical performance as the 709 but it eliminated most of the major disadvantages and application problems of its predecessor. The LM101 was more stable than the 709 and was easier to frequency compensate. Further, the inputs and outputs were protected from overloads: and the amplifier would not latch up even in a situation where the common-mode range was exceeded. Although the 101 circuitry was more complex than 709, the size of die was smaller, about 168×168 µm; in comparison the 709 was 188×188 µm. Talbert established better process control and used more advanced photolithography at National than earlier at Fairchild. Widlar eliminated base resistors, using instead pinch resistors, and used extensively active collector loads.

An additional offshoot of the LM101 was the beginning of cooperation between two Bob's: Bob Widlar and Bob Dobkin. In the spring of 1969 Bob Dobkin was then with Philbrick/Nexus Research in Dedham, MA. Widlar said that: "*I am getting letters from the kid at the East who constantly criticized my designs or offered ways to improve them. After a few letters I got ticked of this guy and I called him.*" Widlar and Dobkin set up a meeting. Because Widlar always considered a bar as the best place to brainstorm new technical ideas, they met in a bar. The morning after a fruitful technical dis-

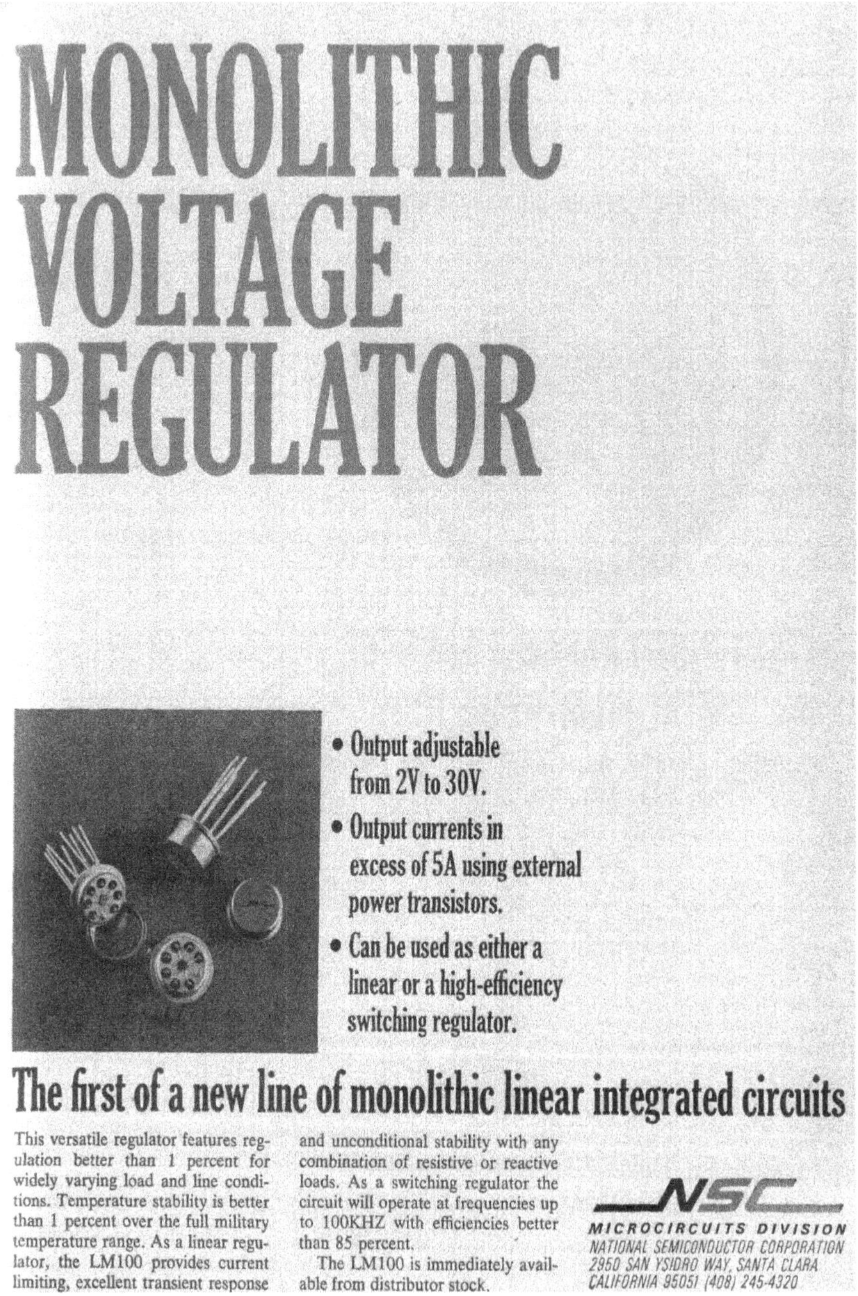

Fig. 9.9. The first advertisement for National's LM100 voltage regulator (September, 1966)

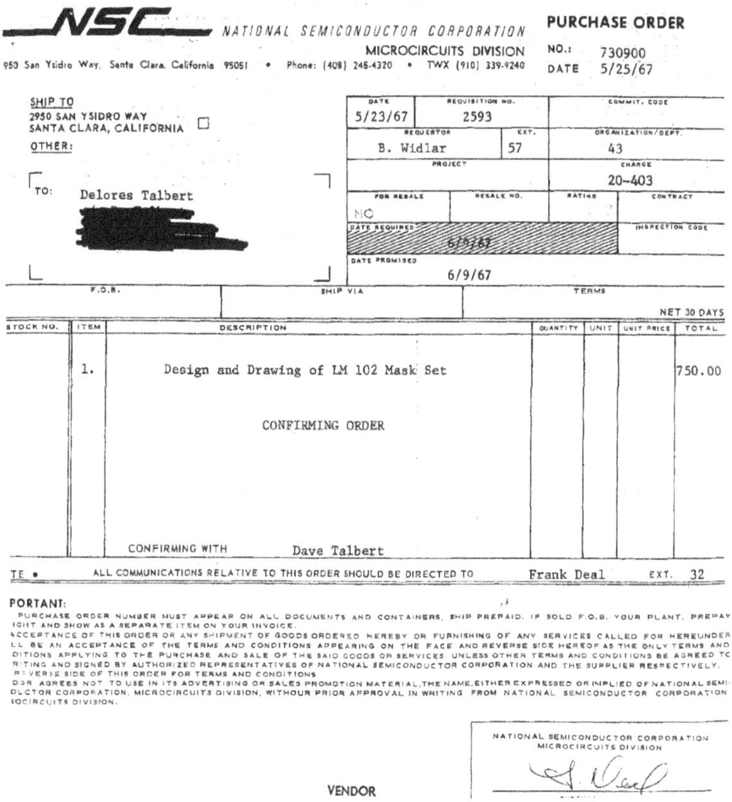

NSC NATIONAL SEMICONDUCTOR CORPORATION
MICROCIRCUITS DIVISION
950 San Ysidro Way, Santa Clara, California 95051 • Phone: (408) 245-4320 • TWX (910) 339-9240

PURCHASE ORDER
NO.: 730900
DATE 5/25/67

SHIP TO
2950 SAN YSIDRO WAY
SANTA CLARA, CALIFORNIA
OTHER:

TO: Delores Talbert
[redacted]

DATE	REQUISITION NO.		COMMIT. CODE
5/23/67	2593		
REQUESTOR		EXT.	ORGANIZATION/DEPT.
B. Widlar		57	43
PROJECT			CHARGE
			20-403
FOR RESALE	RESALE NO.	RATING	CONTRACT
NO			
DATE REQUIRED			INSPECTION CODE
6/9/67			
DATE PROMISED			
6/9/67			

F.O.B.	SHIP VIA	TERMS
		NET 30 DAYS

STOCK NO.	ITEM	DESCRIPTION	QUANTITY	UNIT	UNIT PRICE	TOTAL
	1.	Design and Drawing of LM 102 Mask Set				750.00
		CONFIRMING ORDER				
		CONFIRMING WITH Dave Talbert				

TE • ALL COMMUNICATIONS RELATIVE TO THIS ORDER SHOULD BE DIRECTED TO Frank Deal EXT. 32

PORTANT:
PURCHASE ORDER NUMBER MUST APPEAR ON ALL DOCUMENTS AND CONTAINERS, SHIP PREPAID. IF SOLD F.O.B. YOUR PLANT, PREPAY IGHT AND SHOW AS A SEPARATE ITEM ON YOUR INVOICE.
ACCEPTANCE OF THIS ORDER OR ANY SHIPMENT OF GOODS ORDERED HEREBY OR FURNISHING OF ANY SERVICES CALLED FOR HEREUNDER LL BE AN ACCEPTANCE OF THE TERMS AND CONDITIONS APPEARING ON THE FACE AND REVERSE SIDE HEREOF AS THE ONLY TERMS AND DITIONS APPLYING TO THE PURCHASE AND SALE OF THE SAID GOODS OR SERVICES, UNLESS OTHER TERMS AND CONDITIONS BE AGREED TO RITING AND SIGNED BY AUTHORIZED REPRESENTATIVES OF NATIONAL SEMICONDUCTOR CORPORATION AND THE SUPPLIER RESPECTIVELY.
REVERSE SIDE OF THIS ORDER FOR TERMS AND CONDITIONS
DOR AGREES NOT TO USE IN ITS ADVERTISING OR SALES PROMOTION MATERIAL, THE NAME, EITHER EXPRESSED OR IMPLIED OF NATIONAL SEMI-DUCTOR CORPORATION, MICROCIRCUITS DIVISION, WITHOUR PRIOR APPROVAL IN WRITING FROM NATIONAL SEMICONDUCTOR CORPORATION ROCIRCUITS DIVISION.

VENDOR

NATIONAL SEMICONDUCTOR CORPORATION
MICROCIRCUITS DIVISION

Fig. 9.10. The Purchase Order for design and drawing of LM102. The cost: $750.00

cussion and "tying one on," Bob Dobkin was sick and threw up. The next month "Doby" became an employee of National Semiconductor, working with his "close friend," Bob Widlar.

Beside Dobkin, Mineo Yamatake, Jim Solomon, Carl Nelson, Tom Frederiksen, Bob Hirschfeld joined Widlar's linear team, and eventually headed their own design group at National.

The second op amp Bob designed at National was the LM101A. This device was nominally identical to the LM101, except that the input currents were reduced by more than an order of magnitude. This device for the first time used a collector FET (also called epi-FET.) The collector FET (Fig. 9.14) provides bias for all current sources. All the currents can be determined by resistors in the current sources which have relatively small voltages dropped across them; and there are no resistors connected directly across the supplies. This approach not only minimizes chip area, as well as current drain, but also permits the circuit to operate over a wider range of supply voltages. In the course of work on biasing circuits used in the LM101A, Widlar come up with another new device – Multiple Collector Lateral Transistor Device,

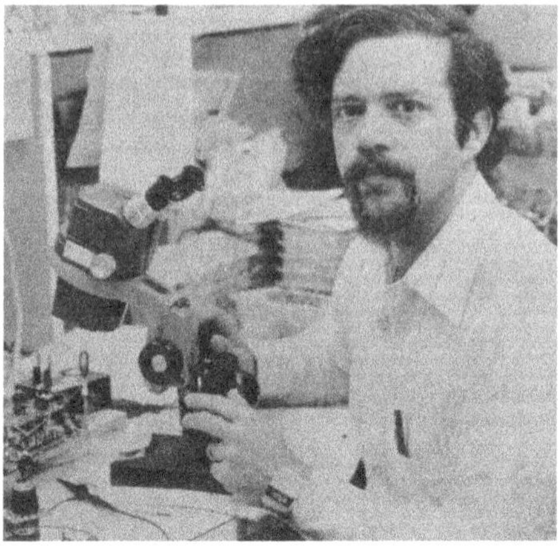

Fig. 9.11. Robert Dobkin at National in 1968

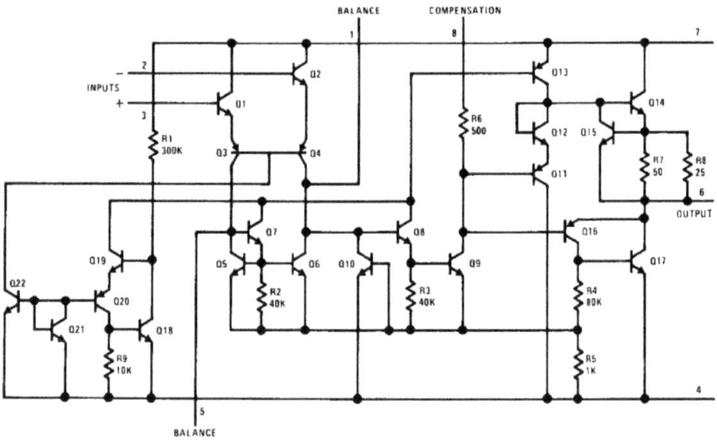

Fig. 9.12. Complete Schematic of the LM101 Operational Amplifier (1967 version)

later in 1968. Widlar recognized that utilization of different current sources is wasteful and requires additional die area, and he suggested a device having multiple collector current output circuits which are isolated from one another. The Multiple Collector Lateral Transistor has multiple separate collector regions diffused into the base region to form multiple PN junctions in the base regions. The junction between each collector region and the base region, in combination with the junction between the emitter region and the base region, provides transistor operation and the sum of the individual collector current is equal to the emitter current less the base current.

Talbert's process started with high resistivity P-type <111> silicon into which a N+ sub-collector was diffused (Mask # 1.) In the next step the high resistivity N-Type epi was grown followed by P+ diffusion isolation diffusion

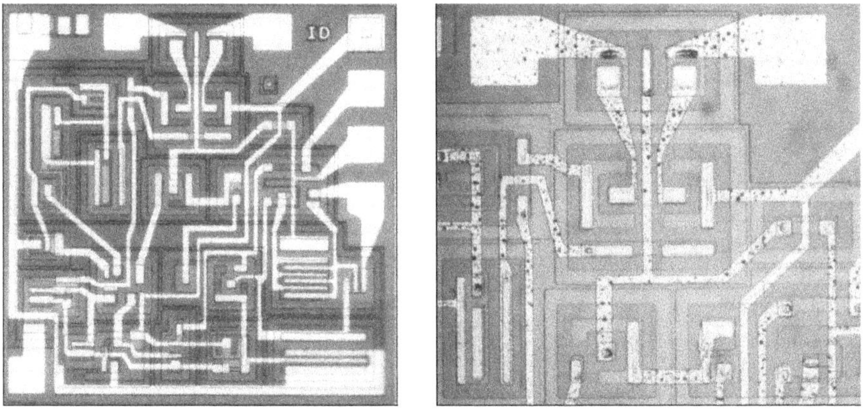

Fig. 9.13. LM101 – Low offset is achieved by making the input transistors identical and locating them close together

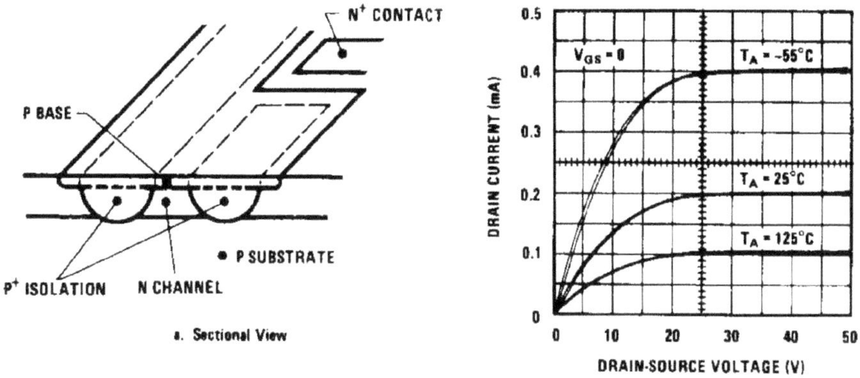

Fig. 9.14. Collector FET (The FET used in the LM101A was 75 μm wide, between $P+$ isolation, and 500μm long)

(Mask # 2). The next masking step is a base diffusion (Mask # 3) and an emitter diffusion (Mask # 4). Contact holes are etched into insulating oxide using Mask # 5. The last mask (Mask # 6) is used to patterned metal interconnects.

The selection of resistivities and thermal budget was centered around high-quality NPN transistors. Lateral PNP transistors had long been known for their low DC current gain. Talbert was able to make PNP's with gain greater than 100 (Fig. 9.16) without a major process modification.

The LM101A was announced in December 1968; Charles Sporck said in "National Semiconductor Essays Op Amp Alps." Bob Widlar's enthusiasm resulted in a small personal change: David Talbert's daughter had head injuries and when Bob come to the hospital and saw her shaven head, he made

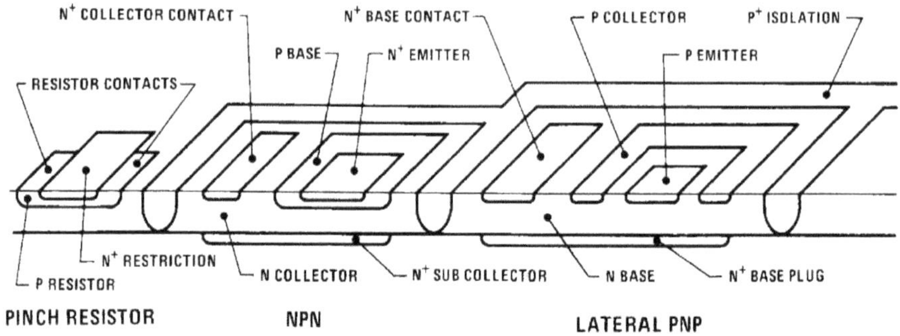

Fig. 9.15. National Semiconductor six masks epitaxial process developed by Dave Talbert in 1968

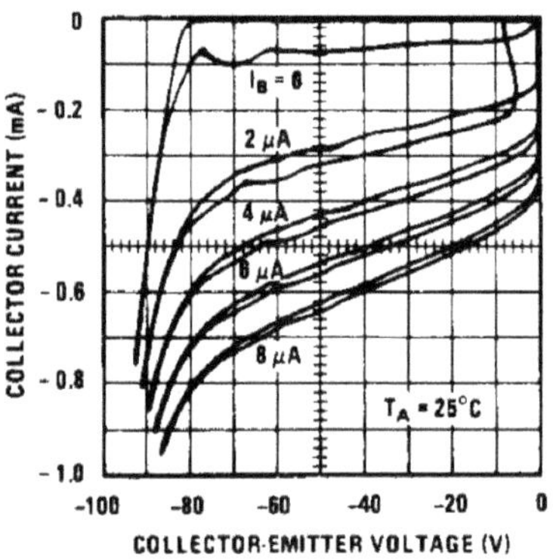

Fig. 9.16. Talbert's Lateral PNP with high current gain

a bet with her that he could grow a beard faster than she could grow hair on head. That was the last day he was ever clean shaven.

In mid 1968 Bob and Dave Talbert developed super beta transistors (U.S. Patent 3,566,218, filed on October 2, 1968). These transistors had very thin bases, and much higher beta than normal devices, although low C-E breakdown. This necessitated some great circuit tricks which was Widlar's expertise.

During Widlar's tenure at National, Bob generated a lot of money for the company. Sporck was willing to tolerate almost everything in Widlar's behavior. When Widlar ended up in jail, Charlie used his strings to get Widlar out. Widlar could work or not work anytime and anywhere. Widlar frequently abused Charlie's protection. Naturally, Pierre Lamond, manager, became less and less capable of contributing to solutions to technical problems, and he suffered most from Widlar's jokes. When Pierre issued an order that engineers could not eat their lunch at their desks, Widlar took his sandwich and enjoyed lunch in the laboratory. When Pierre complained, Widlar pulled out of his

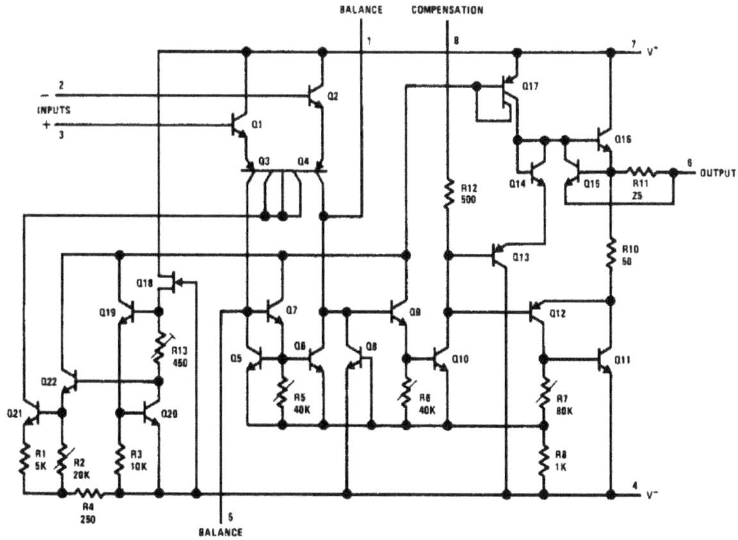

Fig. 9.17. LM101A

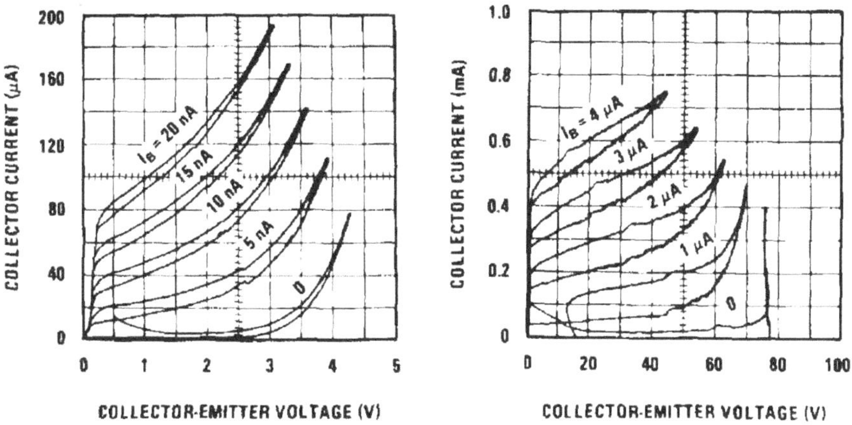

Fig. 9.18. Comparison of Super Beta and Conventional Transistor

pocket Pierre's memo and said "*No food or lunch inside cubicle. It does not say in laboratory.*"

In December 1968 Charles Sporck said "capitalizing on the talents of Robert Widlar, David Talbert and later William Routh, we have the widest range of linear IC's in the business and one of the highest in dollar volume."

In May, 1968, competition in linear IC's business significantly increased and prices were dropping from level of tens of dollars to the level of cents (Fig. 9.19). Charlie Sporck wanted to offset this trend by moving National

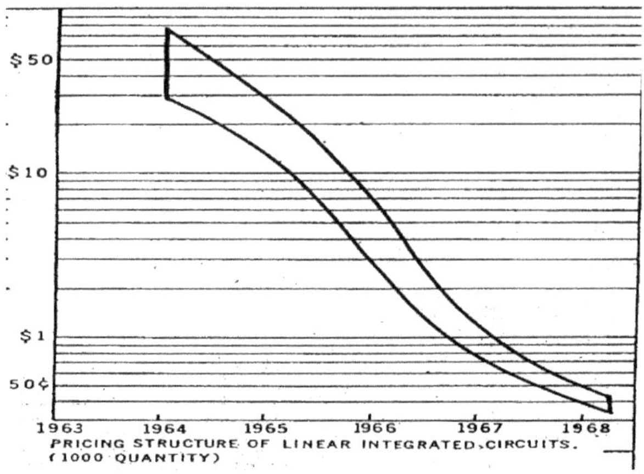

Fig. 9.19. Pricing of Linear Integrated Circuits (1964–1968)

into MOS and bipolar TTL products, in the latter area to second-sourcing Texas Instruments' 54/74 series.

Floyd Kvamme took responsibility for MOS development, and for a while, the duo Widlar – Kvamme kept National in the center of attention. Half of this duo, Kvamme, very polite, quiet and always avoiding any controversy, with the second half Widlar, who did not want to miss any controversy. It could be somehow surprising that two so different personalities could work for more than a decade together. Later, Kvamme said, that one reason why he joined Sporck's defectors was that he wanted to work with Widlar.

The first products the duo announced in February 1969 were a 2048 bit MOS ROM (MM253) and the LM108. The P-MOS device and MOS process was basically the same process which P. Lamond put into production at Fairchild. Fairchild was never comfortable with MOS processing and never actually solved problems related to new and unknown processing.

In the end of the sixties the National was challenged in linear business only by Motorola; however, Motorola had many processing difficulties and National become an industry leader. National become the second largest manufacturer in just about every product category.

The National Semiconductor story is a nice example how a few creative individuals, headed by maverick Charles Sporck, can transform company almost from bankruptcy to a top performing business. National also created an enjoyable working atmosphere which is still remembered by many. National had no problem to attract professionals; then, National was a very informal, technology driven company and National staff was known as "the animals of Silicon Valley."

Sporck belonged to the last generation of managers of semiconductor companies who occasionally showed up on the laboratory floor or in a clean room and talked with operators. At National it was easy to communicate directly with top management. Sporck can easily take a joke and he created an enjoyable working environment which attracted many talented people. The hard work then was also fun.

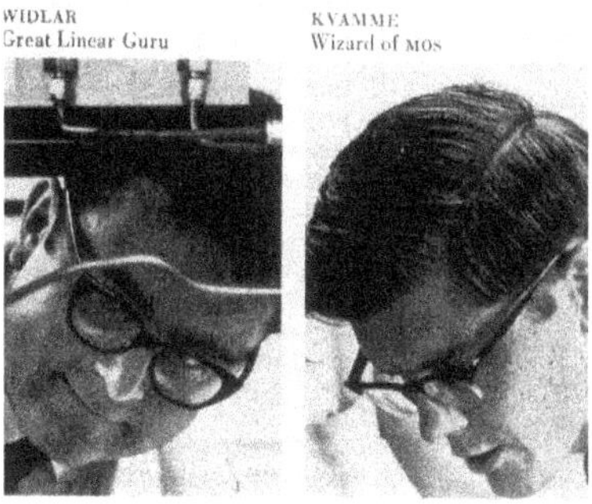

Besides lunch you get all this for $15.

9:00 – 9:30	CIRCUITS AVAILABLE TODAY	*Kvamme*
9:30 – 11:00	VOLTAGE REGULATORS	*Widlar*
11:00 – 12:00	MOS IN MEMORY APPLICATIONS	*Kvamme*
1:00 – 2:30	OP AMPS & VOLTAGE COMPARATORS	*Widlar*
2:30 – 3:00	MOS ANALOG SWITCH APPLICATIONS	*Kvamme*
3:00 – 3:45	COMMUNICATIONS CIRCUITS	*Hirshfeld*
3:45 – 4:15	FUTURE TRENDS FOR LINEAR DEVICES	*Widlar*
4:15 – 5:30	INTEGRATED CIRCUIT CLINIC	*Kalb, Thorkelson, Widlar, Hirshfeld & Kvamme*

Fig. 9.20. National seminars featuring two National stars – Widlar and Kvamme

I keep in my office a framed original Boxer Glows flyer announcing a Widlar Seminar. This is not very unusual. Exceptional is that Bob Widlar, "The industry's champ in linear IC's" conducted a seminar at Madison Square Gardens. The seminar took place on March 23, 1970 and all 500 seats sold out in a two days. Bob was a terrific teacher, and he learned how to get the attention of the public. Well presented technical stuff and humor – this was Widlar's approach. I found one of his notes where he prepares his bullets for a panel moderated by Jerry Eimbinder of EEE. The surprisingly tough Widlar knew that he needed to control himself and "Kindness" was his first bullet (Fig. 9.21.)

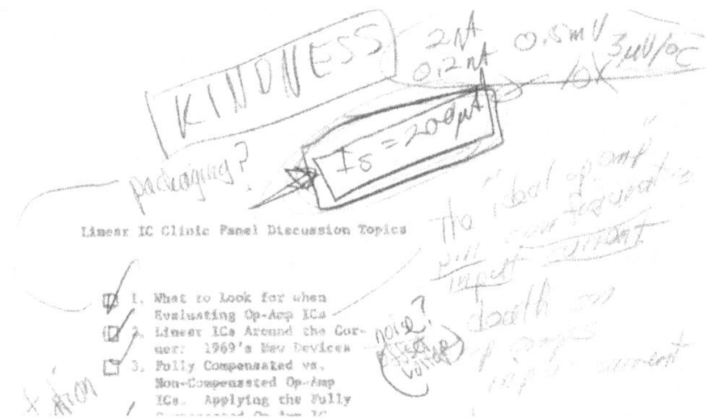

Fig. 9.21. Widlar "bullets" for panel discussion

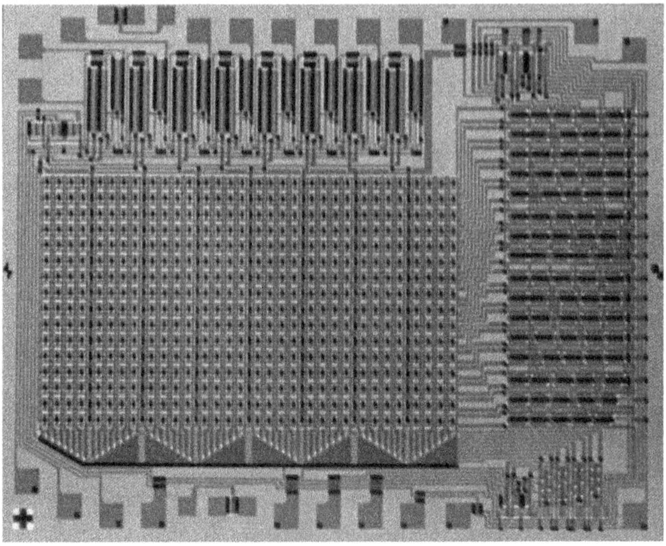

Fig. 9.22. The first National MOS product: P-MOS ROM MM253 (1969)

The reasons why Widlar and Talbert left National are not completely clear. Widlar and Talbert resigned from National on December 21, 1970 at 10:32AM. Although they joined National as a team and resigned simultaneously, Bob said: "*we have no immediate plans, we are buddies and I suppose we will be looking around for something, but how soon we get our tails in gears will depend on how interesting it is.*" It would not have been characteristic for Bob Widlar to just separate from the company and not make a distinguishing mark. During end of 1970's when cash was short for National, Sporck decided not to spend any money on landscaping of the Santa Clara plant. Bob Widlar in order help "*to keep the firm's austerity program by cutting moving expenses*" drove to the farm of his friend Don Wise, and "borrowed" a sheep and tethered it in front of the National facility and then called a San Jose News reporter (Fig. 9.23.)

Pierre Lamond did not consider Widlar's effort funny, and contrary to the common myth that Widlar auctioned the sheep that evening in Wagon Wheels, the sheep was mysteriously "stolen."

What Talbert hoped would be interesting never materialized. Widlar had some 20,000 National shares, which were trading over the $ 50.00 range; he excised his option and said to his friends that "*I don't like the water in Silicon Valley any more.*" He moved south to Puerto Vallarta, Mexico where Widlar hired an old Mexican lady as housekeeper and did not work for a few days. He quit the "*rat race and establishment*" but he never stopped working.

★★★ San Jose News Sat., Dec. 12, 1970 3

Economy Mowing

This is not Mary's little lamb grown up and following her to work. It is a lawnmower. Bob Widlar, director of advanced circuits development, "borrowed" the sheep for the front lawn of National Semiconductor in Santa Clara to help the firm's austerity program by cutting mowing expenses. Widlar admitted it is "putting a lot of gardeners out of work," but notes "at the same time the grass gets cut, it gets fertilized, too."

Fig. 9.23. "Economy Mowing" by Bob Widlar (1970)

Fig. 9.24. Robert J. Widlar on December 12, 1970

Widlar's popularity was enormous. In 1973 Teledyne Semiconductor run advertisement "Godzila Meets The Linear Monster" featuring Bob Widlar swinging the ax and with the following comment:

> *"Godzilla, alias Bob Widlar, is the well known king of the linear IC underworld. Teledyne, on the other hand, is known as the semiconductor and IC producer of monstrous proportions. We compete in just about all areas of IC's. When we started out to do battle in the linear*

market, we come against Godzilla's forces: the 101, 101A, 105, 107, 108, 108A formidable line. You see, Teledyne, though big, is friendly. To oppose such a line would be contrary to our normal cordial, compatible, helpful nature. So the only thing to do is join Godzilla's forces. After all, Teledyne can do it in a very big way. Just to prove how friendly we really are, we'll give you absolutely free one of the above (1 only) IC's ... plus a signed picture post card of Godzilla ... if you send us a note on your company letterhead and tell us why you want one free.

Note: Bob Widlar; inventor of the 709, 101, 105, and 108 does not work for Teledyne Semiconductor. Bob Widlar does not work."

On November 6, 1974 National Semiconductor Corporation and Robert J. Widlar reached an agreement whereby Bob would be developing linear ICs for National. The contract was originally prepared by National's attorney, Bob scratched and changed several paragraphs, and Charlie scratched the attorney's name and affiliation and signed an agreement by himself. Widlar was on the board again.

When driving from Mexico to the North, Widlar had problems each time he passed the border with the border patrol. The guards frequently asked Bob where he was working. "*I do not work*" was Bob's reply. This only irritated the guards more and they searched his car and delayed his pass as much as they could. Bob complained to Charlie Sporck about that, and Charlie suggested that he should have a business card and tell the guard that he was on a business trip. Bohemian Bob Widlar could not have a regular faked business card; he designed a card with skull and crossbones on it, and it stated that Robert J. Widlar was a Road Agent associated with Morgan Associates. The business card did the trick and future border crossing was without troubles.

Fig. 9.25. Business Card of Road Agent – Robert J. Widlar

Bob Widlar's notes are similar to Shockley's notes. He spent hours working on his calculations. In Mexico, and at the time when no numerical modeling of semiconductor devices was available, Widlar was working on the secondary breakdown of bipolar power transistors and he needed to solve continuity equations. Bob solved the problem with only a pencil and slide rule. He was very analytical. When he asked Mineo to run some experiment in the lab he always wrote down what he was expecting. Widlar was a tenacious problem solver and in the same way as Shockley, he was always worried that others were not "doing it right."

In 1981 Widlar, Robert Swanson, and Robert Dobkin co-founded Linear Technology Corporation. The new company started as a second-source for National and other products. Widlar was more interested in designing of new parts and gradually his relationship with Bob Dobkin turn sour. Widlar's position was that the LT1-10, -15, -16, -17, -18, and 20 were based on his invention and were covered under the Exclusive License Agreement. Linear Technology disputed Widlar's claim and during May 1984 to October 1984 both parties communicate only through attorneys. Bob Dobkin was in a difficult situation between Swanson and Widlar. Swanson was a different type of executive than Sporck and fired Widlar. A letter of "Termination of Consulting Agreement and Stock Repurchase" was mailed to Widlar on October 15, 1984 and a letter titled "Confidential Information" with agreement termination was hand delivered to Bob in Puerto Vallarta on October 24, 1984. Linear also requested via their attorney reimbursement for their expenses in connection with the disputed patent application. The request was denied.

I found in Widlar's technical notebooks from 1976 to 1981 Bob Dobkin's signature and they were part of the Agreement. Many of these design ideas given to Dobkin are dated before they left National. Unfortunately the crucial part of the Agreement was not well formulated and each side interpreted its contents differently. Widlar spent an enormous amount of time with the analysis of the breakdown voltage of bipolar power transistors. At the time when no device analysis was available, he solved continuity equations with pencil and slide rule (Fig. 9.26.) Nobody from Linear Technology really saw how much work Widlar put into this product and it is understandable why Widlar considered the crucial part of new device as his work.

Widlar returned to work as a "contractor" for National and the former close friends Dobkin and Widlar did not talk to each other since then.

There are numerous stories about Widlar, some of them with passing time are more than exaggeration. On the other hand there is no doubt that Robert J. Widlar was a terrific performer, genius, and mysterious loner. He established the Linear Integrated Circuits as a business; he put Fairchild into a position of leader, and he did the same with National Semiconductor.

He kept an ax in the corner of his office, which prompted the story that he worked off frustrations by driving into the country and chopping wood for three or four hours. Actually the house he was renting had a fireplace, and the ax was used to replenish his supply of firewood. The best story about the ax is different. Regis McKenna, marketing manager at National, was preparing a data sheet for the print shop. The printer guy was in Regis's office, when he called Bob. Bob just returned from lunch in a good mood and brought his draft stapled, and when he sat he was fumbling, trying to remove the staple out of papers. Suddenly he gave up, ran into his office, returned with his axe and chopped the staple and the desk.

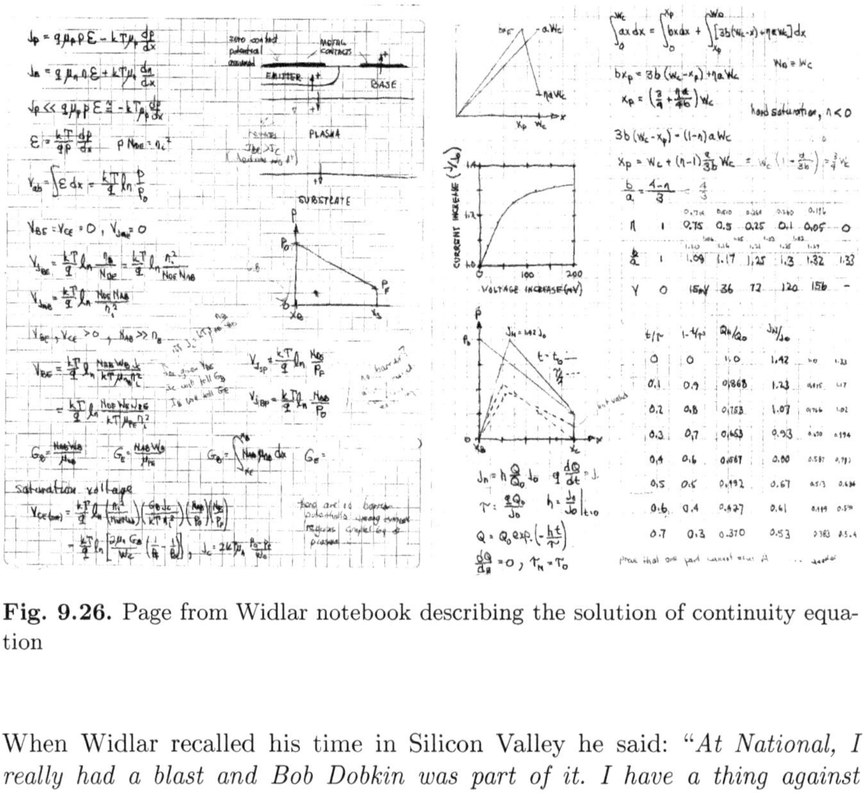

Fig. 9.26. Page from Widlar notebook describing the solution of continuity equation

When Widlar recalled his time in Silicon Valley he said: "*At National, I really had a blast and Bob Dobkin was part of it. I have a thing against paging systems and bells. In the lab we replaced bells on the telephones with lights so the ringing in wouldn't disturb me or the other engineers. However, in order to get some processing equipment into the lab the door was removed and the entrance widened. After a few days of listening to the paging system, right outside our entrance, I decided to do something about it. I went to my car, got an M-84, placed it in the speaker and blew it partly away*". Pierre Lamond heard the noise and come running over." Pierre was really annoyed at Widlar about that, and Widlar was annoyed too because the blast did not completely disable the speaker. The next thing, Widlar had put another M-84 into the speaker. The second charge "did the trick."

In another instance Widlar and Dobkin installed "The Cycle Dropper" circuit into the clock in Pierre's office that caused an unpredictable "time shift" in Pierre scheduling.

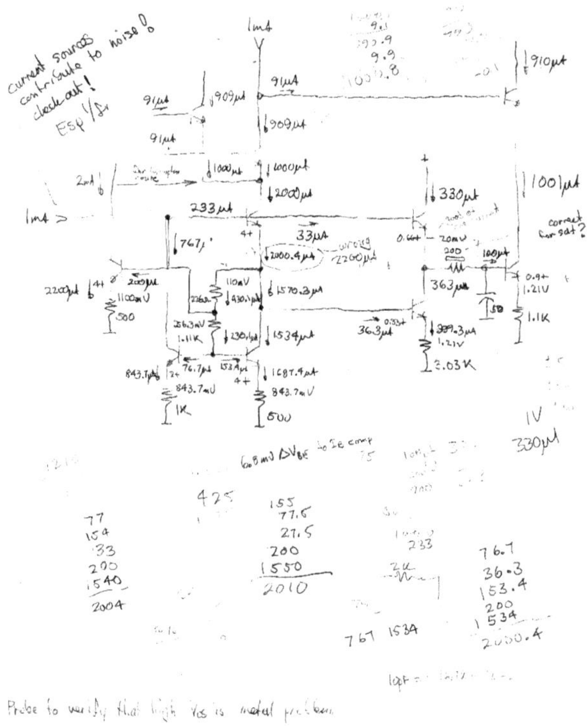

Fig. 9.27. Widlar circuit simulator: paper, pencil, and slide rule (circuit analysis of LT 1016)

Widlar's success did not come without hard work. It is not know that Bob was frequently spending hours in the Stanford University Library reading and studying. People at Fairchild were complaining that Widlar often possessed copy machine and copied all he consider as important. His engineering notebooks are piece of art, very neatly organized and mainly showing the geniality of the Widlar's engineering.

Bob Widlar had also another face: he was passionate about photography and had an impressive collection of Nikons; he had a telescope and was dreaming and looking into stars; he did amazing photographs of eclipses. As he was older, he was for the first time able to keep a relationship with one woman.

Widlar never stopped pushing himself farther and farther. Contrary to popular myth, he did not die jogging on a beach, he was taking a bigger challenge, he was jogging up a hill. His heart, scarred by his early wild life and pre-dispositioned by family genes, collapsed the morning of February 27, 1991.

Fig. 9.28. Robert J. Widlar 1937–1991

Robert J. Widlar accomplished an extraordinary amount of work in his short life. Engineer Widlar never accepted anything less than the best and he underlined each of his monumental accomplishments by his unique and clever humor. Widlar could not live a conventional life; he needed, as many creative people, to live hard.

Interestingly enough, today we are not benefiting from the accomplishments of so called "nice people," who diluted their personality with obedience and hypocrisy, sunken into an already forgotten past. Bob Widlar, who searched for truth and perfection, will live with us forever.

10

The MOS Transistor

The surface controlled transistor has a very bad drift problem. We have been fooling with this problem for a long time and have no hope of an early solution. In fact, I am not sure I have a strong hope of an eventual solution.

Gordon Moore
Fairchild Progress Report, February 15, 1962

Although the MOS devices are still at the research stage because of fabrication problems and incomplete physical understanding, their impact on microelectronics is expected to be significant.

George Warfield, RCA
Electron Device Meeting, October, 1962

The Metal-Oxide-Semiconductor transistor is based on a relatively old idea. Contrary to common myth the very first proposals for field effect devices is not a device patented by J. Lilienfield in 1928 (U.S. Patent 1,900,018, Fig. 10.1) or the idea patented by O. Heil in 1935 (British Patent 439,457). The first experimental observation of the surface and its impact on the electric current was disclosed in the paper "*The action of light on Selenium*" by W. G. Adams and R. E. Day in the Proceeding of Royal Society in 1876.

William Shockley explored this concept at Bell Telephone Laboratories in 1945. In mid 1945 Shockley and Pearson [Phys. Rev. 74, 232 (1948)] envisaged the device which would change the conductivity by modulation of the surface charge by normal electrical field. Several features of this device are the same as what would become the MOS transistor. The device consists of a thin layer of N-type semiconductor placed on an insulating support. In close proximity of semiconductor is a metal plate forming a capacitor as shown in

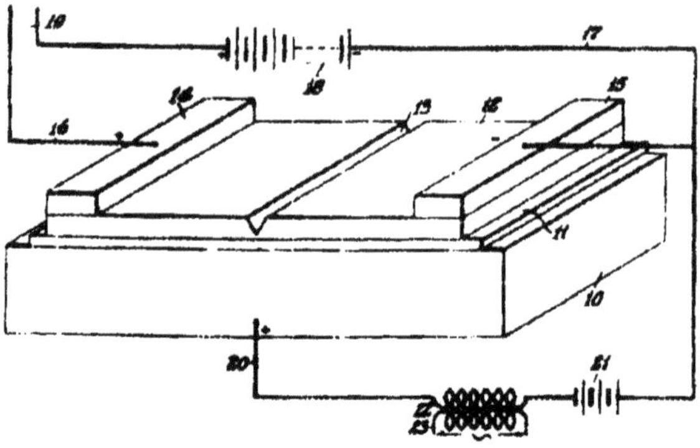

Fig. 10.1. Field effect control device proposed by J. Lilienfield

Fig. 10.2. Shockley assumed that if the metal plate was charged positively, the additional charge on the semiconductor would be represented by an increased concentration of free electrons. Free electrons should contribute to the conductivity.

Numerous experiments were carried out with various semiconductors; however, the degree of modulation was much smaller than expected.

In October of 1945 Shockley brought his calculations to John Bardeen whom he had recently hired. On November 7, 1945 Bardeen confirmed the correctness of Shockley's calculations and was equally puzzled by the discrepancies between theory and experimental data.

At the same time Walter E. Mayerhof who was pursuing his Ph.D. degree at the University of Pennsylvania developed an experimental technique for measurement of the difference in the work functions of metal and semiconductor in contact [2]. Bardeen analyzed Meyerhof's study of the relation between the surface potential and work function of metal point contacts applied to Si and Ge, and he found little correlation. Bardeen correctly recognized that the electronic states on the surface of the semiconductor correlate to surface potential.

On March 19, 1946 Bardeen recorded in his notebook an explanation for the failure of the field-effect structure. He suggested that electrons drawn to the surface of the semiconductor by charging it negatively were not as free to move as were electrons in the bulk. Instead, these added electrons were trapped in the surface states. Once the free electrons are trapped in the surface states, they shield the bulk of the semiconductor from the influence of the charged control plate. Bardeen's suggestion of surface states explained not only the failure of the field-effect device but also the mysteries of the rectifying characteristics of semiconductors contacted by point metal contacts. Based on Bardeen's result Shockley's semiconductor research team abandoned activity on the field-effect device and focused on new experimental conditions and a new observation leading later to the point-contact transistor. The failure of

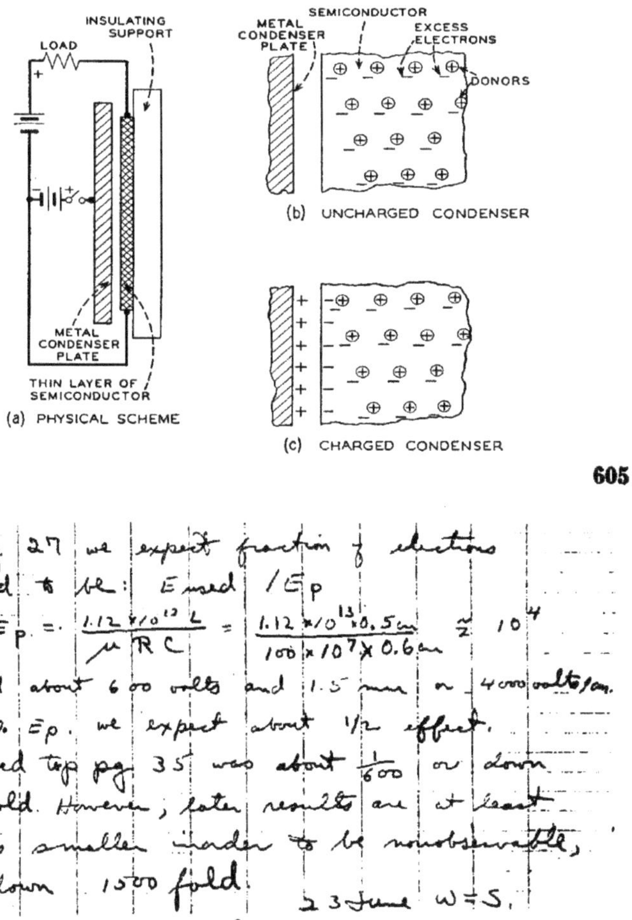

Fig. 10.2. Field-effect device proposed by Shockley on June23, 1945

the field effect device caused the abandonment of further development efforts at BTL. Instead, for a while, Shockley become fully occupied with the theory of minority carriers and PN junction devices.

A. E. Iunovich presented at the All-Union Conference on Non-Equilibrium Processes in Semiconductors, Moscow, March 14–16, 1957 a landmark paper describing measurement and theory of the semiconductor surface layer. A quite unusual British lady, Fedora Berz, of Mullard Research Laboratories in Redhill, Surrey presented her extension of Russian papers at a meeting of the Electrochemical Society in Houston in October 9–13, 1960. The C-V term was not used at that time. Berz choose the unfortunate title of her presentation, "*Field Effect at High Frequency*" similar to the title of the original Russian presentation: "*Variation of field effect in semiconductors on frequency.*" Berz's highly mathematical presentation was not breathtaking, and not many attendees paid attention to her. The one who did was Kurt Lehovec. He was interested in this problem since he saw the World War II era memo of Dr. Heinrich J. Welker analyzing electrical behavior of the field effect device. Welker followed on the work of P. Brauer and noticed that conductivity of Cu_2O could be controlled by the partial pressure of water vapor on the crystal surface[1]. The water caused the hole depletion in Cu_2O. The memo described calculation of the thickness of depletion layer, hole concentration, and current-voltage characteristics.

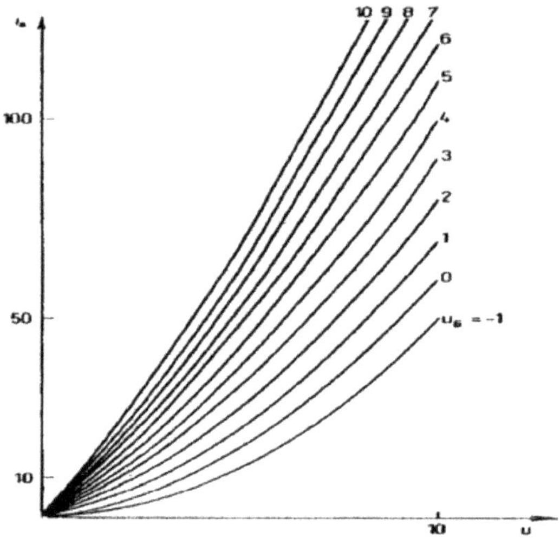

Fig. 10.3. I-V characteristics of Welker's field effect device [1944]

Welker was involved in radar research and he suggested the adoption of semiconductor detectors in cm wave receivers. He worked on this problem with Klaus Clusius at Munich University during 1942 to 1945. Their work, as Welker pointed out, was "*interrupted in May 1945 by circumstances over which I had no control.*" Their patent application was filed on April 7, 1945, but was not awarded until 1973 (DE Patent 980,084). H. Welker was the first to describe correctly the current – voltage characteristics of the FET device in the form as used today (drain current as function of threshold, gate, and drain voltages.)

Lehovec published several papers dealing with capacitance measurement of selenium rectifiers [J. Applied Physics, Vol.20 (1949), p. 123]; however, his interpretation still cannot explain other components to the loss-angle than the resistivity. After a meeting in Houston, Lehovec returned to the old problem

[1] Robert Gibney's inspiration to use electrolyte "to overcome the blocking effect of the surface states" (Brattain and Gibney U.S. Patent 2,524,034) originates in Welker's work.

and at the beginning of 1961 submitted a paper "Field Effect Analysis of Surface States on Silicon" with John Sprague and Alexis Slobodskoy as co-authors to Physical Review. The paper was rejected for publication but was presented at the Stanford Solid State Device Research Conference on June 26–28, 1961 in a session chaired by John Moll. The presentation had the title "The effect of surface states on Si-SiO_2 interface." Finally, in 1963, Physica Status Solidi published the paper "Field Effect Capacitance Analysis of Surface States on Silicon," describing the C-V technique as used today.

At the Houston meeting Henry Theuerer of BTL presented other work which originated in Soviet Union. In 1957, N. N. Sheftal, N. P. Kokorish and A. V. Krasilov published a paper describing the epitaxial growth of silicon and germanium layers [Izvest. Akad. Nauk, SSSR Ser. Fyz., Vol. 21 (1957), p. 140.] This work was followed by an RCA group (E. F. Cave, M. A. Klein, H. Kressel, A. Meyer and B. R. Czorny), IBM group (E. S. Wajda, B. W. Kippenhan, W. H. White, M. J. O'Rourke, J. C. Marinace, R. L. Anderson) and a BTL team, H. C. Theuerer, J. J. Kleimack, H. H. Loar, H. Christensen, J. M. Goldey and Ian M. Ross. The BTL group first announced the epitaxial collector transistor in June 1960 at the IRE Device Research Conference in Pittsburgh. In September 1960 Rheem announced the first epitaxial transistor, RT409. Two months later Motorola introduced the 2N834 at the National Electronics Conference in Chicago. The epi sensation that year completely overshadowed the pioneering work on field effect transistors by Dawon Kahng and Mohamed M. Atalla of Bell Labs.

In the end of the 1950's Atalla's work was partially supported by the Army under a general program to improve transistor reliability. Atalla presented his work at several conferences. The first was the IRE Solid State Device Research Conference, Ohio State University, June 18–20, 1958.The next important paper "Silicon-Silicon Dioxide Surface Device" was presented in 1960 at the IRE Device Research Conference. The paper summarized the work of the authors and their team – E. E. LaBate and E. I. Povilonis who fabricated the device; M. O. Thurston, L. A. D'Asaro, and J. R. Ligenza who developed the diffusion processes, and H. K. Gummel and R. Lindner who characterized the device.

Dawon Kahng was born in Seoul, Korea and studied physics at Seoul University. In 1959 he received a Ph.D. from Ohio State University in Columbus. While at Ohio State University, he was involved in the study of diffusion in oxide. He joined BTL in Murray Hill in 1959. Mohamed M. Atalla, alias Martin or John Atalla, graduated from Cairo University in Egypt and for his master and doctorate degrees he attended Purdue University. Before joining HP Associates in 1962 he was for eleven years with Bell Laboratories at Murray Hill. Atalla found a cool reception of his work at BTL and this consequently led to his resignation from BTL. He joined HP Associates, the obscure semiconductor arm of Hewlett-Packard, and he never published another paper related to semiconductor surface or MOS device.

Fig. 10.4. M. M. Atalla (smoker) as a Director of Semiconductor Research at HP Associates, an affiliate of Hewlett-Packard Company in 1963 (in picture with J. L. Melchor)

The starting material of Kahng's and Atalla's device was the 250 μm thick <111> silicon bar. Two types of devices were fabricated: P-MOS and N-MOS. After the initial wet oxidation at 1200°C for 120 minutes (approximately 10 kÅ of oxide), photoresist technique was used to define the device geometry shown in Fig. 10.5. Source/Drain diffusion was carried out in a closed tube at 1250°C (phosphorus for N-MOS, which were called at that time n-p-n units) and 1300°C (boron for P-MOS). The junctions' depth was varied from 12 μm to 25 μm with surface concentration typically 10^{20} atoms per cm^{-3}. After this step the device was manually "stained" to define the gate area which would be etched in 10:1 nitric acid – HF acid mixture for 20–30 second to clean the gate area. The bars were then cleaned for 15 minutes in boiling DI water, 15 minutes in boiling nitric acid and again in three baths of boiling DI water. Bars were loaded immediately after cleaning into a high-pressure steam oxidation "bomb" designed by J. R. Ligenza, and oxidized at 650°C and 120 atmospheres to grow gate oxide with thickness of 1000 to 2000 Å. Immediately after gate oxidation the aluminum gate was evaporated and residual oxide removed from the back of the sample and S/D contact areas. The individual devices were mounted on the four-lead header (Fig. 10.6.) Gold wire was used to contact source and drain. The gate was contacted by a spring loaded point contact.

Only the P-MOS type of devices were working. As drawn, the channel length was 25 μm; typical L_{eff} was in the range of 20 μm. The channel width W was 250 μm. Threshold voltage ranged from 3 to 6 V.

Several years before Bell Laboratories demonstrated a MOS transistor, Paul K. Weimer and Swedish'-born J. Torkel Wallmark of RCA Laborato-

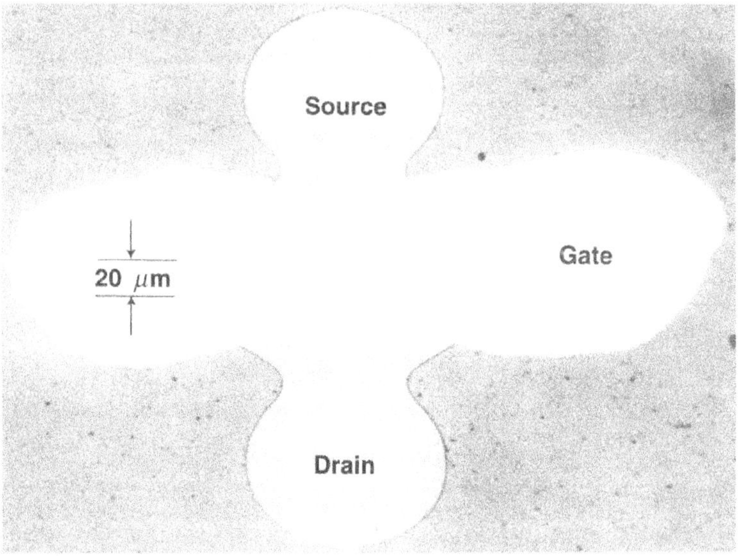

Fig. 10.5. The first MOS transistor demonstrated by Kahng and Atalla in 1960. Although this photograph is in many publications, only a few people know what is in the picture. The photograph does not actually show the complete transistor but rather the device after manual "stain" masking of the gate region (before stripping oxide from the gate region and subsequent gate oxidation)

ries laid out in the end of the 1950's the groundwork for a MOS transistor and integrated circuits made of thin films. Paul Weimer successfully implemented Wallmark's ideas with field-effect transistors made with thin films of cadmium sulfide and cadmium selenide, deposited on insulating substrates.

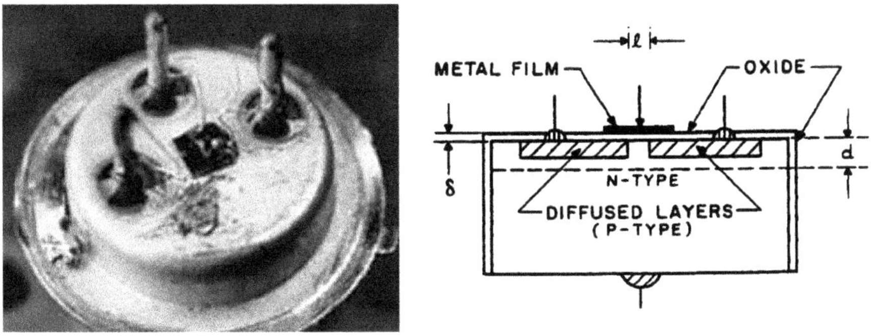

Fig. 10.6. Kahng's and Atalla's MOS device bonded to the header. Note spring loaded point contact to aluminum gate (middle pin) and drawing from D. Kahng, "Silicon-Silicon Dioxide Surface Device", BTL Tech. Memo, MM-61-2821-1, January 16, 1961

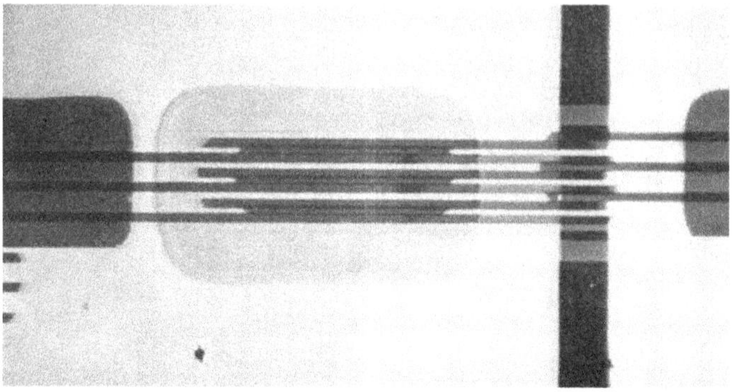

Fig. 10.7. 25 μm thin film transistor designed by P. Weimer and H. Borkan of RCA (1961)

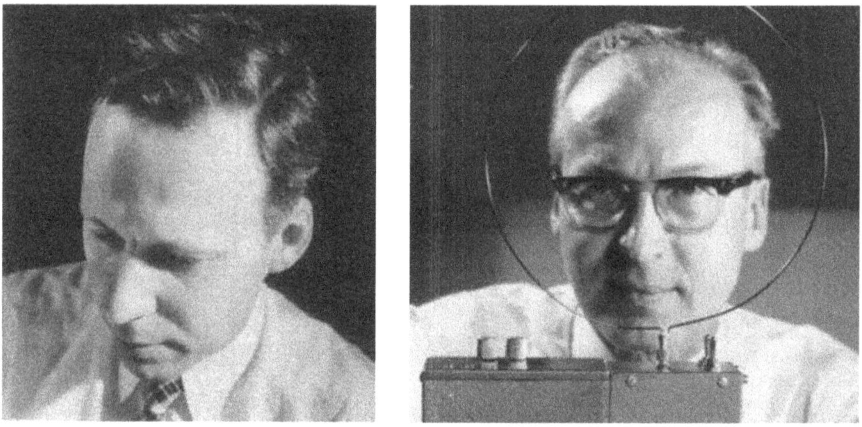

Fig. 10.8. John Torkel Wallmark and Paul K. Weimer in 1966 with RCA TV camera using Weimer's thin film sensor and MOS circuitry. The pictures from the moon in 1968 were made with this camera

The volume of the work and the vision of John Torkel Wallmark was really remarkable. He was working on the concept of integrated electronics from devices to systems. The forty years or so of development and scaling of MOS devices resulted in what we call today the double gate transistor, FinFET, or surrounding gate device. Wallmark filed a patent application describing such a device on February 28, 1957. (Fig. 10.9.)

Torkel was one of few who correctly interpreted Shockley's work. In his paper Shockley elaborated device requirements for device structure and concluded that the variation of the electric field along the channel should be much less then corresponding variation perpendicular to the channel. [W. Shockley, Proc. Inst. Radio Eng., 40 (1952), p. 1365]

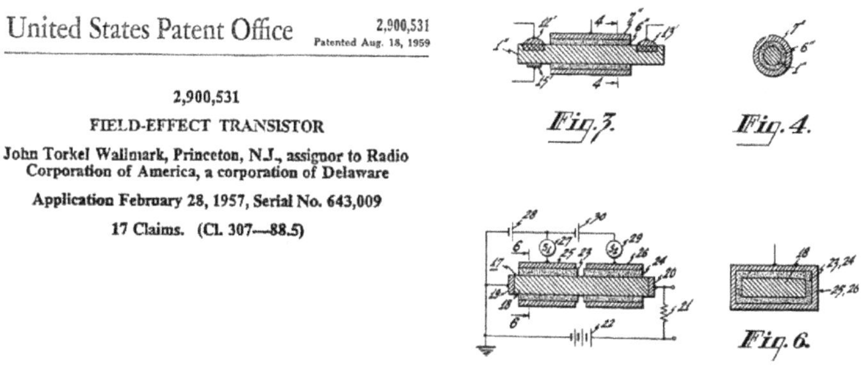

Fig. 10.9. "Surrounding Gate" or "Double Gate" transistor as described in J. T. Walmark's Patent filed on February 28, 1957

The ideas and work developed by Atalla, Kwong, Walmark, and Weimer, bore fruit in the first MOS integrated circuits. RCA Laboratories announced in December of 1962 a major advance in the technology of MOS transistors as a result of the RCA – Air Force Project. Thomas O. Stanley, head of the Integrated Electronics Group at RCA Electronic Research Lab in Princeton, N.J. said: "*we worked on MOS because we wanted to develop a reproducible IC technology to implement J. Wallmark's concept of universal logic gates.*" Wallmark developed this concept when RCA was awarded in 1956 a major contract from the National Security Agency for project "Lighting" with the goal to produce fast computers.

William Webster was attending Atalla's 1960 presentation. Webster made significant contributions to the junction transistor theory at RCA and maintained close informal relationships with many members of BTL. RCA was newer in the telephone business and was not considered a BTL competitor. The idea of integrating multiple electronic components on a single piece of semiconductor was considered by RCA engineers as early as 1953. Webster put T. O. Stanley in charge of the MOS project. He rightly recognized that as MOS technology matured it would be the one most widely used because:

1) The gate oxide, the most critical portion of an MOS transistor, is covered up immediately after it is grown.
2) Even with only one layer of metal, MOS devices may be more dense than their bipolar counterparts.

With the desire to encourage work in silicon at Princeton, Webster took a strong interest in Atalla's work and showed more interest in his work than BTL management. Webster hired Karl Zaininger who was interested in the field of semiconductor surfaces, and after attending the 1960 DRC he immediately began work on the "MOS Diode." In December 1960, working with Charles Mueller and Ethel Moonan, he built in the Somerville RCA transistor plant the "MOS Transistor". The team was ultimately joined by two additional engineers – Steve Hofstein and Fred Heineman, whom Webster hired later that year.

At the 1962 Electron Device Meeting in Washington, D.C. Frederic P. Heiman and Steven R. Hofstein presented a paper with the title "Insulated-Gate Field-Effect Transistor." The device was later described in Frederic P. Heiman's patent application for "Method of Fabricating Semiconductor Device" filed on September 7, 1962.

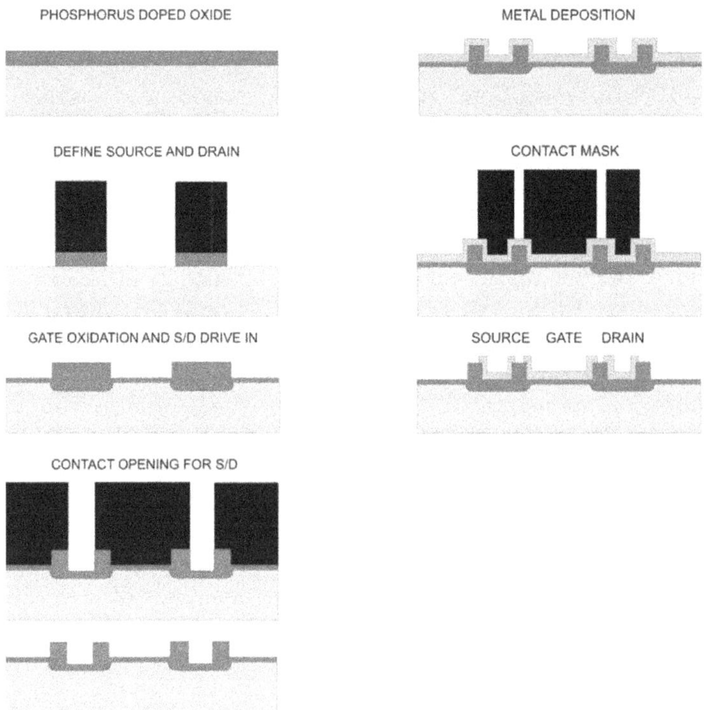

Fig. 10.10. The main processing steps for the MOS transistor developed by Hofstein and Heiman of RCA in the end of 1961

At the time when Hofstein and Heiman worked on the MOS device, the RCA Research Labs were continuing to make integrated circuits with thin films. These two efforts, although parallel, had quite different objectives. The application of the MOS device was going to be components for digital circuits. The application of the thin-film FET, on other hand, was for an electro-optical usage – a solid-state vidicon. After all, in the beginning of the 1960s television was a huge business; computers were not. Since the work on thin film devices started first, many of the techniques now used to make MOS devices were actually developed for the thin-film effort.

The processing steps for the Hofstein and Heiman device are shown in Fig. 10.10. The starting N-type silicon sample with <111> orientation and resistivity 500 Ω-cm was chemically cleaned and followed by the deposition of oxide at 750°C in the mixed vapor of ethyl silicate and trimethyl phosphate. (If trimethyl phosphate is added to the orthosilicate solution, phosphorus-rich oxide is deposited; if trimethyl borate is added, the boron-rich oxide is deposited.) The thickness of doped oxide was approximately 8 μm to 10 μm. After resist masking and patterning the resist was removed from all parts of the device except source and drain regions. The exposed oxide was then removed in diluted HF solution. After the resist was stripped and cleaned, samples were heated at 950 to 1050°C in dry oxygen for 60 minutes to "drive-in" boron to form the source and drain regions. At the same time, silicon dioxide thermally grew on the exposed silicon surface (channel region) with a thickness about 1300 Å. In the next masking step, the openings in the exposed portion of heavily doped source and drain are formed. After aluminum sputtering the final mask is employed to form contacts to the source/drain regions and to the gate. Both types of devices (enhancement-type N-MOS and depletion P-MOS were fabricated and evaluated. If the devices worked, the quite reasonable performance was achieved as may be seen from the I-V characteristics shown in Fig. 10.12 and 10.13. The main problems were high threshold voltage, and stability of the device.

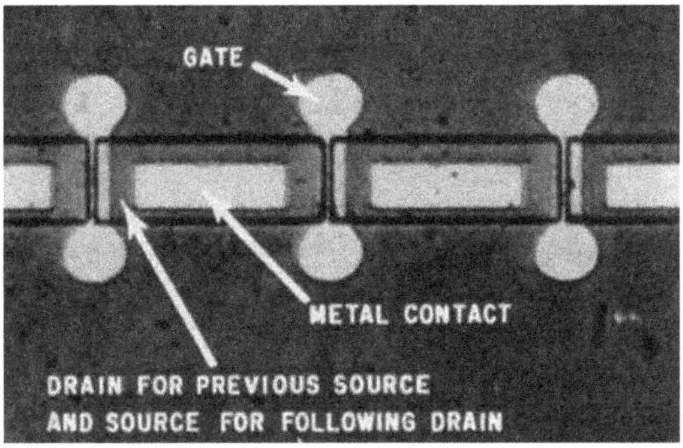

Fig. 10.11. Typical RCA device with channel length $L = 15$ μm and width $W =$ 125 μm and gate oxide thickness 1200 Å. $V_{\mathrm{Dmax}} = 35$ V. (Note the large metal gate misalignment

Dr. William M. Webster, director of RCA Laboratories' electronic research program said that the RCA Semiconductor and Materials Division in Somerville, NJ. will start production of the first unit, TA2330. The new MOS transistor was similar to the thin-film transistor developed earlier by Dr. Paul K. Weimer of RCA. TA2330 was based on the P-MOS enhancement-mode device in the so called "8-neighbor NOR circuit" concept developed by Air Force Labs. The circuit layout consisted of vertical and horizontal lines

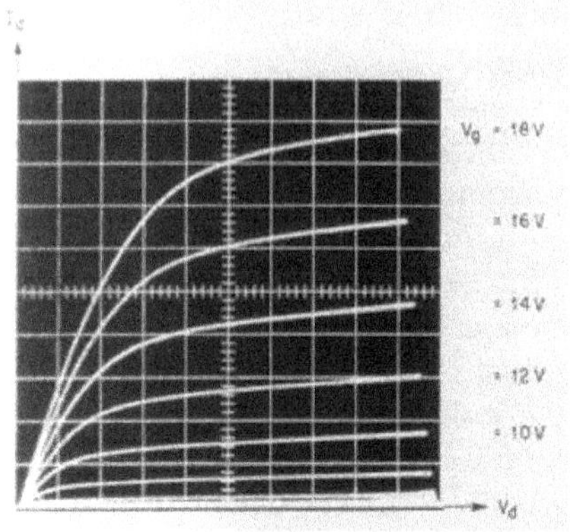

Fig. 10.12. I-V characteristics of enhancement N-MOS (vertical scale 1 mA/div, horizontal scale 2 V/div)

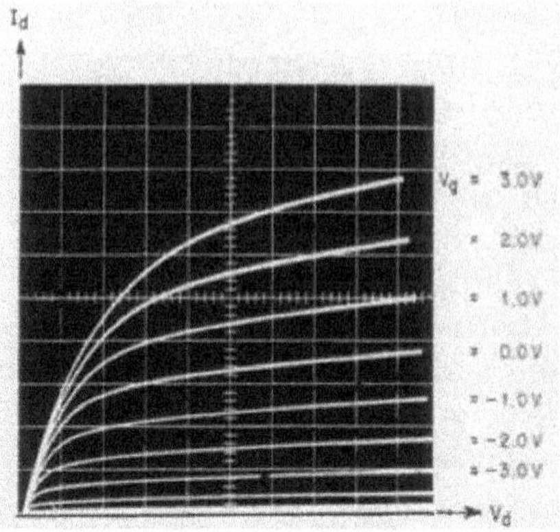

Fig. 10.13. I-V characteristics of depletion P-MOS (vertical scale 2 mA/div, horizontal scale 5 V/div)

with identical circuit elements at each node, which was connected to its eight nearest neighbors. Since signals could be distributed in both directions along the lines, each node becomes, in effect, the 8-input and 9-output NOR gate.

However, the problems ahead were still monumental. It took another year and several technological process variations before the RCA 3N98 and 3N99, the first commercially available MOS transistors, got into production in September 1964.

RCA Computer Division insisted on the highest speed ECL for mainframe logic and used discrete bipolar transistors in its peripherals. They did not see a need for slow logic array. Except for Webster, Stanley, and a few others, very few people in the RCA product group believed useful MOS devices could be built economically and almost all agreed it was not worth the effort. Fortunately, the Laboratories' Applied Research Program, which funded the division's work, kept the effort alive.

The situation turned for the better when the Air Force called for studies on the feasibility of large-scale integration arrays and incorporation of such arrays in computers suitable for Air Force applications. Dr. Webster made a valiant attempt to win the contract, even though RCA's existing skills in the digital IC business were inferior to those of Texas Instruments, Fairchild, and others. RCA had only one unique thing to offer – CMOS. The Air Force awarded three contracts, one to TI for an all-bipolar approach, one to General Micro-electronics for all PMOS transistor approach, and one to RCA, for computers using bipolar ECL arrays for logic, and CMOS arrays for memory. Only TI and RCA completed the program.

The RCA MOS IC program continued for military and NASA contracts, but from the list of suppliers of commercial integrated circuits, RCA was missing for a long time. It took until November 1965 when Radio Corporation

Fig. 10.14. The very first MOS Integrated Circuit designed by Steven R. Hofstein and Frederic P. Heiman of RCA Electronic Research Laboratories in Princeton, N.J. (Die size 1270 × 1270 μm, December 1962)

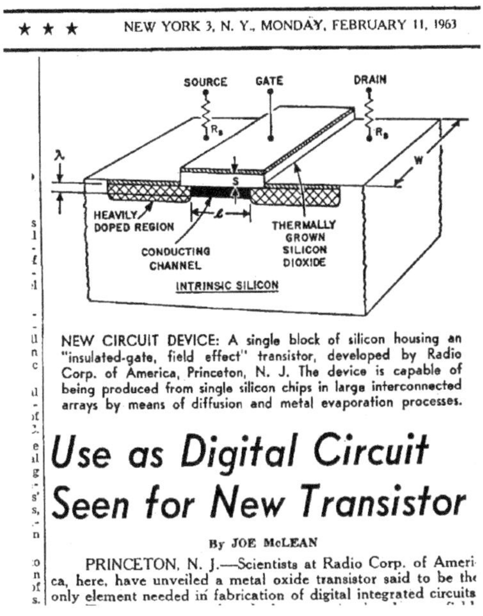
★ ★ ★ NEW YORK 3, N. Y., MONDAY, FEBRUARY 11, 1963

NEW CIRCUIT DEVICE: A single block of silicon housing an "insulated-gate, field effect" transistor, developed by Radio Corp. of America, Princeton, N. J. The device is capable of being produced from single silicon chips in large interconnected arrays by means of diffusion and metal evaporation processes.

Use as Digital Circuit Seen for New Transistor

By JOE McLEAN

PRINCETON, N. J.—Scientists at Radio Corp. of America, here, have unveiled a metal oxide transistor said to be the only element needed in fabrication of digital integrated circuits

Fig. 10.15. RCA Announcement of MOS transistor (February 11, 1963)

Fig. 10.16. Steven Hofstein and Thomas O. Stanley of RCA

of America served notice that as a giant corporation it intended to get into the (at that time burgeoning) IC market. RCA announced availability of a line of seventeen silicon monolithic circuits. Not MOS, but eight ECL and DTL digital and nine linear circuits were immediately available. Digital parts were used in the RCA Spectra 750 computer; RF, IF, video, and audio amplifiers were used in RCA TV sets.

In mid 1968, RCA was the first company to introduce the 4000 series of CMOS digital integrated circuits. The metal gate was later replaced with a polysilicon gate. CD4000 operated in the range of voltages from 3 to 15 V of V_{DD}, and used thick top oxide (field device) for isolation. The threshold voltage might be as low as 0.9 V. Design rules gradually reduced from an original 20 μm to 10 μm. The typical process flow as used by RCA Somerville Fab is shown in Fig. 10.17.

The starting material was N-type <100> silicon with 1–2 Ω-cm resistivity. The first mask was used for the P-well. The concentration of the P-well determined the threshold voltage of N-MOS devices. In subsequent masking, oxidations and patterning the N+ and P+ diffusions for source/drain, guard rings, and resistors were performed. A new oxide with a thickness of approximately 10k Å was grown and removed from all active areas. The gate oxide (typically 1k Å) was grown over active regions and covered by aluminum gate or P-type polysilicon (typically 2k Å thick). In the next steps the gate was patterned and annealed. Subsequent masks defined contacts and after appropriate cleaning the interconnect metal was formed.

The acceptance of CMOS technology was not enthusiastic. All the "big boys" were pushing P-MOS technology. The calculators were the major driving force. Fairchild, Texas Instruments, AMI and Collins Radio Co., created design cell libraries. Fairchild developed "Micromosaic," AMI "Prophet,"

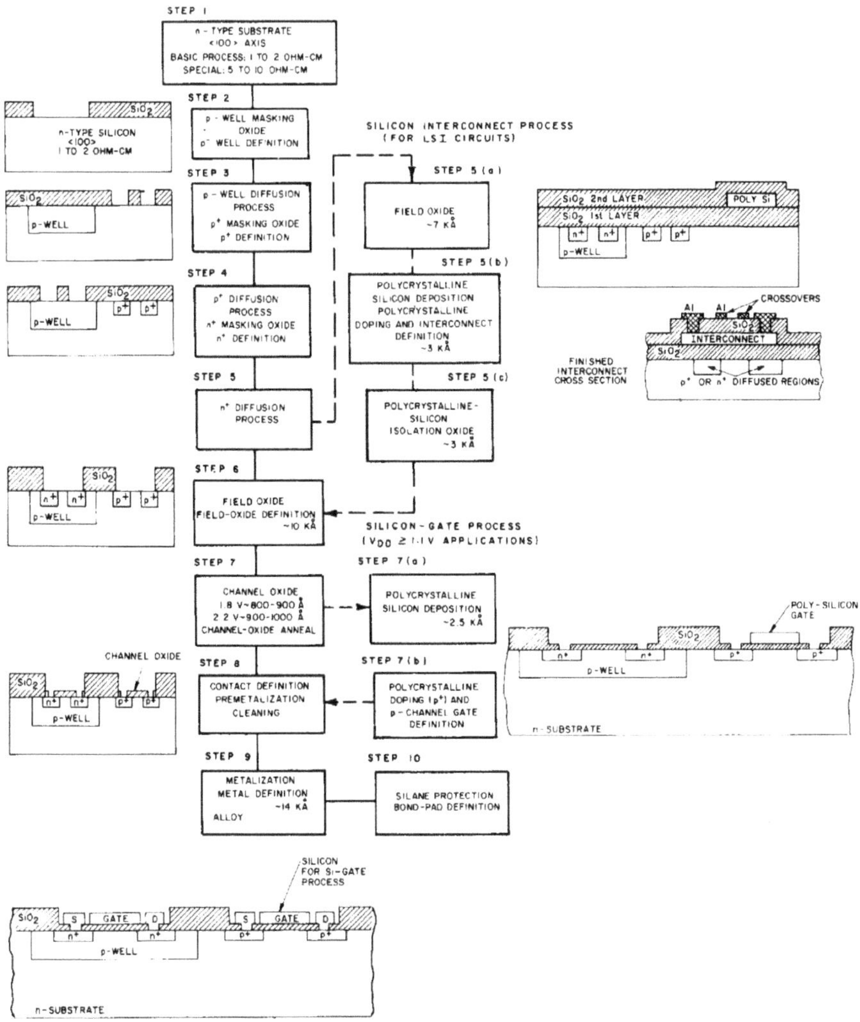

Fig. 10.17. Five-mask RCA COS/MOS process for CD4000 Series of integrated circuit (~ 1970)

Collins "Logicomp," and although such programs were quite simple and needed to run on CDC or IBM mainframe computers, they reduced design time. There was no such work in the CMOS area. In 1968 there were only two vendors of CMOS technology – RCA and a small company, Solid State Scientific Devices, Inc., in Montgomeryville, PA., who developed the first BiCMOS process. The skepticism about CMOS was everywhere, with the prediction that MOS would never operate at 100 MHz (Fig. 10.20). In such an environment you really needed to be brave and confident to stand in front of a customer and defend the CMOS parts. RCA, by emphasizing the mi-

Seven COS/MOS types now available

RCA COS/MOS Type No.	Pkg.	Circuit Function	Price 1-99 Units	Price 1000 Units
CD4000D	DIC	Dual 3-Input NOR Gate plus Inverter	$13.05	$ 9.00
CD4001D	DIC	Quad 2-Input NOR Gate	$13.05	$ 9.00
CD4002D	DIC	Dual 4-Input NOR Gate	$14.50	$10.00
CD4003D	DIC	Dual "D" Type Flip Flop	$20.30	$14.00
CD4004T	TO-5	7-Stage Binary Counter/ Frequency Divider	$31.90	$22.00
CD4005D	DIC	16-Bit NDRO Memory	$23.20	$16.00
CD4007D	DIC	Dual Complementary Transistor Pair plus Inverter	$ 9.90	$ 6.00

Fig. 10.18. RCA kit with 15 CD4000 integrated circuits. The author paid an incredible $75.00 for this kit in June 1969

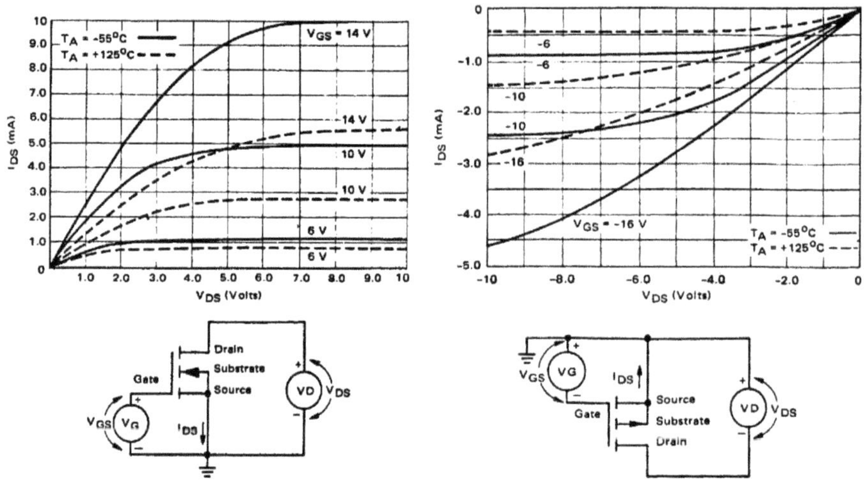

Fig. 10.19. Typical I-V characteristics of N-MOS and P-MOS devices used in early RCA CD 4000 series Integrated Circuits with aluminum gate

crowatt power consumption, prevailed and CD4000 became the pre-cursor of the so-called 54/74 HC CMOS which were produced by every single major semiconductor manufacturer in the nineteen seventies and eighties. HC parts were compatible with bipolar TTL devices, and typically operated from 1.5 to 5.5 V and they used Local Oxidation as isolation.

In August 1962 Fairchild Semiconductor hired a very bright, young, mulish, unyielding, obstinate, and persistent Ph.D. from the University of Utah in Salt Lake City – Frank M. Wanlass.

Wanlass was assigned to the solid state physics group headed by C. T. Sah who, however, was located in Illinois. "Tom" Sah often presented impressive lectures during his visits in Palo Alto, but was not involved in the day-to-day operations of the group. If Gordon Moore's time permited, he managed the group. Frank Wanlass took advantage of the poorly assigned responsibilities

The crystal ball is not altogether pellucid at this viewing. Assuming that MOS makes the scene—as many prophets assert it will—it is still not certain whether single-channel or complementary MOS will win out. And within single-channel technology, the n-type devices have not yet come into their own, despite their potentially superior performance.

In the foreseeable future, for ultra-high-speed scratch-pad operation, bipolars can't be touched by anything. All the experts seem agreed on that.

One specialist in systems designed for Government end-use, George Podraza of the Bunker-Ramo Corp., Canoga Park, Calif., concedes the point, but he qualifies it as follows:

"There's nothing exceptionally fast about MOS memories. Bipolars are 10 to 1000 times better in speed-power product figure of merit." But Podraza adds that this is true only at high frequencies because capacitance is less critical as the frequency is lowered.

Dr. Friedolf Smits, scientist at Bell Telephone Laboratories, Murray Hill, N.J., predicted at the Solid-State Circuits Conference in Philadelphia in February that MOSFETs would never operate at 100 MHz.

Fig. 10.20. The very first wrong prediction of MOS capabilities presented at 1968 Solid-State Circuit Conference in Philadelphia. [Electronic Design, April 1, 1968]

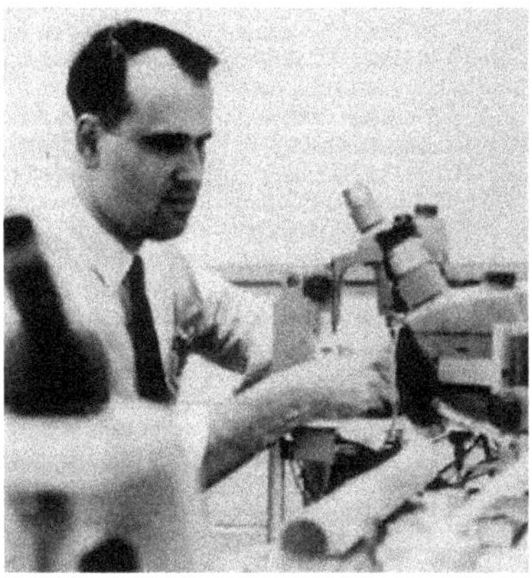

Fig. 10.21. Frank M. Wanlass in 1964 at GMe

in the physics group and defined his own priorities. He wanted to develop field effect devices. Wanlass's interest in FET devices originated from his work at Utah when he was studying for his doctorate in solid-state physics and when he become familiar with the work of RCA and especially the work of Paul Weimer on FET devices and Michael E. Szekely on FET logic circuits.

F. Wanlass and K. Eaton's development work was part of the Fairchild Project # 142 "Surface Controlled Devices." In less than six months Wanlass designed several discrete N-MOS and P-MOS devices and simple MOS integrated circuits with P-MOS devices. All N-MOS devices were not functional; only some P-MOS devices were functional. In the Fairchild's Progress Report from February 1, 1963 Wanlass reported the integrated flip-flop circuit and simple amplifier (Fig. 10.22). Fairchild's Planar process was used for 5 and 10 μm long P-MOS device with width 250 μm and 1000 Å thick gate oxide. Aluminum gate was deposited in vacuum immediately after gate oxidation. Threshold voltage was not defined the way we know today; Wanlass reported threshold voltage −5.5 V as gate voltage for 10 μA drain current at $V_{DS} = V_{GS}$. V. G. Redi and C. Bittman were involved in evaluation of devices while fabrication of devices was done by M. Papkoff and J. Kelly. The main problem of all Wanlass' devices was the same as the problem of RCA devices – instability and dependence of the threshold and the pinch-off voltage on the aluminum alloying process (Fig. 10.23.)

Wanlass presented his idea of the CMOS device in a paper co-authored by C. T. Sah at the 1963 Solid-State Circuits Conference. The presentation outlined the concept of CMOS with very limited experimental data; however, the main features of CMOS were correctly identified. The presentation concluded: "*... because 1) both standby power density will be extremely low and 2) switching power density can be high, it should be possible to construct field-effect triode logic circuitry with a very high packing density.*" The second Wanlass paper, presented at the WESCON Show and Convention in San Francisco in August 1963, summarized the results of his work on FET devices at Fairchild. The P-MOS transistor, process, and performance of logic circuitry were described. Wanlass patented the idea of CMOS in a patent application "Low Stand-By Power Complementary Field Effect Circuitry." filed on June 18, 1963 (U.S. Patent #3,356,858) just a few days before he quit his Fairchild job.

Wanlass left Fairchild because Fairchild announced that there was no "*immediate plan to use a new technology*" before he had any experimental data and had a chance to complete the project. Wanlass was characterized by his peers as "highly impatient;" however, he was a very creative person never comfortable in a large organization. He wants to traverse his way and he does not like detailed theoretical studies. He did not trust the work of technicians, and he preferred to do the work in the lab himself, including device fabrication. Although Wanlass had a degree in physics, he was an excellent experimenter. He built electronic circuits and he designed and built electrometers based on the high input impedance of MOS transistors which had the same instability problems as Sah's tetrode. While Grove and Sah preferred to use the MOS capacitor and the CV technique to characterize oxides and semiconductor surfaces, Wanlass consided the efforts of Sah's group as overemphasis on studying MOS physics instead of developing and produc-

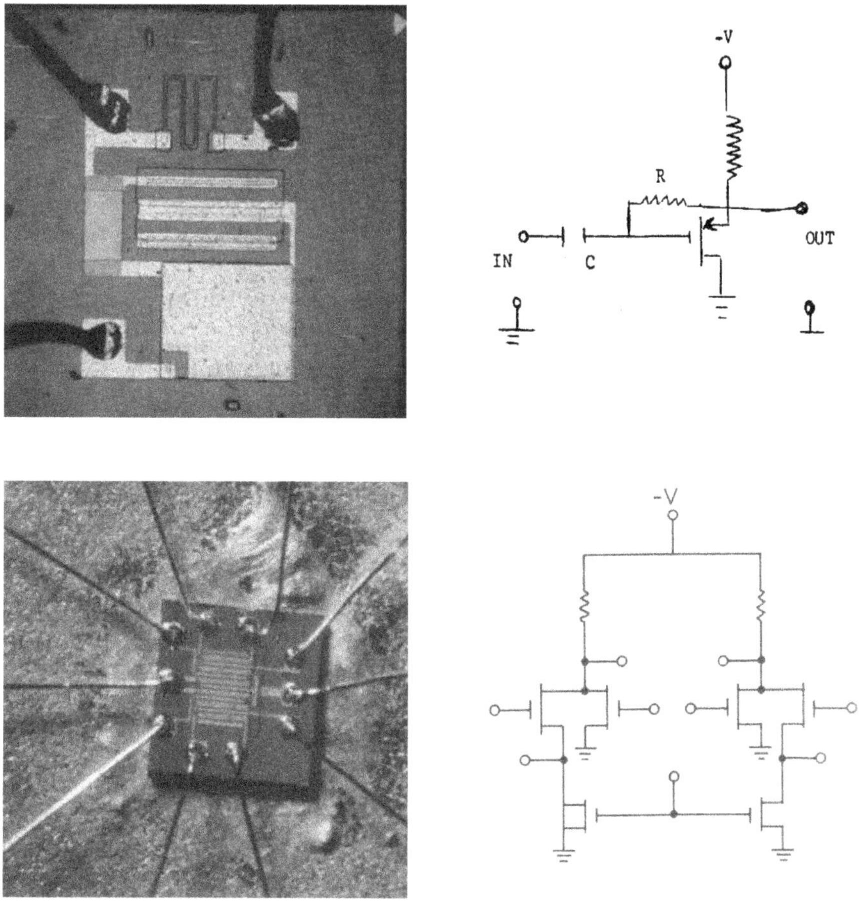

Fig. 10.22. Simple amplifier and circuit for verification of a half-adder, shift register stage, and flip flop as designed by Frank M. Wanlass at Fairchild [February 1963]

ing MOS products. Wanlass generally worked with packaged MOS transistors. His typical oxide stability test was to put a transistor on the Tektronix curve tracer, heat the transistor and measure the shift in I-V characteristics. Wanlass' goal was to develop technology, not the theory of MOS integrated circuits.

Gordon Moore became very suspicious of any device where surface effects might play any role. Moore's attitude originated from several incidents in the early Fairchild years. In 1960 Fairchild was forced to stop production temporarily of the 2N1740 PNP mesa transistor because of problems with inversion layers. In August 1961 G. Moore wrote: "*The inversion layers that appear on pnp mesa transistor are the most elusive things to study we have found*". The work of C. T. Sah on the so-called semiconductor tetrode, which

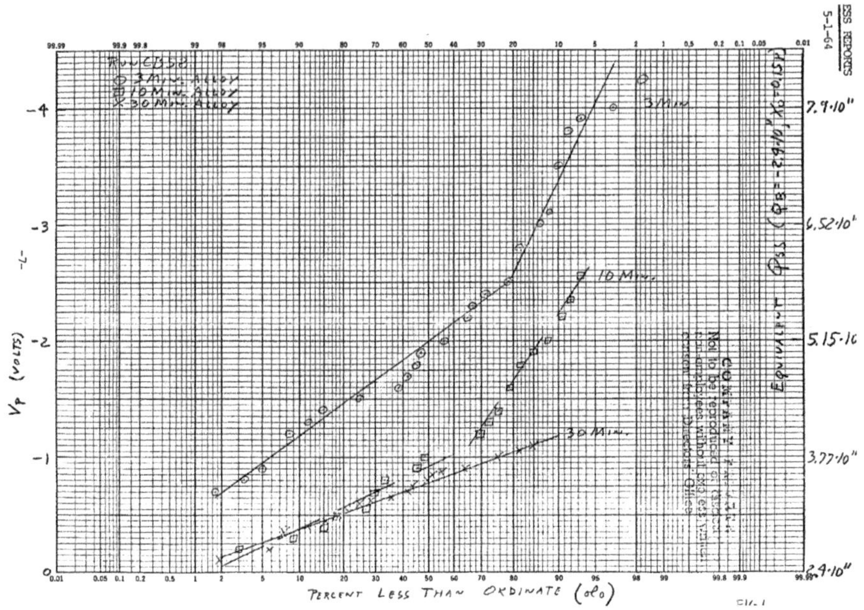

Fig. 10.23. Dependence of V_P on aluminum alloying (3, 10 and 30 min) for one of the earlier batches of Wanlass's MOS transistors (~ February 1963)

Although it was reported that complementary n- and p-type devices have been constructed, a spokesman for Fairchild said the devices are still in R&D, with no production scheduled.

Fig. 10.24. Fairchild announcement from March 1, 1963

he investigated at Fairchild during 1961 and 1962, resulted in even bigger frustration. The semiconductor tetrode was basically a bipolar transistor where the surface potential of the base-emitter diode was controlled by an additional electrode. At that time a lack of understanding of surface effects, relatively primitive processing techniques along with poor quality of silicon substrates hampered Sah's work and G. Moore terminated Sah's work in this field.

Wanlass and Moore also argued about the name for the new field effect device. Wanlass (and latter also Bruce Deal) used the name P-MOST or N-MOST for "*field-effect hole (electron) conducting metal-semiconductor transistor.*" Moore insisted on the name "*n- and p-channel metal-oxide-semiconductor field-effect triodes.*" This term was used in the Lecture Series which AGARD's Avionics Panel organized for several NATO members in June 1963. One of the lectures was prepared by Wanlass, but presented by

G. Moore[2]. Contrary to internal Fairchild memos, Moore's give an up-beat presentation about CMOS advantages, even though the last sentence of the presentation was "*integration of the complementary structures has not yet been accomplished.*" An additional interesting fact about this presentation is the list of references. To my knowledge this is the only presentation made by G. Moore where Shockley's works are listed in references.

In March 1963 Bruce E. Deal, unhappy with Raytheon's takeover of Rheem, joined Sah's group. Bruce was allowed take with him his assistant Maija Sklar, who become the first female member of Fairchild's R&D group. In June 1963, Ed Snow, who knew Wanlass and who was finishing his thesis at the University of Utah, joined the group. Snow had been working on diffusion of ions in glasses, which were similar to oxides. Another new member of the group was Andy Grove who just completed his Ph.D study at UC Berkeley. Grove, who was more a theoretician in those days, was responsible for Fairchild Project # 114 "Diffusion Research." Grove with A. Roder programmed an IBM650 computer and solved impurity distribution during the epi growth.

After Wanlass' departure, Bruce Deal and Ed Snow were driving Fairchild's R&D work on instabilities in the thermal oxides. D. Hilbiber from Fairchild's Applications department designed the Fairchild C-V meter and the group presented in 1964 the IEEE Solid State Device Research Conference in Boulder, Colorado the paper "Investigation of thermally oxidized silicon surfaces using metal-oxide-semiconductor structures."

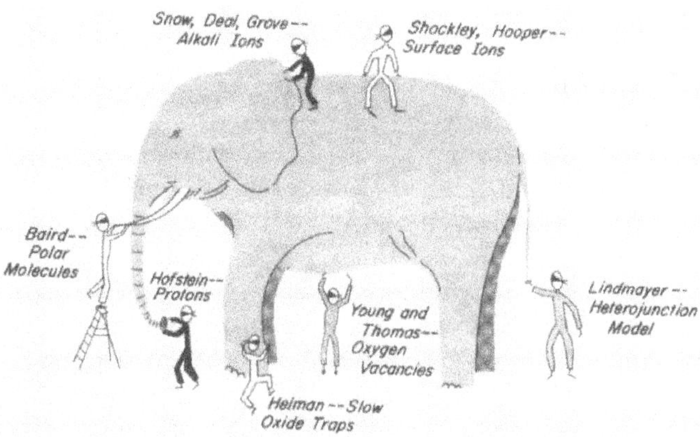

Fig. 10.25. Robert P. Donovan version of the poem "Blind men and the elephant"

[2] G. E. Moore, C. T. Sah, F. M. Wanlass, Metal-Oxide-Semiconductor Field-Effect Devices for Micropower Logic Circuitry, AGARD, 1963

The instabilities in the thermal oxides of MOS transistors were actually due to mobile ions, like sodium, lithium, potassium, etc. Several laboratories were working on this problem such as IBM, RCA, Shockley Lab (Clevite), Bell Labs, Philco and others. There was strong animosity among the various groups. Bruce Deal often described the situation and competition between RCA and Andy Grove who "*tended to be very volatile in his discussions and opinions.*"

Bob Donovan of the Research Triangle Institute presented at the Electrochemical Society Meeting in 1967 his version of John Godfrey Saxe's famous poem "The blind man and the elephant." In the drawing Donovan showed each of the blind men to be one of the groups that were working on the oxide instability problem. Andy Grove had heard about Donovan's elephant and asked Bruce, who knew Donovan, to send him the picture. When Grove found that he was shown on the head of the elephant and one of his arch-rivals Hungarian Lindmayer was shown on the tail, "*you could hear Grove's shout of glee all over the whole valley area.*"

Fairchild's April 1964 reorganization with Bob Noyce as a group vice-president resulted in pressures that they must get something new on the market. Because they had been very successful with the Planar process, the marketing group come up with Planar II to indicate that they now had stable MOS. Planar II really had nothing to do with Hoerni's Planar process, other than the fact that it was used in a Planar process. After enormous effort of the production group headed by P. Lamond to bring MOS instability under control, Fairchild introduced in November 1964 the FI-100 MOSFET. With a confusing statement about "locking of surface charge" (Fig. 10.26) the FT-100 did not became a commercial success, but was successful enough to attract funding from the Air Force (Contract AF 33 (647) 9033) to continue Fairchild's MOS work.

Early MOS circuits were very susceptible to damage from electrostatic discharge (ESD). Subsequent generations were thus equipped with sophisticated protection circuitry that helped absorb electric charges with no damage to the fragile gate oxides and pn-junctions. Still, antistatic handling precautions continued to be enforced to prevent excessive energies from building up.

Conversely, early generations such as the 4000 series that used aluminum as gate material were extremely tolerant to supply voltage variations and operated anywhere from 3 to 18 volts DC. Later generations used polycrystalline silicon ("polysilicon") as gate material. For many years, CMOS logic has been designed to operate from the then industry-standard of 5 V imposed by TTL. Starting in the mid 1990's, it proved necessary to downscale the supply voltage along with the geometric dimensions to maintain sustainable electric fields and to improve energy efficiency. Modern CMOS circuits operate from voltages as low as 1 V.

Although RCA Labs and Fairchild made the first MOS integrated circuits, their companies were not the first in the market. That distinction is held by

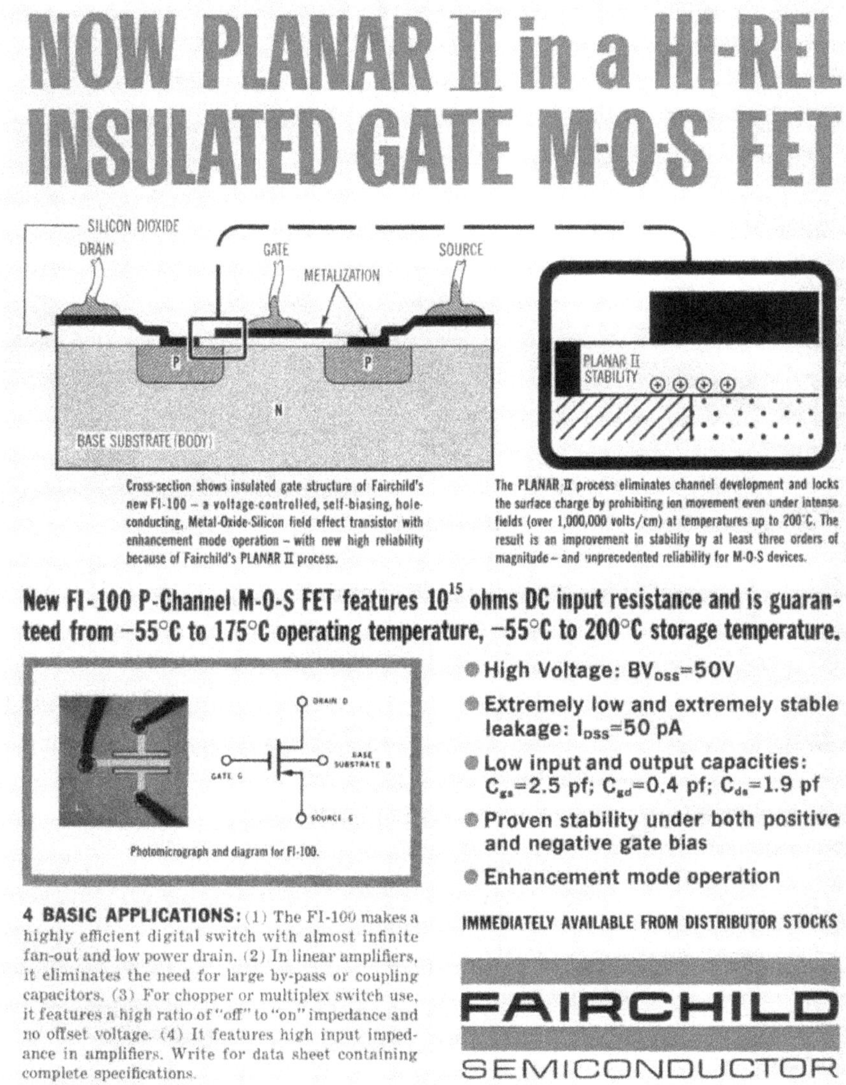

Fig. 10.26. Fairchild's FI-100 ad released on November 1964

General Micro-electronics and General Instrument, which offered the first commercial MOS integrated circuit in late 1964.

General Instrument Corporation based in Hicksville, N.Y., was one of the early companies working on germanium and silicon devices. As early as 1963, the company pioneered and set the MOS technology as the company mainstream. One of the main problems with the RCA and Fairchild MOS process was gate oxide formation. The gate oxide was grown during the drive-in diffusion of source/drain dopants. In the next step oxide was covered by resist which was used for patterning of source/drain areas. Any defect in resist resulted in a pinhole in the gate oxide, which would short the gate metal and channel. General Instruments introduced a thick oxide process (MTOS – Metal Thick Oxide Semiconductor) in 1965 (Fig. 10.27.) Gate oxide thickness was greater than the thickness of oxide which needed to be removed from the source/drain region. As a result, any possible resist defect caused only thinning of gate oxide, not an electrical short.

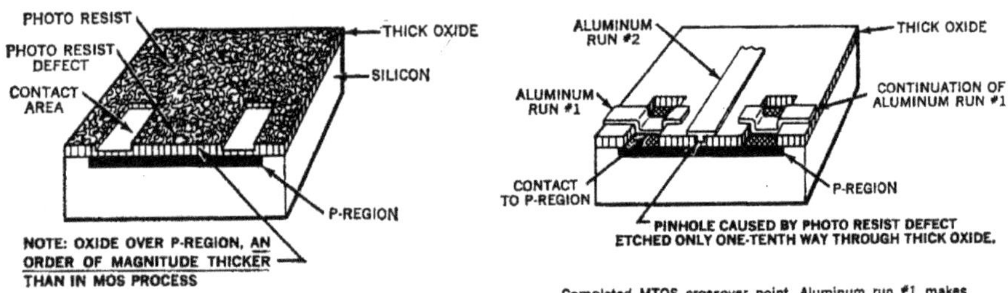

MTOS crossover point prior to etching of thin oxide from contact areas. Note photo resist defect which will allow etchant to attack the oxide.

Completed MTOS crossover point. Aluminum run #1 makes electrical contact to the conducting P-region which crosses underneath aluminum run #2. Contact is then made to the continuation of run #1. The two runs are separated by the thick oxide insulation layer.

Fig. 10.27. General Instrument's second generation" of MTOS devices (1965)

General Instrument launched in the end of 1965 a massive campaign promoting GI as a leader in LSI. The GI MEM family consisted of an 8, 12, 16, 21, 25 Bit shift register, an OR/NOR gate, RS flip-flop, and 8 Bit adder. The most complex circuit was the "Digital Differential Analyzer MEM 5021". At that time the term "Large Scale Integration" was used for components with a die size as large as 1800 × 2200 μm with approximately 250 transistors. (Fig. 10.29.)

General Instrument's entry into the MOS business was eased by luring away several key General Micro-electronics employees, including Dr. Frank Wanlass, who become director of MOS engineering, and whom the GI drum promoted as "*the father of MOS*". Other key men who defected from GMe included Dr. Lee Seely, director of MOS operations and Robert Pace, associate director of MOS research and engineering. In order to accommodate these heavy-weight engineers, GI opened a facility in Utah.

Later GI had second sourcing arrangements with Ferranti, and also a plant in Glenrothes, Scotland.

The report "Electronic Calculator Markets and Suppliers", of 1974, says "GI is believed to be shipping a quarter million (calculator integrated) circuits per month, and has supplied nearly all British manufacturers (CT5000 series)." In addition GI chips were used by Monroe, Victor, Commodore, Royal and Singer in the USA, as well as Sanyo in Japan. Before the Japanese took over the calculator market, General Instrument was the third biggest

Fig. 10.28. General Instrument advertisement announcing the Digital Differential Analyzer (1965)

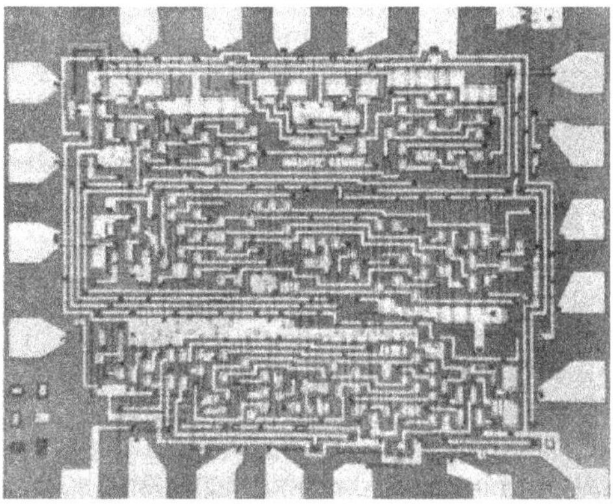

Fig. 10.29. General Instruments MTOS Differential Analyzer MEM 5021

calculator chip vendor behind Texas Instruments and Rockwell. From 1974 the company steadily declined and was active until 1997, when it split into General Semiconductor (power semiconductors) and NextLevel Systems (the cable TV division, which took the GI name with it and was later purchased by Motorola).

The fast, but brief, success of General Instrument is attributed to GI's president Moses Shapiro. Shapiro was a kind of "Renaissance Man" with the philosophy that the educated man needed to maintain broad knowledge, not only in the narrow field related to his profession. In his luncheon presentation "The Engineer – the Whole Man," during the IRE Electron Device Meeting 1962, (which was unfortunately attended by only a few engineers,) Shapiro briefly commented on the semiconductor business and attributed some of the present problems in the industry to an irresponsible financial community and also to marketing gimmicks. In the rest of his presentation Shapiro displayed a keen understanding of the engineering community. Although a lawyer himself, Shapiro urged the engineers to broaden their education and gain an appreciation for the arts, classic literature, music, and poetry. Can the reader imagine similar thoughts from today's business leaders?

On June 31, 1963 Phil Ferguson[3], head of Fairchild's device development department. Howard Bobb, marketing manager; and Robert H. Norman resigned and with Colonel A. C. Lowell formed a new company. Lowell retired from the Navy on August 1, 1963. Pyle-National Co., Chicago based electrical and electronic equipment manufacturer, acknowledged a financial interest in GMe and announced that GMe began in October 1963 (four months after incorporating!) the production of its first family of IC's.

Strangely enough, Colonel Lowell was a Government employee who had monitored the RCA contract and had taken to heart RCA's descriptions of the virtues of MOS transistor arrays. He observed how RCA struggled to make N-MOS devices stable, and virtually ignored the basically more stable, but slower P-MOS devices. Arthur Lowell used to be the Marine Corps' colonel in charge of the avionics division of the Navy's Bureau of Weapons. He was responsible for the planning of Integrated Light Avionics System (ILAS) and Integrated Helicopter Avionics System (IHAS).

The Colonel had also another virtue – he would promise to customers all they wanted, no matter if it was feasible or not. He was technically very weak, but had the ability to sell anything. The drawback was that such transactions could be sold only once. It did not take a long time for people to not take Colonel Lowell seriously. Colonel Lowell's behavior actually contributed

[3] There is an interesting note about Ferguson's departure from Fairchild. Bob Polata was hired by Ferguson, just a few months before Ferguson left Fairchild. When Ferguson decided to leave he assumed in his naiveté that he should notify management about his planned resignation, so someone new could be assigned to work with Ferguson and make the transition smooth. Polata and Ferguson were walking through the hallway in the Palo Alto facility when they suddenly approached Gordon Moore. Ferguson asked Moore if he could set up a meeting with him. Moore asked what the purpose of meeting was. When Ferguson answered, Gordon Moore escorted Ferguson to the entrance gate and Ferguson was not allowed to return to his desk for his belongings.

to delayed acceptance of MOS technology, because he foolishly discouraged many customers with his false promises.

But back in 1963 Colonel A. C. Lowell, president, and general manager of GMe was a new star in the semiconductor business and he announced development of the device family called Picologic. Lowell said that the first such device would be a 20-bit serial shift register, the PL200, which contained 120 transistors on the die of size 1050 × 1050 microns, priced $250.00 per unit. The circuit was designed by Frank M. Wanlass who was a senior member of the technical staff. GME began production of its first family in the second week of October 1963, exactly twelve weeks after moving into empty office space. Contrary to skeptical Fairchild's Gordon Moore, GMe recognized that MOS technology would become a major factor in the electronics industry. They argued that "*even based on the cheapest discrete device types from any country, we will be able to compete with reductions up to 50 per cent, compared with discrete devices.*" This and Colonel Lowell's former connections secured an important contract from the National Security Agency and NASA, and on October 17, 1963 GME offered seven MOS devices: a gate ($19.80), a dual gate, an adder, a half-adder, a buffer, an expander gate and a register ($ 51.00).

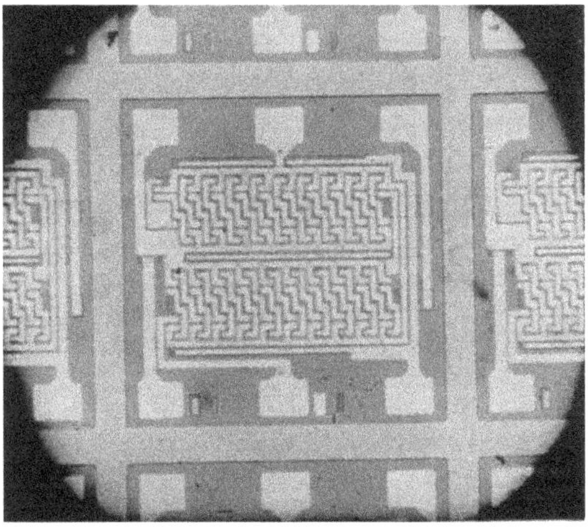

Fig. 10.30. World's first commercially available MOS integrated circuit. The General Micro-electronics 20-bit serial shift register contained 120 P-MOS devices designed by Frank Wanlass and was introduced at the Western Electronics Show and Convention in August, 1964

The 3900 calculator of Victor Comptometer Corporation had been the hope and despair of General Micro-electronics. The initial design and construction of the prototype moved right on schedule after the project was begun in September 1964. By October 1965, every circuit worked and the assembled machine performed well. Unfortunately, Lowell's unsustainable management,

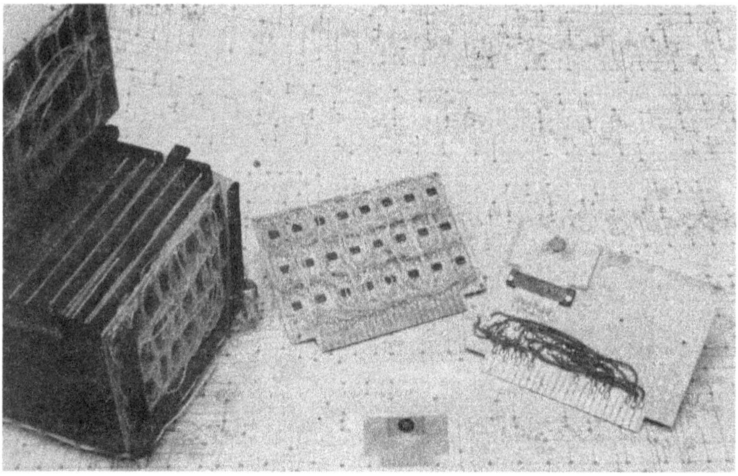

Fig. 10.31. Breadboard version of GMe electronics calculator (1964) (left). In the middle is a typical logic card with 48 MOS integrated circuits which was substituted with a single integrated circuit (right). This digital system was designed by J. Donald Trotter, just two years after his graduation from Stanford University in 1961

and the ambitions of that small company to do many various things at the same time bore bitter fruit. General Micro-electronics was acquired by Philco-Ford. Philco-Ford's Microelectronics Division introduced with great fanfare the first completely electronic desk-top calculator in November of 1965 at the annual show of the Business Equipment Manufacturers Association.

General Micro-electronics was running a pilot-line operation during 1965. The switch from pilot-line to mass-production became a nightmare. Employment tripled during six months and floor space was increased from 30,000

Fig. 10.32. 44-lead GMe package for complex MOS ICs (1964)

square feet to 100,000. The company, which promised the calculator at a price low enough to make them competitive with conventional mechanical machines, found itself spending money at a rate it could not sustain. To make the situation worse, two founders of GMe, Howard S. Bobb and Robert Norman resigned on April, 30 1966. Colonel Arthur Lowell was technically on leave on absence, but no longer had job responsibilities.

The yield of some circuits was "*substantially less than 1%.*" On one or two, yields were zero; they simply could not be produced in high volume. J. Philip Ferguson, one of the founders of General Micro-electronics, who become president of Philco-Ford's Microelectronics Division, conceded that serious production problems delayed the calculator's debut.

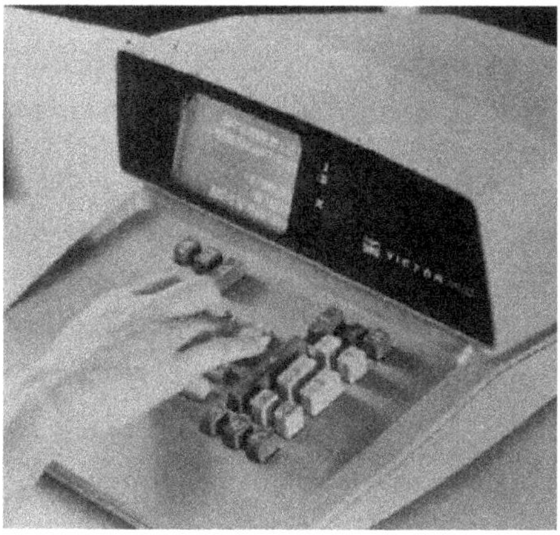

Fig. 10.33. The first completely electronic desk calculator, the Model 3900 from Victor Comptometer Corporation

Fig. 10.34. *"To take big gains you take big risks"* J. Philip Ferguson [1967]

"The MOS was in infancy and both system designers and circuit designers were going to school on the calculator learning the technology," Ferguson said: The original technology used the P-MOS process with a typical threshold voltage of −5 V and 6 µm design rules with 1 µm alignment error. It turned out that Kodak Photoresist exhibited variations that exceeded 1 µm tolerances. Since technology was so new and engineers at General Microelectronics had no computer-aided design tool to help lay out masks, all circuits were drawn by hand in a trial-and-error basis. To deal with this reality the original design was loosened to 10 µm design rules.

Another necessary design alteration involved the ratio between load resistor and inverter. A typical switching function was realized by using inverter or gate devices with a load resistor. Since both are MOS transistors, the design of the switching function depends on the transconductance ratio between the two. The resistance of the transistor depends on the width of the

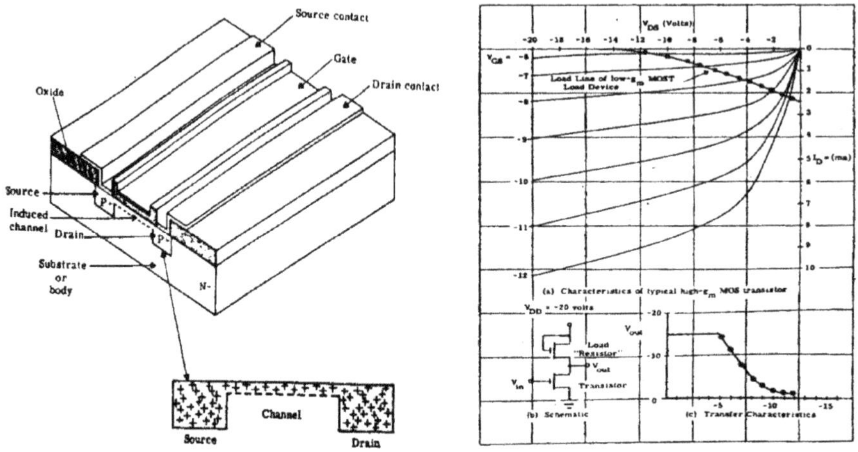

Fig. 10.35. Structure and typical I-V characteristics of GMe P-MOS transistor

P diffused S/D regions. Initially, the designers chose a 7:1 transconductance ratio so that 14 V supply would put 2 V across the output of one transistor. But misalignment may alter the width of each transistor, changing the ratio to 4:1. This will put 3.5 V instead 2 V across the transistor.

The testing of the parts turned out to be another area in which General Micro-electronics engineers grossly underestimated the circuit complexity.

After experiencing great difficulty in making the 29 extremely complex MOS integrated circuits, and after many false starts, the calculator went into production in October 1967. But Victor complained constantly that production problems had kept shipments of parts below its requirements.

In May 1968 both companies negotiated the end of the contract, and Philco had already shut down its MOS production line, which was converted for production of diodes and rectifiers for the automotive industry. At the same time Philco's MOS R&D work was transferred from Santa Clara to Blue Bell, PA.

During GMe – Philco turmoil young UC Berkeley graduate, Boyd G. Watkins, continue in the work of Alfred J. Gale of Ion Physics Corporation, Burlington, MA, who described the application of Ion Physics implanter for fabrication solid state devices by ion-implantation (U.S. patent 3,434894 filed on October 6, 1955). Watkins developed self-aligned MOS transistor. Watkins started his career by designing airborne navigation systems at Kaiser Aircraft Corporation. In 1963 he joined GMe where he designed devices and circuits. His idea was described in an internal memo to Donald Farina on March 30, 1965. The Watkins memo stated: *"The test vehicle can be used for evaluating the polycrystalline silicon-gate process but optional masks have to be designed and cut. Considering the present state of our polycrystalline film development and the over-loaded conditions in our own mask making area, it was not considered feasible to design and cut these optional masks at present."* Farina used "polycrystalline silicon-gate process" in a GMe funding proposal for a Large Scale Integrated Circuit Array project submitted to the Air Force on August 9, 1965. GMe filed a patent application on September 26, 1966 (now U.S. Patent 2,576,478).

The GMe priority of self-align MOS transistor was challenged in court by Hughes Aircraft Company. On February 22, 1974 the District Court in Delaware ruled that the Hans G. Dill patent 3,544,3999, filed on October 26, 1966 and assigned to Hughes Aircraft, was awarded the priority. The court argued that while GMe was first to conceive the idea, they were last to reduce the device to practice.

Hughes Aircraft hired Swiss born Dill in 1966. Dill developed the self-alignment concept independently of GMe when he was working on insulated gate tetrodes at the Swiss Federal Institute of Technology in Zurich. Dill, after joining Hughes, teamed up with Robert W. Bower and they later applied the idea to self-aligned ion-implanted MOS transistor. Hughes never manufactured any device based on the Dill and Bower ideas, because in their approach implantation was performed after the patterning of metal (aluminum) gate and no high temperature process can be effectively used to anneal out the implantation damage.

The other patent interference hearing awarded R. E. Kerwin of Bell Laboratories thc priority on the invention, because Bell Laboratories were able to establish from Kerwin's notebook that idea of self-aligned ion-implanted devices was conceived in February 1966 and the working device was measured at Bell Laboratories in May 1966.

Regardless of the failure of General Micro-electronics, the future of the MOS technology was determined. *"The true potential of MOS technology is in very complex circuits, but be prepared that most large steps forward take longer than you think"* says Philip Ferguson, one of the visionaries who put MOS technology on the roadmap.

In July 1966 Howard Bobb and Warren Wheeler formed American Microsystems, Inc., and started construction of a 25,000-square-feet plant in Santa Clara. Howard Bobb was president of the new company and also head of the marketing operations. Mr. Bobb explained that "*AMI is not going to compete with established companies, but going to fill one of the gaps existing in the microcircuit fields – custom MOS.*"

The posture towards MOS of large companies – Texas Instruments, Fairchild, Motorola and Westinghouse – had ranged from negative to lukewarm, but with no parts offered. General Micro-electronics (Philco), General Instruments and American Microsystems firmly established MOS technology and in 1966 almost completely controlled the MOS market. Motorola, an almost always wrong company, introduced at WESCON (Los Angeles, August 21–25, 1966) a bipolar memory array containing 524 components on 3×3 mm die. The Motorola spokesman said decisively "*you cannot get that kind of den-*

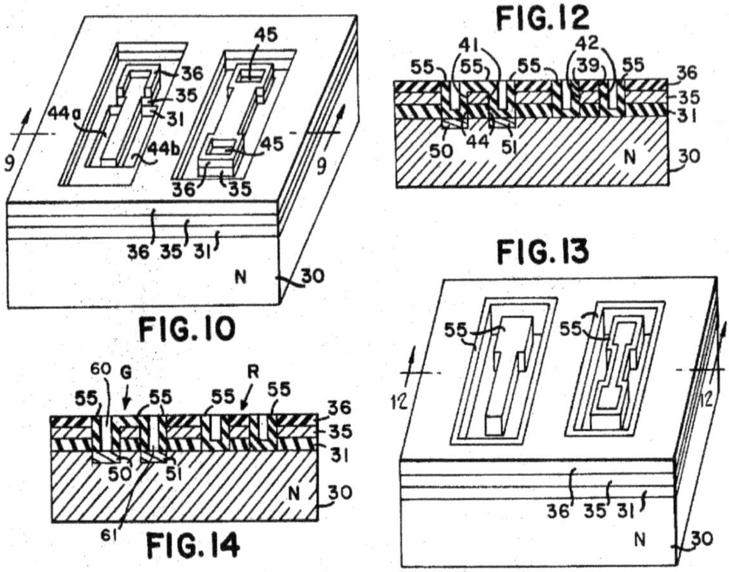

Fig. 10.36. The self-aligned MOS transistor as conceived by B. G. Watkins of GMe (1965)

sity with MOS." Others already knew that such a claim was nonsense: RCA had agreed to purchase $25 million annually in MOS circuits from newly established American Microsystems. This was substantially more than half of the total projected AMI output. In a very emotional atmosphere during WESCON'S MOS session where a Swiss financier concluded his analysis with statement "*the greatest enemy of the business is hope*," Howard Bobb looked upon all the MOS doves benignly and with a smile said *"they don't know, they just don't know!"*

What "they" do not know was the mortal work of group at Sperry Rand Research Corporation, who developed et the end of 1966 the electrically erasable and programmable nonvolatile memory cell (E^2PROM). H. A. Richard Wegener, the Sperry Rand lab's chief device developer, and his team when working on ionic drift in gate dielectric discovered at the end of 1965 that a nitride process does not depend on reaction with the silicon surface, so the interface can be better controlled. They produced very stable transistor with a channel effective length of 10 μm (gate diameter 125 μm) and nitride thickness 1000 Å and typical threshold voltage just under 5 V. (Fig. 10.37.)

During electrical testing of this device they found that threshold voltage can be set to −12 V or −2 V applying approximately 25–30 V between the substrate of the transistor and the gate. The amount of threshold voltage change during programming depends on the duration and the amplitude.

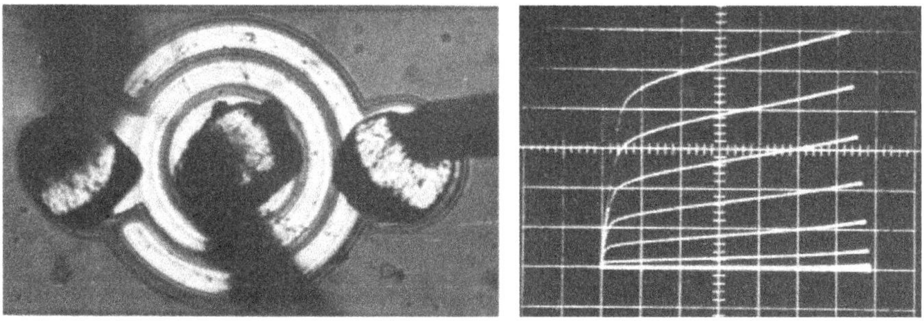

Fig. 10.37. The Sperry Rand Corporation's MOS transistor with nitride gate insulator (1965)

The shift direction of the threshold voltage change depends on the polarity of the applied voltage. Wegener found instantly an application for device with such originally undesirable behavior and invented the non-volatile memory cell. The data retention was specified as *"a few days."* Wegener and his group described the world's first E^2PROM at 1967 Electron Device Meeting (Fig. 10.38.)

11.4 THE VARIABLE THRESHOLD TRANSISTOR, A NEW ELECTRICALLY-ALTERABLE, NON-DESTRUCTIVE READ-ONLY STORAGE DE-VICE, H. A. R. Wegener, A. J. Lincoln, H. C. Pao, M. R. O'Connell, and R. E. Oleksiak, Sperry Rand Research Center. Sudbury, Mass.; H. Lawrence, Sperry Semiconductor, Norwalk, Conn.

A memory element has been developed that has the structure of a typical silicon planar p-channel enhancement insulated-gate field-effect transistor (IGFET). The information is stored by setting the threshold voltage of the 1GFET to a high or low value. Interrogation is accomplished by applying a gate voltage intermediate to the threshold voltage extremes. Current will flow between source and drain if the recorded threshold voltage is less negative than the interrogating gate voltage; none will flow if the recorded threshold voltage is more negative. A high (negative) threshold voltage is written by applying a pulse of -50 v or higher for a duration of 1 msec or less between gate and substrate. A low threshold voltage is similarly obtained with positive 50 v pulse. The persistence of stored information has been demonstrated for periods of at least several months.

The storage mechanism will be discussed and experimental data that describe the write-in, read-out, and storage characteristics of this device will be presented. The results of its application, in various experimental circuits, including a one-by-four electrically-alterable read-only memory, will be given in conclusion.

Fig. 10.38. An abstract of Wegener's presentation describing the Sperry Rand E^2PROM at 1967 IEDM in Washington, D.C.

The second towering work Howard Bobb was referring was switching circuit techniques used in MOS circuits. In mid-1966 several research groups initiated new circuit techniques which enabled the low-power digital circuitry to work from very low frequencies up to a limiting frequency. Frank Wanlass and his co-workers described their approach in [2]. Boyd Watkins of GMe reported a similar work in [1]. This concept, was by far, perfected by Lee L. Boysel, the manager of the MOS circuit design group at Fairchild in Mountain View.

When Boysel developed key components, arithmetic logic unit, ROM, and RAM, and had a proof that his concept would work, he presented Late News at 1968 IEDM and left Fairchild.

The following LATE NEWS papers have been accepted for presentation at the 1967 International Electron Devices Meeting:

2.7 MOS/LSI 8-Bit Parallel Accumulator. L. Boysel, J. Kelley and R. Cole, Fairchild Semiconductor, Mountain View, Calif.

The impact of Large Scale Integration on future computing systems will have great economic significance only if some degree of standardization at the subsystem or I.E.C. level can be achieved. The arithmetic unit, when considered bit by bit is, like the memory, repetitive in nature and becomes suit-able for standard circuit arrays.

This paper describes a section of such an arithmetic unit fabricated monolithically using P-channel enhancement mode MOS transistors.

The accumulator is an eight-bit, fully parallel slice with series connections for varying word bit lengths in eight-bit increments. It consists of approximately 600 MOS transistors representing a functional complexity of 200 gates.

The circuit may be divided into four sections:

1) a loading or buffer register for temporary storage of operands.
2) the parallel adder/subtracter performs addition and subtraction of negative numbers directly.
3) the accumulator register shifts right or left and/or accumulates, and
4) the push-pull output buffers.

The circuit was fabricated using standard single metal layer MOS processing techniques with the addition of a low temperature dielectric deposition to prevent field inversion and improve operating characteristics. The chip size is 116 by 86 mils. Devices were assembled by conventional techniques in a 36 pin dual-in-line package.

Evaluation of circuit is currently being performed. Initial results indicate an eight-bit carry thru time of the order of 1p sec, at a power level of between 150 and 200 mw.

Fig. 10.39. An abstract of Boysel's presentation at 1968 IEDM in Washingthon, D.C.

In February 1969, Boysel and Jack Faith incorporated Four Phase Systems. Corning Glass Works provided besides funding of $2 million, also, safety net for Four Phase employees with an agreement that if the start-up failed, all engineers will be transferred to Signetics, which was also owned by Corning Glass. In a matter of months, the key persons of Fairchild's MOS design group joined Four Phase. The business plan of Four Phase called for the design of MOS based computers using a reduced number of parts in order to avoid the problems experienced by GMe. Four Phase Systems was a design house only. Robert Cole, Boysel's friend from Fairchild, also left and started the foundry company Cartesian, Inc., "using" the MOS process that had been used at a Mountain View Fairchild manufacturing facility.

The Four Phase circuit concept could be explained, with for example, a 2-input NAND gate and an inverter shown in Fig. 10.40. The ϕ_1, ϕ_2, ϕ_3, and ϕ_4 clocks need to be non-overlapping. Considering the gate during the ϕ_1 clock high time (precharge time) the output C precharges up to $V(\phi_1)–V_{th}$. During the next quarter clock cycle (the sample time), when ϕ_1 is low and ϕ_2 is high, C either stays high (if A or B are low) or C gets discharged low (if A and B are high). The A and B inputs must be stable throughout this sample time and the output C becomes valid during this time – and therefore a 1 gate output can't drive another 1 gate's inputs. A 2 gate precharges on ϕ_2 and samples on ϕ_4. Therefore, 1 gates can drive 2 gates and/or 3 gates; 2 gates can drive only 3 gates, 3 gates can drive 4 gates and/or 1 gates, 4 gates can drive only 1 gates. Of course there are some difficulties, the main one being that the gate output is dynamic. This means that its state is held on the capacitance at the gate output. But the output track can cross clock lines and other gate outputs, all of which can change the charge on the capacitor. In order that the gate output voltage remains at some safe 0 or 1 level during the cycle the amount of change has to be calculated and, if necessary, additional (diffusion) capacitance has to be added to the output node.

Four Phase Systems introduced their 8-bit computer in September, 1970 as AL1 Computer System. Although the system became very successful the Four Phase Systems marketing avoided the term microprocessor. The term *microprocessor* was first introduced in 1968 by Viatron Computer System, Inc. Viatron was established as a spin-off from MITRE in 1967 and, with out-of-proportion humbug, introduced "distributed data processing" based on their Viatron' System 21. System 21 was a set of electronic data processors, displays, keyboards, and tape sub-assemblies that could be assembled as a "desktop microprocessor" connected to the network or as a terminal for large computers. Viatron's president and board chairman, Edward M. Bennett, was promising by mid-1970 monthly production of 5000 to 6000 machines monthly! Such numbers were based on the assumption that System 21 may be leased for a low monthly fee as a home terminal to access mainframe computers. The Computer Research Bureau Newsletter from October 21, 1968 promised to investors *"Viatron Blitzkrieg the Industry."* Viatron had all money they wanted, but they were not able to deliver. In March 1971 Viatron filed for Chapter 11 bankruptcy. Boysel and the Four Phase System marketing department knew that marketing a new computer system with the term microprocessor after Viatron's "blitzkrieg" was suicide.

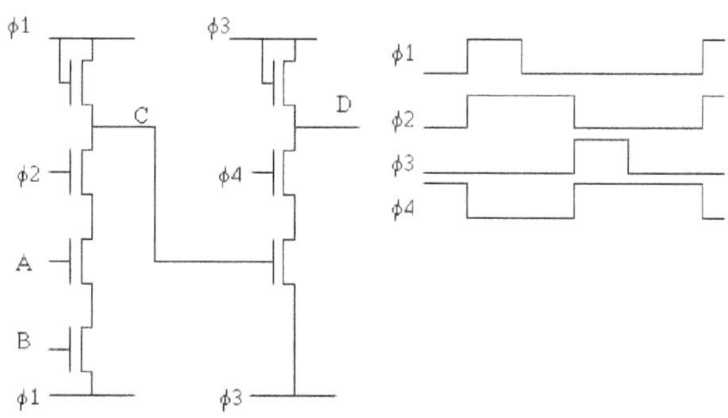

Fig. 10.40. Fairchild (and Four Phase Systems) concept of multiphase circuit technique

Boysel's well designed AL1 system was a major technical breakthrough and it is considered by many as the first microprocessor design. However, because the company was more of a computer than a semiconductor business, the growth of Four Phase Systems was relatively slow and required constant rising of capital to keep the company going. The company grew to almost \$180 million by the end of 1979 and was sold to Motorola in 1981.

Calculator chips and computer components were a major driver of MOS technology at the end of the nineteen sixties. In July 1970, AMI developed the 6-chip set that contained all the MOS circuitry for an arithmetic calculator. AMI was quickly followed by Texas Instruments and North American Rockwell. With average retail price \$100.00 per unit, John McNichol of The Electronic Engineer magazine estimated the calculator market to be by the end of 1972, more than \$50 million (Fig. 10.41.)

John McNichol, New Directions Editor

"Talk about an electronic calculator to the average layman, and he'll crap in his pants."

Indelicate? Perhaps, but this sentiment expressed by a very canny electronics executive, agrees with many other knowledgeable electronic industry executives.

Nonetheless, a new market of imposing size and seemingly easy reach for entrepreneurial electronic engineers—a million and one-half units within the next 18 months is the best estimate—has appeared within the last year—the personal electronic calculator.

Fig. 10.41. Calculator market prediction by J. McNichol of The Electronic Engineer Magazine

Up to 1970 the integrated circuit's innovation and business was almost completely a U.S. phenomenon. One of the decisions which changed this situation and gave rise to Japanese competition and later complete dominance was the decision of Autonetics, a division of the North American Rockwell Corporation, to develop a MOS array for a 3-pound calculator produced by Japan's Hayakawa Electric Co. in January 1969.

Autonetics, which profited highly from programs financed by the government, showed at the Fall Joint Computer Conference in 1968 an avionics computer with only MOS large-scale integrated arrays in central processor, control unit and memory. The computer consumed 10 W; the processor had 24 arrays ranging in complexity from 142 MOS transistors to 1053. The multiplication operation time was 108 μsec, divide time 112 μsec and 4 msec memory cycle. Other instructions took 8 μsec. This work was now transferred to Japan. When commenting on the deal with Hayakawa (later re-named Sharp), Autonetics president S. Fred Eyestone said *"they're fine people to deal with."* A two year contract to deliver 2 million MOS LSI arrays for Hayakawa's QT-8D calculator was worth \$30 million. The fine print of the contract said that after 1971 Hayakawa would produce the circuits itself in Japan under the terms of a technical assistance agreement.

NAR Forms Separate MOS Unit

By NEIL KELLEY

LOS ANGELES — North American Rockwell last week spun out its MOS operation from the Autonetics division in the wake of a new $30 million add-on contract from Sharp Corp. of Japan.

The new profit center will be known as North American Rockwell Microelectronics, Inc., and will be headed by Sam Carlson, who directed Autonetics' Products division.

The new company is expected to play a major role in helping Sharp (formerly Hayakawa Electric) become a self-sufficient producer of MOS circuits which it uses extensively in desk calculators and other products.

The contract, which had been anticipated in the industry is a follow-on from one awarded last March for a similar amount and which was described as a "technology transfer." Sharp presently assembles and tests MOS circuits shipped from Autonetics' Anaheim, Calif. facility.

Autonetics spokesmen insisted that wafer fabrication had not yet been negotiated for transfer to Sharp.

Right Move

Conceding that Sharp is going to start MOS manufacture sometime in the future, officers of the new Autonetics MOS company said that the Japanese firm is making the move in the right way by entering the field in steps, starting with assembly and test.

(Sharp sales for fiscal 1970 were $420 million. The firm is heavy in electronics and consumer products and claims 35 per cent of the world calculator market. Sharp employs 25,000.)

Harold Edge, NAR Microelectronics Co. vice-president of operations who did much of the negotiation in Japan, claimed Sharp would be making MOS mostly for in-house consumption.

Industry sources outside Autonetics have largely discounted the possibility of any Japanese company becoming highly competitive with U.S. makers except within Japan.

Sharp originally came to Autonetics for the MOS/LSI circuits after Japanese makers could not fulfill Sharp's needs for the desk top calculator. Mr. Edge said no Japanese firm is now producing MOS in quantity.

Autonetics spokesmen preferred not to discuss technology transfer, stating the topic was a "very sensitive" area with the Japanese Trade Ministry. It was explained the country is trying to establish an image of self sufficiency.

Autonetics last week was likewise trying to establish the image of a semiconductor manufacturer with its semi-autonomous Microelectronics Co.

Sharp's competitors would benefit from this information.

Three other companies, Caxom and Sanicom from Japan and Victor from the U.S. have recently announced plans to build calculators with MOS from Texas Instruments, General Instrument and Fairchild Semiconductor respectively.

Autonetics said production rates for the five devices had resulted in total output of 160,000 units in June of this year. The firm expects to have production up to 200,000 units next month.

At its press conference last week, Autonetics named officers of the newly formed Microelectronics company.

Mr. Carlson, who was vice-president and general manager of the Autonetic products division, becomes president of the Microelectronics Company. Charles Kovac, former marketing manager for MOS, becomes vice-president for marketing in the new unit. John Milberg, factory manager, becomes vice-president of engineering; Mr. Edge, manager of business operations becomes vice-president of operations.

Nova Market Manager

CAMBRIDGE, Mass. — Stan Harris has been named marketing manager, Nova Devices, Inc., subsidiary of Analog Devices, Inc., here. Mr. Harris was eastern area marketing manager for Computer Microtechnology, Inc.

Fig. 10.42. Electronic News July 6,1970. Note the paragraph in the middle of the article. In modern society, intelligent people may come up with plausible reasons to justify anything, to do nothing, or to do all they want to do

By January 1973 the U.S. market was dominated by 20 Japanese calculator makers who took over from U.S. manufacturers. By 1999 Rockwell was out of semiconductor manufacturing. "*Fine people*" demonstrated that we have a problem when it comes to competing.

The calculator chips also resolved the long-standing hassle over integrated circuits between Texas Instruments and Japanese trade officials. Ministry of International Trade and Industry (MITI) warned Japanese producers not to export IC hardware until licenses had been arrange with Fairchild Camera & Instruments, which held the basic U.S. patents on the planar process. With the exception of Matsushita Electronics and Sony, Japan's major IC producers had acquired rights to Fairchild's patents. But TI, which also owns basic IC patents, has been holding off on licenses because its bid to set up a wholly owned subsidiary had been turned down by MITI. Hayakawa executives insisted they were looking for a desk calculator market in the U.S. rather than a run-in with TI. The company was ready for mass production of its calculator and needed the large American market as an outlet. This dispute was resolved when TI was allow to set up a joint venture in Japan with Sony to manufacture integrated circuits and allowed the granting of licenses to allow integrated circuits to be used in calculators.

The failure of General Instruments and General Micro-electronics in the mid nineteen sixties should be seen also from a different point of view. At the time when they initiated MOS technology, thermal oxidation and thermal oxidation equipment were rather primitive. The unpredictable charge in oxide and charge at the interface severely limited the production yield. The quality of the wafers did not meet the stringent requirements for MOS processing. Table 10.1. shows quality evaluation of silicon wafers. The quality of resist and cleaning processes were still in their infancy.

Table 10.1. Quality evaluation of silicon wafers

	1960	**1965**	**1970**
Wafer Diameter	0.5–1″	1–2″	2–3″
Dislocation density	50,000	10,000	1,000
Surface Finish	0.5 μm	0.25 μm	Damage Free

With many MOS transistors on a chip, a parasitic transistors formed over field region may limit the functionality of devices. Such a parasitic transistor is formed between adjacent P+ regions when a high voltage metal line crosses them. Unless the field oxide under this line is thick enough, the high voltage inverts the surface of the N-substrate and turns on the parasitic transistor. This parasitic transistor operates like a conventional device, except it has a thicker gate oxide. Since voltages were often above 30 V, the field oxide must be at least 1.5 μm thick. To grow this thickness of oxide thermally required a significant thermal budget. No other robust technology was available at that time.

Despite these difficulties there were engineers who recognized the advantage of the MOS technology. By 1969, there were over twenty companies developing MOS technology. The pocket calculator, envisioned electronic memories, and electronic wristwatches using quartz crystal frequency control, were fueling the MOS market which rose to $80 million in 1970. Table 10.2 summarizes the status of MOS semiconductor manufacturers during period 1968 to 1970.

In 1965 the Bell Laboratories group of J. C. Sarace, R. E. Kerwin, D. L. Klein and R. Edwards began work on a self-aligned silicon-gate MOS process. J. C. Sarace and his team correctly recognized that if they introduce silicon nitride between the thermally grown and the deposited oxide, the thermal oxide-silicon nitride combination would provide a barrier against sodium and others contaminants.

The processing flow of Bell Labs MOS process consisted of the following steps:

1) Initial oxidation of silicon substrates (steam, 1050°C/180 sec, oxide thickness ~ 600 Å)
2) Nitride deposition (silane and ammonia, 1000°C, nitride thickness ~ 400 Å)
3) Thick oxide deposition (tetraethoxy silane, 550°C)
4) Mask – Gate Area Definition
5) Wet etch silicon oxide
6) Strip resist
7) Blanket Silicon evaporation
8) Mask – Source/Drain Definition
9) Silicon Gate etch (Source/Drain Area, HF-HNO_3-HAc-I_2 mixture))
10) Silicon oxide etch (Source/Drain Area, ammonium bifluoride, silicon films serve as mask)
11) Strip resist
12) Mask – Contact Pads Definition
13) Remaining poly-silicon etch
14) Nitride and oxide etch (Source/Drain area, hot phosphoric acid and ammonium bifluoride)
15) Source/Drain Diffusion (deposition of 6000 Å ~ 10 Ω/sq film and Drive-In)
16) Strip diffusion glass
17) Metallization of entire surface
18) Mask – Interconnect Definition
19) Etch Aluminum

The key features of the process were:

1) The Si-SiO2 interface was established at the beginning of the processing flow with the best pre-cleaning techniques known at that time.
2) Silicon nitride layer served as a barrier against mobile ions and as a hard mask
3) Silicon gate electrode defined the Source/Drain channel edges

Both P-MOS and N-MOS devices were fabricated. Threshold voltages for an equivalent oxide thickness 1000 Å and channel concentration of ~ 1E16 cm^{-3} were −2.6 V for P-MOS and 1.35 V for N-MOS (V_D=10 V.)

Table 10.2. Market share of U.S. MOS and CMOS manufacturers 1968–1970

Company	**1968**	**1969**	**1970**	**Technology**	**Note**
American Micro-Systems, Inc.	37.6 [%]	27.5 [%]	**21.9 [%]**	P-MOS <111>,<100>	Leader in Custom Circuit
Autonetics (North American Rockwell Corpotration)		1.7	**21.9**	P-MOS <111>	
National Semicondcutor	2.1	10.7	**13.1**	P-MOS <111>, <100> Poly Gate	Best Low Cost production
General Instrument Corporation	26.9	17.7	**10.9**	P-MOS <111> / Nitride / CMOS	
Texas Instruments	10.7	10.7	**8.7**	P-MOS <111>,<100> / CMOS / Implant	
Hughes Aircraft Company	2.1	1.4	**3.3**	P-MOS <111> / Nitride / Poly Gate / CMOS / Implant	
Intel		0.4	**3.3**	P-MOS <111>, <100> Poly Gate	Second Best Yield (~10 %)
Fairchild	11.8	5	**3.3**	P-MOS <111> / Poly Gate / CMOS	
Advanced Memory System, Inc.			**2.2**	P-MOS <111>	
Electronics Arrays, Inc.			**2.2**	P-MOS <111>	
Motorola			**2.2**	P-MOS <100> / Poly Gate / CMOS / Implant	
Philco-Ford (General Micro-Electronics)			**1.6**	P-MOS <111>,<100>	
Signetics			**1.6**	P-MOS <111> / Nitride / Poly Gate / Implant	
Mostek			**1.3**	P-MOS <111> / Implant	The Most Significant Innovation (Design)
RCA			**1.1**	P-MOS <111> / CMOS	
Ragen Semiconductor			**0.8**	CMOS	The First MOS A/D Converter
Solid State Scientific Corporation			**0.2**	CMOS	
Cogar Corporation				N-MOS	Samples - 1970
Computer Microtechnology Inc.				P-MOS <111>	Samples - 1970
MOS Technology, Inc.				P-MOS <111>	Samples - 1970
Semiconductor Electronic Memories, Inc.				CMOS	Samples - 1970
Total Industry Revenues ($M)	**9**	**28**	**80**		

J. C. Sarace, R. E. Kerwin, D. L. Klein and R. Edwards presented results of their work on a self-aligned silicon-gate process at the Metallurgical Society (AIME) Meeting in New York in August 1967. The AIEM meeting was attended by two Fairchild engineers. Contrary to Bell Laboratories, who apparently failed to recognize some of the most significant advantages of Kerwin's and Serace's structure, Fairchild MOS Device development group headed by Leslie Vadasz abandoned the Fairchild MOS process originally developed by Wanlass, (later tuned by Bruce Deal,) and launched a new silicon-gate research-and-development effort. The motivation of Vadasz' group stems

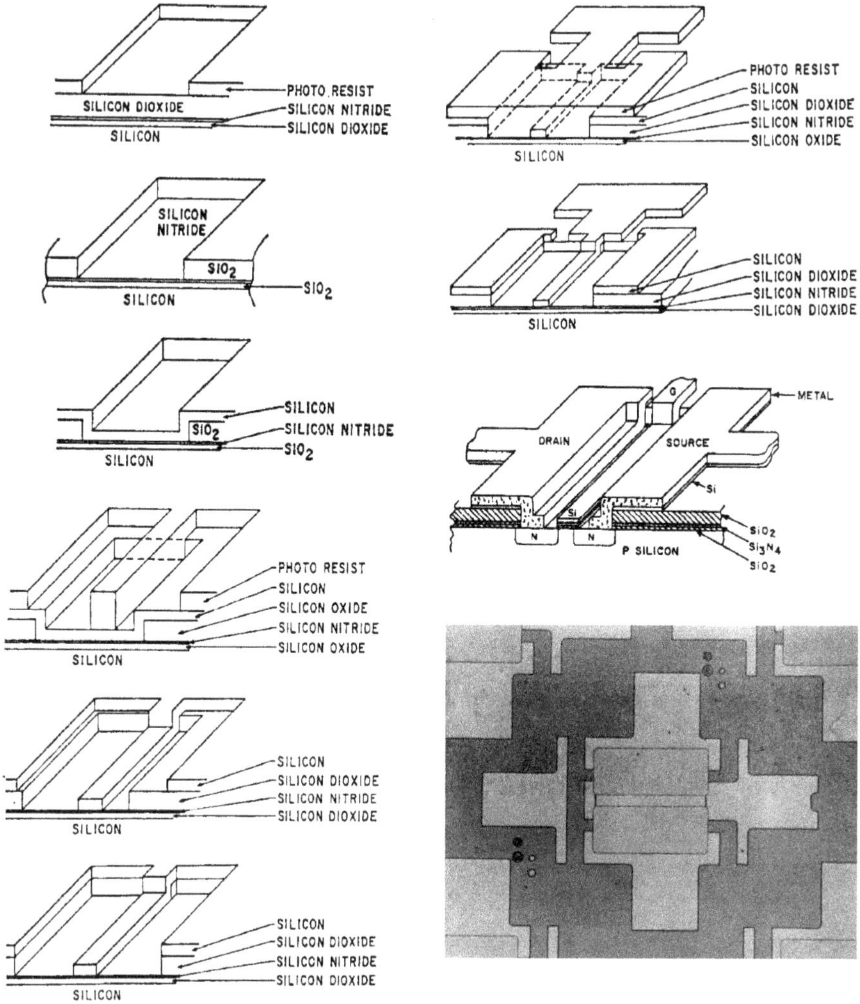

Fig. 10.43. Self-aligned MOS process with silicon oxide/silicon nitride gate developed by Bell Laboratories (1965–1967)

from several factors. The technique reduced threshold voltage by almost 500 mV, it tripled speed, and it had density higher by 50 %. The new technique was equal in terms of complexity – the number of masking steps was the same. The reason why silicon gate technique had not been thought of earlier can be attributed largely to the fact that aluminum has been the least troublesome part of MOS structure. Advances in the technology originated in large part from work on the gate dielectric and Si-SiO_2 interface. Moreover, polysilicon cannot be wire bonded, and this contributed to the delay in considering the poly gate.

Hungarian-born Leslie L. Vadasz received his B.S.E.E. degree from McGill University, Montreal, Canada, in 1961. The same year he joined Transitron where his work dealt with silicon process and device development. He joined Fairchild in 1964 and since 1965 he headed the MOS Device development group. The second native Hungarian, Thomas Klein graduated in physics from Aberdeen University, Scotland in 1960. He started work on integrated circuits at the Mullard Research Laboratories in Redhill, U.K. He joined Fairchild in 1966. When still with Mullard Research, Klein developed tetrachlosilane epi process (British Patent Application No. 19744/64) which was later use at Fairchild for silicon gate process. Federico Faggin, native of Italy, graduated from the University of Padua, Italy in 1965 and after a brief stint with Olivetti, he joined Fairchild in February 1968.

Fig. 10.44. Federico Faggin (left) and Thomas Klein at Fairchild (June 1968)

Vadasz' group used the Fairchild 3705 P-MOS aluminum-gate 8-bit multiplexer subsystem (which contained a 8-bit decoder, and input multiplexer switches as a test vehicle and developed the Fairchild version of a silicon-gate self-aligned MOS process (Fig. 10.45.) The electrical performance exceeded expectation.

Fairchild's Klein, Faggin, and Vadasz presented results of their implementation of Sarace's process at the Electron Device Meeting, Washington, October 1968 [3]. At that time, Leslie Vadasz was already with Intel and headed Intel's MOS design group.

The Intel intention was to be the manufacturer of proprietary memory products for mainframe computers using both bipolar and MOS technologies. The initial effort focused on bipolar parts 3101 and 3301 introduced to the market at the end of 1969 (Fig. 10.46.)

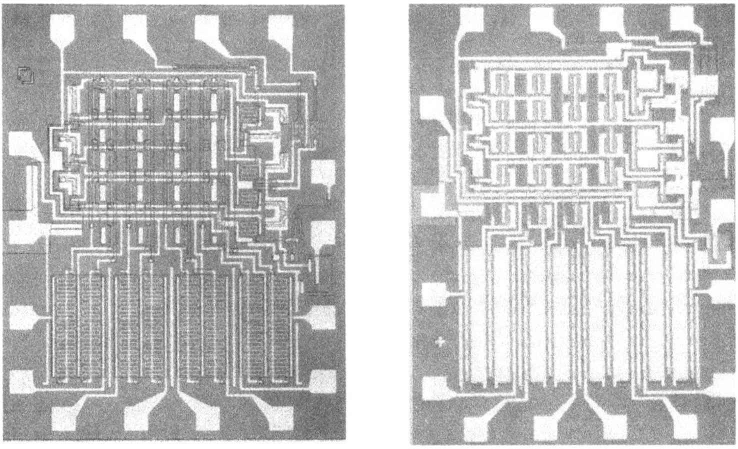

Fig. 10.45. Conventional aluminum gate version of Fairchild's 3705 (left) and silicon gate version 3708. (October 1968)

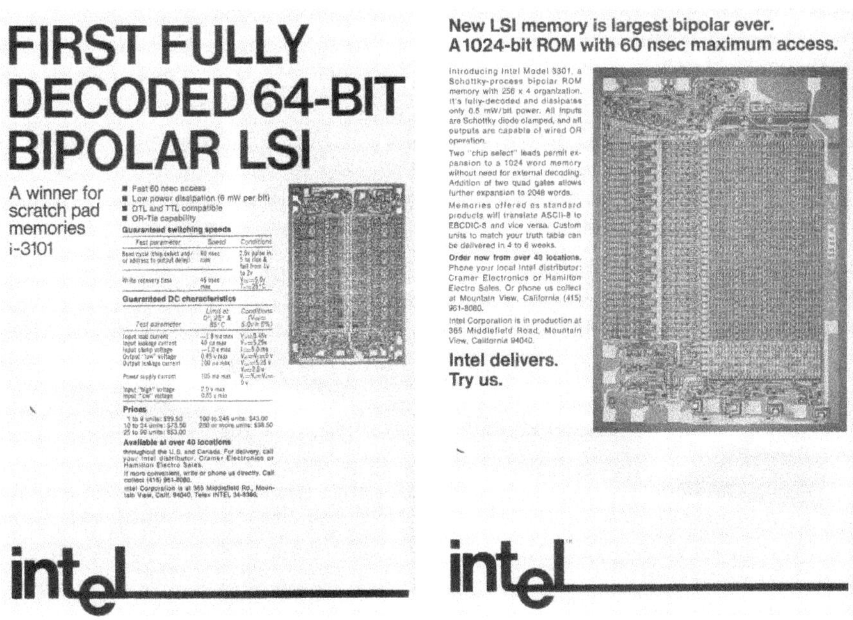

Fig. 10.46. Intel's first bipolar memories (1969)

In search of new products, three months after Intel's incorporation, in November 1968, Bob Noyce visited Tadashi Sasaki at the Hayakawa Electric Industry Co. Ltd.[4] in Nara and offered Intel's manufacturing facilities to Hayakawa.

[4] Hayakawa Electric Industry Co. Ltd. re-named to Sharp Corporation in 1970

Sasaki did not want to turn down Noyce's request and asked Rockwell if Intel, a struggling start-up, could produce a small volume of Haykawa's parts. Rockwell's president Fred Eyestone in one heart beat responded *". . . . of course not. We have an exclusive agreement, and we are not going to change it because some start-up's go under.*" Sasaki directed Noyce to Yoshio Kojima. Kojima who was in charge of the joint project of Electro-Technical Industries and Nippon Calculating Machines Company to develop the chip set for a low-cost printing calculator with the brand name Busicom. In April 1969, Intel signed an agreement to develop a set of calculator chips for Kojima's project. This agreement did not specify the number of chips in the chip set but made an arrangement for three of Kojima's engineers (Shima, Masuda, Takayama) to work at Intel. Kojima would also pay the development cost $100k keeping the rights to manufacturing.

In July Marcian E. "Ted" Hoff, Jr., a young Stanford graduate, was managing Intel's Application Department and was assigned to review the Nippon Calculating Machines Company's design. The design consisted of a programmable decimal computer with ROM-based macro-instructions and shift register data memory, partitioned into seven different chips. Because the Japanese design would require all of Intel resources that they had at that time, Hoff started considering alternative solutions. During July and August 1969, Hoff proposed a general-purpose computer that could be programmed to perform calculator functions. Hoff's basic idea was good but a detail system concept was missing. Hoff did not at that time solve problems with keyboard, display, and printer control. The other major problem was binary to decimal and decimal to binary conversions. In 1969, the maximum size of ROM memory was limited to 2k bits per chip. This was not sufficient for a 4-bit binary processor. Intel's management, however, pulled Hoff from the calculator project to another contract. Les Vadasz, head of Intel's MOS design team took over the Hayakawa contract. Vadasz had, also, responsibility for the main Intel principal business – memory development. Memory i1101 and i1103, pre-occupied Vadasz's group and almost nothing was done on the calculator project.

When the situation with the calculator project become critical and threatened to embarrass Noyce and Intel, Les Vadasz reluctantly hired the most qualified persons available. Stan Mazor joined Intel in September 1969 and Federico Faggin in April 1970. Faggin was disappointed with the slow acceptance of Thomas Klein and his work at Fairchild and called Vadasz and asked him if he had a job opening. Vadasz and Faggin had several conflicts with each other while Vadasz was still at Fairchild, so the decision did not come easily.

Faggin had hands on Fairchild P-MOS self-aligned technology using amorphous silicon gates prepared in an epi system by decomposition of tetrachlosilane in a hydrogen atmosphere at high temperature ($\sim 1200°C$). The process was a subject of a Fairchild patent application filed by Klein and Faggin on

December 16, 1968. The amorphous silicon deposited by atmospheric chemical vapor deposition was quite a novelty in 1968. The original work, started by Thomas Klein, was motivated by a major problem of Bell Laboratories' MOS devices. Everybody who designed a MOS transistor in the way as described by Sarace and Kerwin's group learned very quickly, that sputtered silicon would crack at the edges of thick oxide. For this reason Bell Laboratories could build only discrete transistors (see photomicrograph of device in Fig. 10.43), not an integrated circuit.

By hiring Faggin, Intel became "co-owner" of Fairchild MOS technology, the best MOS technology available in 1969. In a 1994 interview [5] Shima pointed out "*Originally Intel hired many, many people from Fairchild. They brought in lots and lots of Fairchild internal confidential documents. I had many such documents in my cabinet when I developed the 4004.*" Not surprisingly, none of Andy Grove's columns complaining about stealing somebody else's confidential properties have appeared in newspapers.

Faggin re-vitalized Intel's calculator project and after eleven months of working heroism, Faggin, Shima and Mazor completed the Busicom calculator. Although Intel eventually had gained the right to sell the 4004 microprocessor, the management of the company and the marketing department did not recognize the potential of microprocessors.

A major change in the position of Intel's Marketing Department occurred when Intel hired a new marketing director, Ed Gelbach. Gelbach was previously with Texas Instruments and was familiar with the work of Gary Boone and others on a microprocessor under way at Texas Instruments. The reason why Intel was pushed into the microprocessor business was Gelbach's vision. Because of these changes, Intel's advertising agency Bonfield Associates placed an advertisement in Electronics News, November 15, 1971 announcing "A *new era of integrated electronics.*"[5]

At the end of 1969, William Regitz of Honeywell was looking for a semiconductor company to share in the development of a novel DRAM cell developed by himself and one of his coworkers. Intel was very interested in the technology and started a development program that initially produced the i1102. Although working parts were produced, there were problems with the 1102 and the 1102 never made it to the market. Bill Regitz, who worked for Honywell at that time, joined Intel in 1971, and with then hippie, Joel Karp, they look at Hoffs' 3-transistor DRAM cell. Leslie Vadasz and Joel Karp de-

---

[5] TI has stated in company literature that it invented the single-chip microprocessor in 1970. The company applied for a patent in 1971 with the invention being credited to Gary W. Boone and Michael Cochran (the patent was issued in February 1978). The company described their invention as a "microcomputer – the first integrated circuit with all the elements of a complete computer on a single chip of silicon." TI advertised the integrated circuit developed for the Datapoint terminal in the *Electronics* magazine with the caption "*CPU on a chip*" in June 1971. This was the first announcement of a single-chip microprocessor. TI had functional problems with the chip and never marketed it.

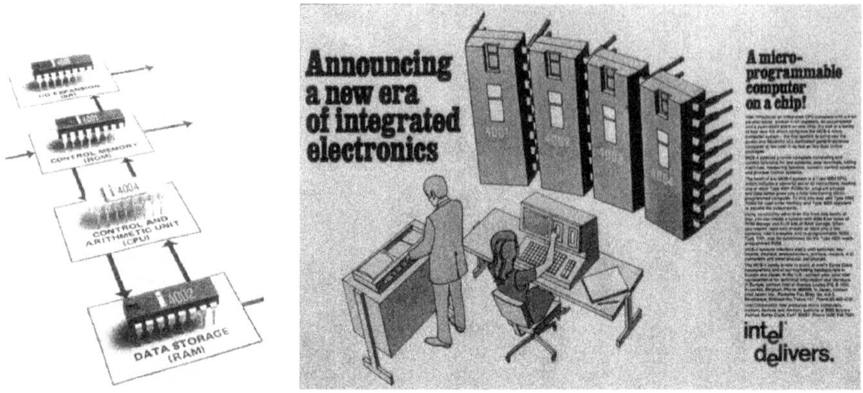

Fig. 10.47. "A new era of integrated electronics" announced by Intel in Electronics News, November 15, 1971

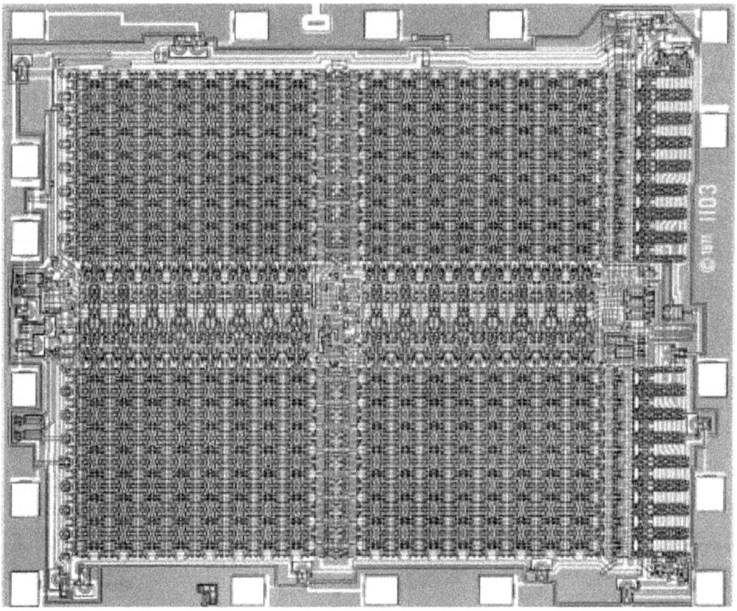

Fig. 10.48. Intel's 1,024-bit dynamic RAM i1103 using silicon-gate P-MOS technology

veloped the preliminary circuit design and the final chip design was assign to Bob Abbott. The resulting product was the i1103 and was introduced to the market in October 1970. The part originally had yield issues (typically one or two working dies on a 2″ wafer) and John Reed, the product engineer had to make several revisions to the part before "good" dice and performance were achieved. The i1103 was manufactured on a 6-mask silicon-gate P-MOS process with 8 μm minimum features. The resulting product had a 2,400 μm^2 memory cell size, a die size just under 10 mm^2 and sold for around $21.

Intel's 256-bit 1101, a six-transistor P-MOS SRAM, could have been the first semiconductor memory to achieve a degree of success in the marketplace. But, it was not. It was expensive – 3 cents per bit. Now, the Intel 1103 became a desperate hope to prove that they could deliver semiconductor memory at the price of ferrite cores. The 1103, introduced in late 1970, was a fully decoded, 1,024-bit dynamic RAM using silicon-gate P-MOS technology. Packaged in 18-pin ceramic or plastic DIP, the 1103 sold at first for about a penny a bit, matching core's price. But, then core prices cratered to 0.3 cent a bit. Intel's prices declined further. Intel boasted that chip, whose core equivalent occupied about a square foot and weighed almost a pound, was the core killer, and Intel was right.

The 1103 knelled the death of core; however, core did not suffer an instant demise. Core memory had proven success, while semiconductor memories could not quite be trusted. Core was nonvolatile while 1103 need to be refreshed every 2 ms. Further, core was far down the learning curve so semiconductor memories would never approach the cost of core. Perhaps most important, the 1103 used MOS technology, a very shaky technology in 1970. However, the 1103 had an advantage in speed. Access time was down to a brisk 300 ns and full cycle time was 580 ns while fast core was in the millisecond range.

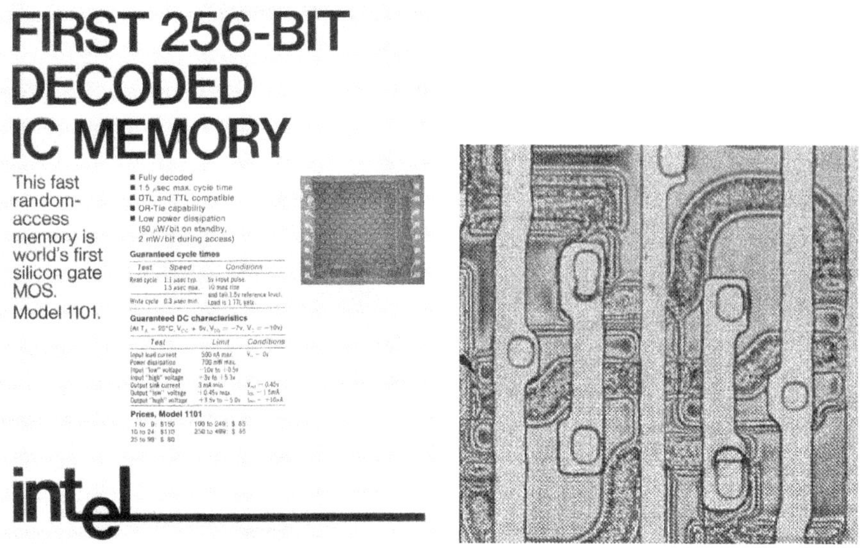

Fig. 10.49. Intel 1101 Ad (October 1969) and detail of memory cell. The technology was developed at Intel by Leslie Vadasz and Tom A. Rove

Fortunately, there were companies like Honeywell, who had been looking into solid-state memories. Engineers there saw little likelihood that a six-transistor cell like that in the 256-bit 1101 would ever make economic sense, but the 1103, with its three-transistors cell, looked promising.

Intel was not the only company competing for Hayakawa business. The other company in the race was Mostek. In 1968, at Texas Instruments, L. J. Sevin created a 256-bit experimental MOS RAM, but he had to fight a management that felt that the way to beat cores was with thin-film memory. Texas Instruments' management had a strong bias toward bipolar and "*MOS couldn't be trusted.*" The attitude of Texas Instruments could result in only one outcome: in mid 1969 L. J. Sevin, R. L. Petritz and L. E. Sharif separated from Texas Instruments. New Business Resources, a Dallas venture capital partnership formed by Richard Petritz and Richard Hanschen, financed the start up, MOSTEK Corporation, based in Dallas. New Business Resources also negotiated an agreement between Mostek and Sprague Electric, which included financial backing by Sprague and a provision for sharing research and development efforts to advance the MOS technology and manufacturing in Sprague's Worcester facility.

In a few weeks, other TI engineers followed Mostek founders: Vincent Prothro, Robert Palmer, Robert Proebsting, Berry Cash, and Vern McKenny.

Kenneth G. Aubuchon of Hughes Research Laboratories described in June 1969 [4] the use of ion implantation to set the threshold voltage of MOS transistors. It happened that Sprague Electric was trying to use ion-implantation to make precision resistors and hybrid circuits on ceramic substrates. They had done excellent research work there. In addition, they built an implanter from scratch. The machine that they had, was completely fabricated at Sprague or at other machine shops. What happened next, was that Bob Palmer of Mostek who was familiar with Aubuchon's work was actually on a trip to Sprague Electric's research facilities in North Adams, Massachusetts. He wanted to learn something about tantalum oxide and he run into Sprague's hand-built ion implantation system. He had never seen an ion implantation system and was asking them what they were doing. Dr. Ken Manchester and Dr. John Macdougall were very open and eager to share what they were doing with Palmer. It turned out that the way they measured the dose was by measuring the threshold voltage shift in a crude MOS device. Palmer got very excited and telephoned L. J. Sevin, "*L. J. you won't believe what these guys are doing up here.*" Palmer, with help of Sprague's engineers, implemented ion-implantation into production technology and commercial integrated circuits.

The Sprague implanter had no automation and all wafer transport was done by hand. The wafers needed to be moved into position by hand. By the spring of 1970, Palmer, Manchester, and Macdougall were doing the first tests to see if in fact they could get a beam, and if they could separate the boron from the other ions, and if they could actually implant wafers. At the end, implantation worked! Mostek needed more implanters, and they asked

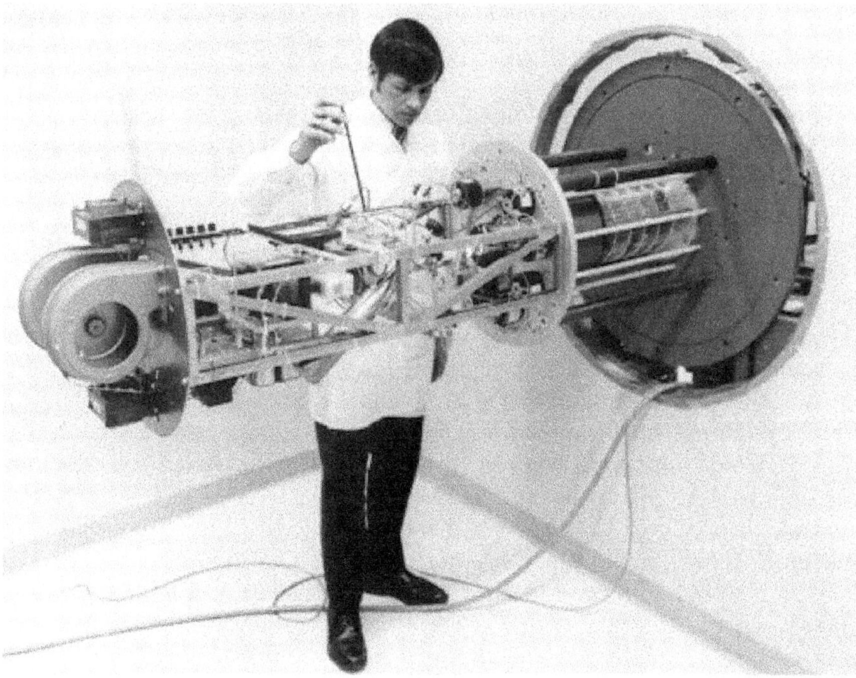

Fig. 10.50. Bob Palmer with Sprague's ion implanter (1970)

Fig. 10.51. Extrion Corporation ion-implanter co-developed with Mostek and Sprague

Peter Rose who formed Extrion Corporation for help. It was only a question of time when the semiconductor industry could "*handle this baby ... with one hand*" (Fig. 10.51).

Mostek's development of ion implantation processing led to several product introductions. Ion implantation allowed, for the first time, the use of depletion load elements in circuits. Depletion load elements were introduced in the MK4006 1 K DRAM and the MK4007 256 bit static RAM simultaneously. In 1971, Mostek had in mass production a single chip calculator MK6010.

L. J. Sevin had the uncanny ability to attract and keep good people. He was very critical and at the same time he encouraged his engineers. Sevin not only created the most technical semiconductor company, but with his colorful personality and sense of humor he had a major impact on Mostek culture.

In 1978, L. J. Sevin and one of his engineers, Steve Goings, in 1978, paid Prof. Per Birch Hansen a visit at USC. L. J. introduced Steve around and said, "*Steve is a pioneer. You can always tell who the pioneers are. They are the ones with all the arrows in their backs.*"

Mostek predicted that VLSI technology soon would make it possible to put an IBM/360 mainframe computer on a single chip! So, L. J. wanted to know if it would be a good idea for Mostek to develop such a powerful chip. Hansen urged that they abandon the IBM emulation and create a microprocessor that could work effectively in a multiprocessor architecture and provide for more direct support of high-level programming languages. While Hansen was explaining all of that, L. J. Sevin looked immensely bored. When Hansen finished, L. J. turned to Goings and said: "*I think we ought to build his machine!*" At the beginning of the project, L. J. Sevin told Hansen that Mostek was funding a dozen research projects that gambled on future computer technology. He expected most of them to fail. During a dinner at a Dallas restaurant, L. J. asked if Hansen would be interested in starting a software research center for Mostek. "*Can I do it anywhere in the world?*" Hansen asked. "*We hope you will do it in Texas,*" he said, "*but, if you prefer, you can also build it in Denmark.*" Hansen said he would need to think about it before he gave up his tenured university position. L. J. responded with the immortal words, "*Let no man complain to me about the size of his balls!*"

The Mostek culture sharply contrasted with Intel culture. In the mid-1970, Bob Noyce was pre-occupied with his personal situation and new girlfriend Barbara. Distracted Noyce paid gradually less attention to Intel's day-by-day operations and his duties were assumed by András Gróf, better known as Andy Grove. Grove expanded on the theme and imposed himself on others. Grove's paranoid management philosophy has spread like, well, the paranoia. He established a "Late List" which each employee who arrived after 8:00 AM must sign. The list appeared in the lobby of the building. Another chilling story recounts Grove's parting words when microprocessor pioneer Federico Faggin left Intel to found Zilog. "*You will fail in everything you do.*" Grove wrote on the first page of his book [6] "*I believe that the prime responsibility of a manager is to guard constantly against other people's attacks and to inculcate this guardian attitude in the people under his or her management."* To be more explicit, people should work in a state of self-inflicted fear, trust no one, and spread that sense of insecurity to every employee [6]. When several top engineers separated from Intel, Sevin's Mostek had no problem to attract the supercharged engineers and they worked to death.

In May, 1970 Mostek's Berry Cash, vice-president for marketing, visited the Tokyo offices of the Nippon Calculating Machine Company. One month before, Canon, Inc. had introduced a "pocket-sized" calculator, which used three circuits and weighed 1.8 pounds. Nippon wanted to introduce a model that was smaller, lighter and less expensive to produce, and, Nippon wanted Mostek to put the calculator's circuit on a single-chip. Nippon agreed to let Mostek work on the idea, and signed a contract, which called for Mostek to demonstrate the circuits by mid-November 1970!

By the time the second design passed the initial tests and was ready for production, October had arrived. Time was literally running out for Mostek and a team headed by Dave Leonard. The team was working mostly at night, sleeping mornings. By the time the chip was ready for production, November had arrived and the design team decided to speed up the process by putting Richard Petty, the layout engineer, with Rubylith foils and other plans on a plane for California where he would hand carry the plans to a contractor for mask production. Petty would then fly to the Sprague production facility in Worcester, Mass., where the Mostek chip was to be made.

In mid-November it was time for Petty to come home with the final product. About 10 one evening, they set up for testing and waited for Petty to walk in the door. About 10 to 15 people had gathered, including Lee Kepley, a young test engineer who had been asked to help with the test. Petty walked in with the finished product and the testers hooked the chip to a calculator. "*We punched the 'Clear' button and it cleared,*" Kepley remember. "*We added and it added; we subtracted and it subtracted; and then everyone started plugging numbers into it-adding, subtracting, multiplying and dividing. The whole thing was incredible.*" The chip did not include a clock generator or display drivers. It was manufactured using a high-threshold, P-channel MOS process operating from −12 and −24 volts because that was compatible with the power supply in the calculator. The chip was also used in the Busicom LE-120A "Handy", the first hand-held calculator. Mostek went on to produce chips for other calculator manufacturers and became the major supplier of integrated circuits for H-P's scientific calculators. Hewlett-Packard soon became Mostek's largest customer.

---

[6] Fortunately for Grove, paranoid-schizophrenia authorities such Ira Glick and Donald MacGregor helped Grove to overcome the past, which he calls today "*running wild.*" "*Paranoia is a mental illness, not a survival technique. I'd never use that word to describe people who are hyper-aware in business. Paranoid people have a mental disorder,*" commented MacGregor.

Mostek, the industry infant, come up (without any "Late List") with the design of the decade, and set the electronic world on its ear and beat the giants of the industry in both the United States and Japan. The company became a leader in technology, having one of the broadest and most advanced product of nineteen seventies.

Fig. 10.52. The photo was taken upon the first testing in November, 1970 of the world's first single-chip calculator circuit. (Far left Gaynel Lockhart (logic designer); front left is Jim Imai behind Gordon Byrd (company comptroller); far back is Berry Cash, (Vice-President); between Berry and Jim is L. J. Sevin, (President); then Richard Petty (layout design); Dave Leonard (chip circuit designer); Bob Crawford (Engineering Manager); Lee Kepley (product engineering); and Louay Sharif, (Vice-President.)

Despite the dominance of P-MOS technology at the end of the nineteen sixties the N-MOS technology was superior for both logic and memory applications. This argument was based on the higher carrier mobility of N-MOS device and on desirable threshold voltages of active and field devices obtainable by substrate biasing. The IBM group from T. J. Watson Research Center, Yorktown Heights, recognized this fact very early, and in a series of presentations during 1967 and 1968 demonstrated that N-MOS offers a factor 2 to 3.4 mobility advantage (depending on the crystal orientation). The IBM group was not the only one who reached this conclusion. Yasuo Tarui, the Chief of the Semiconductor Section of Nippon Electric Company, Ltd. , Tokyo, surprised American semiconductor manufacturers at the February 1969 ISSCC Conference with 40 nsec 144-bit (16 word – 9 bits) N-MOS Memory. However, at that time, the arrogant American semiconductor manufacturers did not see the importance of such work.

George Cogar, a very capable engineer and aviator, started his career without formal training at Univac, where he eventually became a senior architecture designer. In 1964, he became one of the founders of the Mohawk Data Sciences Corporation, a start-up company based in Utica, NY.

Cogar invented an "intelligent terminal" which was an early forerunner of the modern personal computer. The original purpose of such terminals was batch communications between a mainframe computer and the so-called Remote Job Entry (RJE) terminals. The RJE terminals supported a limited number of data formats, and the main application was "word processing." The early terminals were quite simple and limited by the available memory. Cogar realized that functionality of the terminals could be significantly improved if they used emerging semiconductor memories. Cogar wanted to move Mohawk into semiconductor manufacturing. The Cogar philosophy was that the computer company should develop its own customized integrated circuits. Mohawk System /7, /36, and /38 were very successful with sales of over $9 million by the end of 1967, and there was no mood to dilute this business and go into risky and very competitive semiconductor manufacturing.

George Cogar ran into conflict with other founders over the future direction of the company. When he saw that he was not able to realize his ideas at Mohawk Data Sciences, a determined Cogar left the company and in January 1968 incorporated Cogar Corporation. Cogar's business plan called for "*the industry's most automated facility,*" and because Cogar had a well established business record, he had no difficulties raising money for new start-up.

In November 1968, Robert Markle and Ray Pacararo formed Cogar Technology Division of the Cogar Corporation based in Wappingers Falls. Robert Markle was responsible for IBM's SLT (Solid Logic Technology) technology and in 1967 he became responsible for IBM's MOS program in East Fishkill. Pacararo worked for and reported to Markle. Markle became Cogar's Director of Technology Division and Pacararo became Engineering Manager. President of the company, George Cogar, was a strong leader who was respected by his employees. He knew every employee by name, and he was able to motivate his engineers by his personal charisma.

Cogar had very little experience with semiconductor manufacturing, and his original plan assumed that Cogar Corporation would license technology from already established manufacturers. Cogar signed cross licensing for General Instrument's thick oxide technology, and had cross-licensing agreements with Texas Instruments and Fairchild. However, after detailed evaluation of GI devices, Cogar's engineering team decided to establish N-MOS technology for their new products.

Markle and Pacararo lured a group of seventy or so former IBM engineers to Cogar, and they formed two groups: one focused on the design of advanced manufacturing equipment, and the second on N-MOS devices. IBM originally paid little attention to this new start-up. IBM's MOS program had no clear goal, and they decided to follow the crowd. In the spring of 1970,

IBM initiated a P-MOS development program with about twenty-five people involved.

The change occurred when Cogar Corporation published in September 1969 the first information and preliminary data for their new product. IBM sued Cogar Corporation and all former IBM employees who joined Cogar. The charges had weak merit and some Cogar young employees were "guilty" because they left IBM and joined different employers. IBM's attempt to stop Cogar backfired. IBM was sued by the Department of Justice for antitrust violations and in March 1970, both Cogar and IBM settled their dispute out of court. IBM obtained the right of one-time examination of Cogar's facility and Cogar obtained the right to IBM patents.

2.1 A 1024 BIT N-CHANNEL MOS READ–WRITE MEMORY CHIP, W. Krolikowski, W. Brown, R. Dries, R. Foote, D. Lund, R. Plimley, J. Reuter, J. Sandhu, K. Scow and J. Tuttle, Cogar Corp., Wappingers Falls, N.Y.

The Cogar 1024-bit MOS read–write memory chip is fully decoded, has a 100-ns access time, and is only 125 × 125 mils in size. This small size is made possible by the use of a 4-device memory cell and dimensional tolerances as small as 0.15 mil. The chips are fabricated from a 2.25-inch-diameter wafer that consists of a low-resistivity p-type substrate covered by a high-resistivity p-type epitaxial layer. This epitaxial layer provides not only for precise control of surface impurity concentration, but also provides means for a reach-through type of gate protection device. In addition, the low-resistivity substrate tends to minimize unwanted noise voltages due to pulse currents flowing through the substrate.

The basic element is a 500-Å n-channel aluminum-PSG-SiO_2 MOS device. The PSG (phosphorus-silicate glass) layer is used to ensure drift-free and reliable operation of the memory.

The chip is hermetically sealed using r-f sputtered SiO_2, which protects the MOS memory but does not degrade the quality of the MOS devices. The finished chips use a Cr-Cu-Au-PbSn terminal metallurgy, and are flip-chip mounted on ceramic substrates.

Fig. 10.53. An Abstract of Walter Krolikowski's presentation at IEDM 1970

Markle and Pacararo built an impressive facility and their device group produced an even more impressive feat, the world's first 1kb N-MOS RAM. W. Krolikowski, who took Shockley's engineering classes at Stanford and who worked at IBM on bipolar devices, headed a team of engineers, W. Brown, R. Dries, F. Foote, D. Lund, R. Plimley, J. Reuter, J. Sandhu, K. Scow and J. Tuttle. Their work over the next eighteen months resulted in technology that jumped at least one decade ahead over competitors. Walter Krolikowski described the new device at the 1970 IEDM Conference (Fig. 10.53). The crowd booed Walt's presentation because they were told "N-MOS was unstable and could not work."

The memory used P/P+ epi substrate, which was biased (–6 V). Threshold voltage was 1.25 V and minimum channel length was 5 μm. The memory array was supported by MOS and bipolar circuits (drive and sense circuits) with TTL compatible outputs. The access time was 100 nsec with cycle time 400 nsec. All circuits were designed for a modular approach, which allowed assembling many possible sizes and organizations of final memory cards (Fig.10.55).

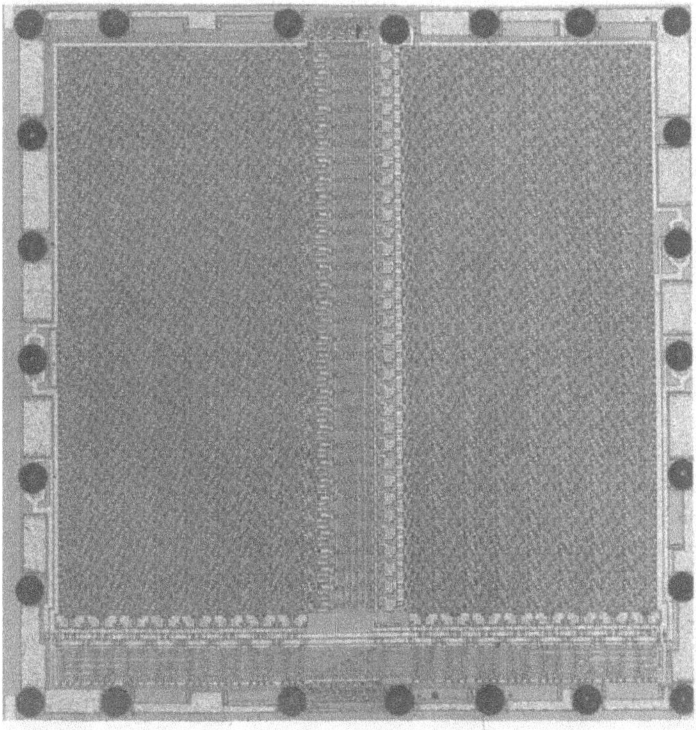

Fig. 10.54. The world's first N-MOS 1 k memory with channel length 5 μm and threshold voltage 1.25 V (Cogar Corporation, 1969)

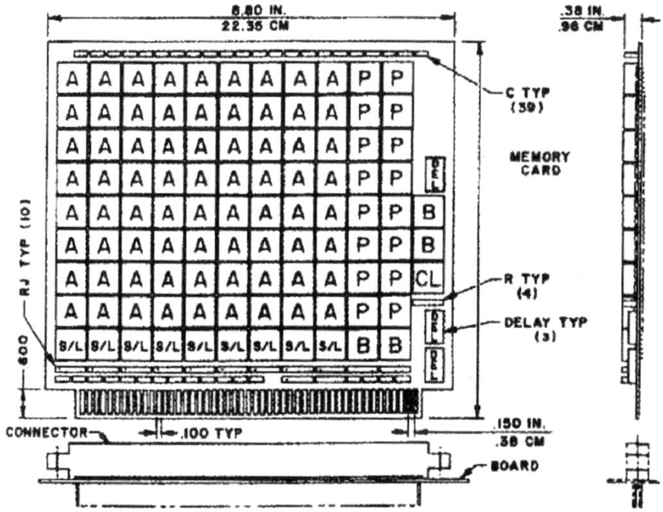

Fig. 10.55. Modular construction of Cogar's memory card (1970). "A" notation indicates memory array, "B" buffers, "P" decoding, "CL" and "DEL" timing and "S/L" sense amplifiers and latches

Individual cards could be assembled into modules, nicknamed "bomb". D. Lund introduced a 32,384 words (36 bit) module at the 1970 Fall Joint Computer Conference. The module consisted of eight cards with a packaging density of 2 millions bits/cu. ft.

In the middle of 1971, Cogar got production fully under control with a yield unprecedented for the type of circuits they were manufacturing. For a moment, Cogar's hopes rose, but only for a while. Good luck is essential to any start-up operation, as in the case of Fairchild, when mesa transistors began failing during the reliability tests performed by Autonetics. Cogar's memory components started to fail when they were assembled into memory cards. The faulty soldering compound, which Cogar purchased from an outside vendor, resulted in unpredictable cracking of soldered joints, resulting in the reliability problems. The panicked customer, Potter Instruments, cancelled its contract with Cogar. With no time and no customer, Cogar's Technology division struggled until April 1972 when the Technology Division closed operations and was available for bids; Singer eventually acquired Cogar's System Division in April 1973. Singer was producing an original Cogar terminal as a model 1500 for some time.

The failure of Cogar Corporation is a typical example that the product, which was ahead of its time, might not always succeed. If the timing for the application of revolutionary products is not right, the success of a start-up always depends on chance and the luck. Cogar was trying to sell a complete memory system for computers; they were not marketing individual memory components like Intel or Fairchild. They not only limited their own access to market, but they also needed to overcome the "not invented here" complex of all possible computer manufacturers which could have used Cogar's memory. With Cogar's overemphasis on the technical aspect of the product and almost no marketing ability, the fate of the company was doomed from the beginning.

However, the Cogar Corporation proved that:

1) N-MOS devices, when applied to memories, could be two or three times faster than P-MOS counterpart with access time less than 100 nsec (typical access time for P-MOS based memories was in the range 200–400 nsec.)
2) N-MOS is two to four times denser
3) N-MOS threshold voltage is generally lower than P-MOS (1 to 5 V for N-MOS, 10 V for P-MOS)

As a result, by the end of 1971, TI, Fairchild, Motorola, Ragen, Standard Microsystems, and Electronics Arrays were putting a large share of their resources into this new technology. IBM and Intel followed the crowd a year later.

A self-educated engineer, who triggered N-MOS technology, and who liked to do things in big style, a self-made entrepreneur, and passionate aviator with his own private airport and runway to his house, George Cogar, was killed in an airplane crash in September 2, 1983 in British Columbia.

Although IBM refused to re-hire anybody associated with the Cogar Corporation for ten years, they in the mid of nineteen-seventies completely adopted Cogar's memory approach and launched one of the largest efforts in semiconductor history, the "N-MOS Technology Program" with the goal to replace magnetic memory with transistor-based memory.

The key contribution to the future MOS technology was made by Robert Dennard who was a manager of a small group at IBM's T. J. Watson Research Center reporting to Dale Critchlow. Other members of Dennard's group were Fritz Gaensslen and Larry Kuhn. Critchlow and Dennard decided to take an existing 8 μm technology and shrank it to 1 μm. They observed that if the electric field was kept constant, the performance of the scaled device could be maintained. Subsequently, Dennard and Gaensslen derived a constant electric field scaling theory and analyzed its limitations. The remarkable result was that if the electric field was kept constant, nearly every other transistor characteristic of scaled MOS devices would improve! With the help of process engineer Hwa Yu, the group demonstrated devices with 1 μm design rules and presented the paper "*Design of micron MOS switching devices*" on the IEDM in 1972.

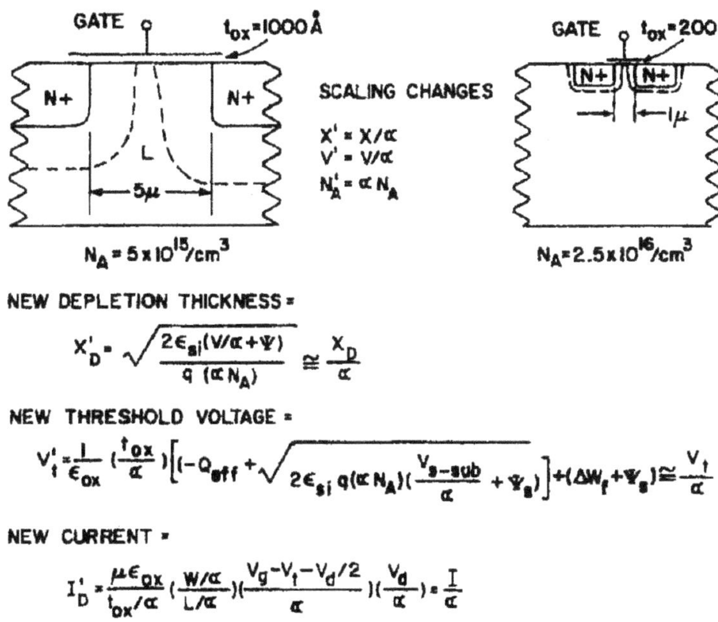

Fig. 10.56. Slide presented by Robert Dennard at 1972 IEDM Conference

Dennard's group was among the first to recognize the tremendous potential of downsizing of MOS devices. The scaling theory he and his colleagues reformulated in 1974 essentially observed that MOSFETs would continue to function while all key figures of merit such as layout density, operating speed, and energy efficiency would improve provided geometric dimensions, voltages, and doping concentrations were consistently scaled such as to maintain the same electric field. The three key scaling factors are the thickness of the insulator between the gate and the underlying silicon; the channel length (which is related to the minimum feature size); and the power-supply voltage, which, when applied to the gate, turns the transistor on.

After the presentation, many of the attendees expressed a lot of reservations. Although it was not generally recognized at the time, scaling theory would eventually end the supremacy of bipolar ICs. Bipolar device did not scale the way MOSFET technology did, and by the early nineties MOSFET technology became the dominant high-density-high-speed technology.

Dennard's seminal work is based on well-established semiconductor physics theory, and as such, the scaling theory became a fundamental driving force in the development of integrated circuits. MOS scaling, new devices, materials, designs and the hard work of thousands of engineers brought us to the "silicon age."

References

[1] B. G. Watkins, A Low Power Multiphase Circuit Technique, IEEE J. of Solid-State Circuits, SC-2(1967), p. 213

[2] L. Cohen, R. Rubinstein, F. Wanlass, MTOS four phase clock systems, 1967 NEREM Record, November 1967

[3] F. Faggin, T. Klein, and L. Vadasz, "Insulated gate field effect transistor integrated circuits with silicon gates," *IEDM Tech. Dig.*, p. 22, October 1968.

[4] K. G. Aubuchon, The use of Ion-Implantation to set the Threshold Voltage of MOS Transistors, Colloque International sur Proprietes et Utilisation des Structures M.I.S. Grenoble , June 17–20, 1969 (Centre d'Etudes Nucleaires de Grenoble)

[5] W. Asprey, Interview with M. Shima, May 17, 1994, IEEE History Center

[6] A. S. Grove, Only the Paranoid Survive: How to Exploit the Crisis Points That Challenge Every Company *(Currency/Doubleday, 1996).*

Epilogue

No one remembers even a few years ago! George Orwell. No memory! Who is that guy?

Michael Crichton, State of Fear, 2004

During my professional career, I met fifty-six people who claimed that they invented the integrated circuit or were part of the invention. Yet, the editor of the publishing house I originally intended to employ wrote after reading my manuscript, *"some prominent people in the center of widely publicized events at those times you described were absent."* The key word in his argument is "widely publicized events." Widely publicized events may be implanted or may be simply different from reality. In new terminology developed by our leaders, the "widely publicized" event is termed "spin" which is a code name for a flat lie.

A success story invariably creates a multitude of "halo" effects, i.e. those who were instrumental in the development are quick to recognize their own contributions, usually placing it completely out of proportion to reality, while those who were not originally enthusiastic for the idea have changed, or been forced to change their evaluation and now are supporters. Indeed, some of the latter group have since become members of the original crusade that initially favored the now successful and accepted alternative.

I refused to "trim" my arguments to fit "widely publicized events." As I emphasized in the Prologue, my goal in this book is to offer an alternative view of "widely publicized events." Science and engineering discoveries have a much bigger impact on society than, for example, any president or dictator. It is only natural that many want the general public to gain knowledge of their makeup, market conditions, or the nation's mood. One of the weaknesses of democracy is that all have the right to remain stupid. The key sources used in my argumentation, such as patents, legal transcripts, and government documents are in the public domain, and it is incumbent upon readers to ponder and reach their own conclusions and benefit from another advantage of democracy, to know.

Outwardly, we take knowledge very seriously, using the expression "Knowledge is Power." This is the main reason why our society is quite uncomfortable with the pursuit of knowledge. For example, politicians and other naïve people believe that hard-line politicians and their policy won the Cold War. This is a very simplistic view, and only people who know nothing about the situation in the former East Block agree with such claims. Advances in microelectronics and technology won the Cold War. Microelectronics revolutionized every single aspect of manufacturing, banking, medicine, communications, the military, the government, everything. The East Block economies were very simply collapsing not because of the shouting in front of the wall, but because they were not able to modernize anything without state-of-the-art microelectronics. Knowledge is what made the difference!

The work of scientists and engineers described in this book had a profound impact on society. It was not only their professional work but also their personal responsibility. It is not well known that Jean Hoerni and his company could have made a lot of money by producing mine fuses. He refused to participate in such an enterprise. On the other hand, Tom Longo made national defense his priority. It is difficult to find any military airplane, which is not using Longo's electronics. Gordon Moore called himself revolutionary during the 1968 street riots "*we do the real revolution, not the crowd in the street.*" Moore was right. They created a revolution!

Anyone who knows anything about engineering would agree that engineers play critical, ubiquitous roles in sustaining our nation's international competitiveness, in maintaining our standard of living, in ensuring a strong national security, in improving our health, and in protecting public safety.

When I started work on the manuscript of this book, I could not comprehend the dramatic change of the U.S. semiconductor industry in the new millennium. The Wright brothers built their airplane in the barn and on the field. My father built from scratch a motorcycle in his garage. I can produce a point contact transistor in my living room. Nobody, no one, however, can construct multimillion transistors microprocessor or a nano-device without huge investments into manufacturing infrastructure. As high-tech manufacturing is moving abroad, the infrastructure for R&D is not going to be available to the U.S. engineers. Still, to use the previously introduced Sidney L. Siegel terminology, there are too many "*financial operators*" who argue that a company can survive if it just does marketing here. Unfortunately, such argumentation cannot withstand scrutiny. Older readers may remember RCA TV sets. There was nothing more American than Radio Corporation of America. RCA led, by far, TV and broadcasting innovation. After a series of wrong managerial decisions, they decided that very expensive soldering of discrete components in TV sets could be "outsourced." Gradually, they outsourced manufacturing of more complex TV parts. In just few years, the brand RCA disappeared.

There is no domestic company that produces TV sets in America. Is it too difficult to extend this trend to semiconductors? Life may be better without TV, but can other important issues such as our defense, airspace, advanced medicine etc. be guaranteed without our leadership in microelectronics?

When Bob Noyce and his compatriots flew to Washington in 1987, they were able to argue, *"semiconductor engineering is vital to our defense industry."* When the chairman of Intel Corporation, Craig Barrett, flew to Washington in 2004, he found that the U.S. has a whole series of complacencies. "*It is complacent on its economic development platform. It is complacent on its infrastructure platform. It is complacent on the whole issue of promoting research and development. So, you go down the list – education, infrastructure, research and development – and the U.S. is basically complacent. In fact, we have been having this great argument in the press about off-shoring, or off-shore outsourcing. The press in general, the politicians in general, have not picked up the issue that you need to be competitive. The fact is that the U.S. is pulling further behind from an infrastructure standpoint and the dismal aspect of the U.S. education system. It is very difficult to go to Washington, D.C., and discuss those three aspects of competitiveness with anybody.*" Semiconductors are pervasive and an important source of productivity in the modern economy. Their rapid technological evolution – characterized by continuously increasing productivity and contemporaneously decreasing cost – is a source of growth throughout the economy, both in emerging industries and in more traditional industrial sectors. A significant element of the strong performance of the U.S. economy in the last decades is rooted in investment in and subsequent application of information technologies, which are ultimately driven by advances in semiconductor technology. Semiconductors also play a crucial role in ensuring our national security. For more than three decades, the semiconductors maintain the highest value-added manufacturing.

In the late 1990s and early 2000s, there was a significant loss in global market share by many U.S. industries to Far East producers. Extensive literature in the social sciences has focused on the decline in U.S. industry during this period. One influential study reported that U.S. manufacturers in general had lost the ability to compete internationally, and that the U.S. industrial weakness was part of a longer period of decline.[7]

In January 2005 during World Economic Forum held in Davos, Switzerland Bill Gates said "*I'm short the dollar. The dollar is going down.*" Microsoft founder seemed to say I am betting on America's decline and putting money on China's rise. Bill Gates' friend, Warren Buffett, shared an even more pessimistic picture and had been badmouthing America's fiscal policy. China held $191 billion in U.S. debt in 2004. At this moment China is still willing to play banker to America, but one day Americans may discover abruptly that the value of their assets has vanished. Gates and Buffet may

[7] Robert M. Solow, Michael Dertouzos, and Richard Lester, *Made in America,* MIT Press, Cambridge, MA, 1989.

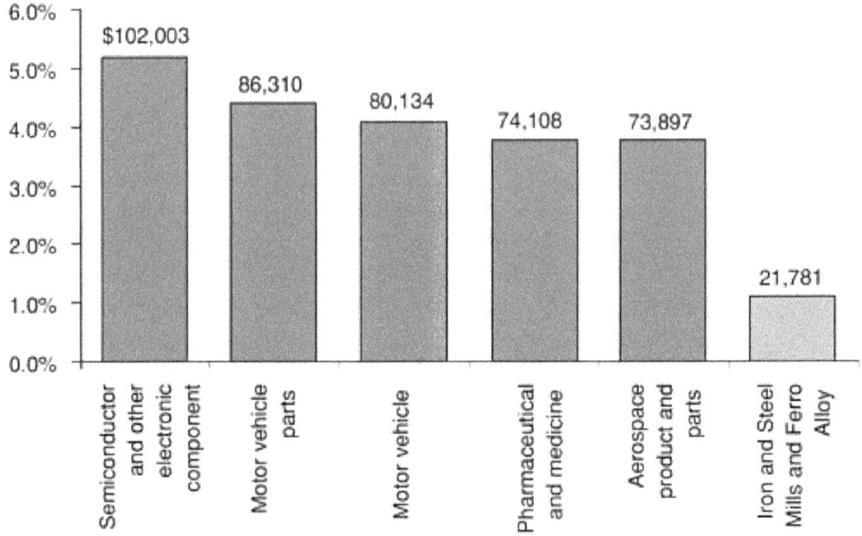

Fig. 1. Largest Five Value-Added Manufacturing Industries Compared with Other Major Sectors Source: US Census Bureau, *Annual Survey of Manufacturers*

be doing Americans a big favor by waking us up to a fact so cold that it may freeze our standard of living. The fact that the Chinese economy has been on a 20-year growth spurt is unprecedented in history. The U.S.-China Economic and Security Review Commission links the loss of 1.5 million U.S. jobs directly to our growing trade deficit with China. Far from taking only low-wage jobs, China is emerging in semiconductors. The Chinese, of course, did not create the U.S. deficit. To get a piece of the Chinese market, U.S. companies spend billions to help build a Chinese economy that can produce goods at prices U.S. manufacturers cannot match. Too often, they help Chinese competitors who usurp U.S. jobs and technology. The Chinese government "helped" by engineering a system in which companies can acquire high technology at exceedingly low costs. Often, without fear of consequence, they simply steal it. Our government is not able to face such challenges and therefore is doing nothing.

The engineers and their achievements described in this book are examples of what an intellectual society could accomplish. The situation is very different today. Electronic Engineering Times and Beacon Technology Partners LLC conducted in summer 2005 "The Engineering Psychographic Survey." The survey with a margin of error of $\pm 1.7\%$ determined that "the creators of technology aren't feeling very appreciated these days." Only 10% think the U.S. will remain the world's tech leader. This is a rather grim prediction for the nation who claims the myth of exceptionality, and expands our preferred way of making other nations be like us. Some answers in this survey assert that the engineering is a poor career choice in these days.

To blame our poor educational system is only part of the problem. The greatest innovations are usually created by the best minds. For engineers like Shockley, Widlar or Hoerni, technical work was glamorous. They were born engineers; they enjoyed the satisfaction in focusing on a problem, educating themselves, mustering the knowledge, sense of pride and accomplishment. They did not care if they were appreciated or not. There will be always "engineers" who feel they have to put up with stuff they do not like at work for the pay. They would always complain that attorneys and medical doctors make more money. Once they have enough money, they suddenly pack it in because they no longer need to put up with that stuff. They are not engineers. Good engineers are born engineers; they are not working for money. School will not make you a good engineer. Engineering is an attitude!

The advent of the transistor and the integrated circuit, and as a consequence, the emergence of the modern computer, telecommunications, and satellite technologies has fundamentally changed the structure of the American economy. Many said that this is perhaps a once-in-a-century event. They are wrong! The transistor, integrated circuit, computer or satellite, are examples of what can be accomplished in a society which does not limit itself by "politically correct" nonsense.

The strength and exceptionality of America was based on unique beliefs of individualism. The politically correct society, in the same way as any dogmatic society, needs to hide the incompetence of their political, clerical, and business leaders. The concept of the team and the suppressing of individuality is the easiest way for such hiding. The only problem is that this is not a good recipe for the longevity of society. In spite of assurances from our political leaders that we were a nation uniquely blessed by God as a "city on a hill," in its brief stint of being the only remaining superpower, we did not demonstrate the myth of exceptionalism. In the short run, we will remain a reasonable economy for several decades leveraging the inertia of the famous economic concepts created by the individualistic creators described in this book. In the long range, the historic rules of life cycles of imperial nations will decimate any society which does not appreciate knowledge, creativity, science, and technology.

It has been said by many, "*the author never finishes a book.*" I am sure that so many things could have been done better and described more elegantly. I am abandoning the book for a while, because of the publisher's deadline. Neverthless, I hope that the information I have provided, may be useful to some.

Name Index

GPSR Compliance
The European Union's (EU) General Product Safety Regulation (GPSR) is a set of rules that requires consumer products to be safe and our obligations to ensure this.

If you have any concerns about our products, you can contact us on

ProductSafety@springernature.com

In case Publisher is established outside the EU, the EU authorized representative is:

Springer Nature Customer Service Center GmbH
Europaplatz 3
69115 Heidelberg, Germany

www.ingramcontent.com/pod-product-compliance
Ingram Content Group UK Ltd.
Pitfield, Milton Keynes, MK11 3LW, UK
UKHW022215051026
15624UKWH00001B/180
9783540342571